VOLUME ONE HUNDRED AND TWENTY FOUR

INTERNATIONAL REVIEW OF NEUROBIOLOGY

Nicotine Use in Mental Illness and Neurological Disorders

INTERNATIONAL REVIEW OF NEUROBIOLOGY

VOLUME 124

SERIES EDITORS

R. ADRON HARRIS
Waggoner Center for Alcohol and Drug Addiction Research
The University of Texas at Austin
Austin, Texas, USA

PETER JENNER
Division of Pharmacology and Therapeutics
GKT School of Biomedical Sciences
King's College, London, UK

VOLUME ONE HUNDRED AND TWENTY FOUR

INTERNATIONAL REVIEW OF NEUROBIOLOGY

Nicotine Use in Mental Illness and Neurological Disorders

Edited by

MARIELLA DE BIASI

Department of Psychiatry
University of Pennsylvania Perelman School of Medicine
Philadelphia, PA, USA

ELSEVIER

AMSTERDAM • BOSTON • HEIDELBERG • LONDON
NEW YORK • OXFORD • PARIS • SAN DIEGO
SAN FRANCISCO • SINGAPORE • SYDNEY • TOKYO
Academic Press is an imprint of Elsevier

Academic Press is an imprint of Elsevier
225 Wyman Street, Waltham, MA 02451, USA
525 B Street, Suite 1800, San Diego, CA 92101-4495, USA
The Boulevard, Langford Lane, Kidlington, Oxford OX5 1GB, UK
125 London Wall, London, EC2Y 5AS, UK

First edition 2015

Notices
Knowledge and best practice in this field are constantly changing. As new research and experience broaden our understanding, changes in research methods, professional practices, or medical treatment may become necessary.

Practitioners and researchers must always rely on their own experience and knowledge in evaluating and using any information, methods, compounds, or experiments described herein. In using such information or methods they should be mindful of their own safety and the safety of others, including parties for whom they have a professional responsibility.

To the fullest extent of the law, neither the Publisher nor the authors, contributors, or editors, assume any liability for any injury and/or damage to persons or property as a matter of products liability, negligence or otherwise, or from any use or operation of any methods, products, instructions, or ideas contained in the material herein.

ISBN: 978-0-12-801583-4
ISSN: 0074-7742

For information on all Academic Press publications visit our website at http://store.elsevier.com/

CONTENTS

Part 5
Alzheimer's disease

CONTRIBUTORS

Janet Audrain-McGovern
Department of Psychiatry, Perelman School of Medicine, University of Pennsylvania, Philadelphia, Pennsylvania, USA

Daniel Bertrand
HiQScreen Sàrl, Geneva, Switzerland

Tanuja Bordia
Center for Health Sciences, SRI International, Menlo Park, California, USA

Corinne Cather
Center for Addiction Medicine, Massachusetts General Hospital, and Department of Psychiatry, Harvard Medical School, Boston, Massachusetts, USA

John A. Dani
Department of Neuroscience, Mahoney Institute for Neurosciences, Perelman School of Medicine, University of Pennsylvania, Philadelphia, Pennsylvania, USA

A. Eden Evins
Center for Addiction Medicine, Massachusetts General Hospital, and Department of Psychiatry, Harvard Medical School, Boston, Massachusetts, USA

Robert E. Featherstone
Translational Neuroscience Program, Department of Psychiatry, University of Pennsylvania, Philadelphia, Pennsylvania, USA

Miranda L. Fisher
Department of Drug Discovery and Biomedical Sciences, South Carolina College of Pharmacy, University of South Carolina, Columbia, South Carolina, USA

Thomas J. Gould
Temple University, Philadelphia, Pennsylvania, USA

Jaakko Kaprio
Department of Public Health, Faculty of Medicine; National Institute for Health and Welfare, and Institute for Molecular Medicine Finland FIMM, University of Helsinki, Helsinki, Finland

Munir Gunes Kutlu
Temple University, Philadelphia, Pennsylvania, USA

Adam M. Leventhal
Departments of Preventive Medicine and Psychology, Keck School of Medicine, University of Southern California, Los Angeles, California, USA

Anu Loukola
Department of Public Health, Faculty of Medicine, University of Helsinki, Helsinki, Finland

Eric J. Nestler
Fishberg Department of Neuroscience and Friedman Brain Institute, Icahn School of Medicine at Mount Sinai, New York, USA

Vinay Parikh
Temple University, Philadelphia, Pennsylvania, USA

Xiomara A. Perez
Center for Health Sciences, SRI International, Menlo Park, California, USA

Maryka Quik
Center for Health Sciences, SRI International, Menlo Park, California, USA

Jianxin Shen
Department of Physiology, Shantou University Medical College, Shantou, Guangdong, China

Steven J. Siegel
Translational Neuroscience Program, Department of Psychiatry, University of Pennsylvania, Philadelphia, Pennsylvania, USA

David R. Strong
Department of Family Medicine and Public Health, University of California, San Diego, California, USA

Jill R. Turner
Department of Drug Discovery and Biomedical Sciences, South Carolina College of Pharmacy, University of South Carolina, Columbia, South Carolina, USA

Tanya L. Wallace
SRI International, Menlo Park, California, USA

Jie Wu
Department of Physiology, Shantou University Medical College, Shantou, Guangdong, China, and Divisions of Neurology and Neurobiology, Barrow Neurological Institute, St. Joseph's Hospital and Medical Center, Phoenix Arizona, USA

Danhui Zhang
Center for Health Sciences, SRI International, Menlo Park, California, USA

PREFACE

Thanks to the efforts of friends and colleagues who contributed, "Nicotine Use in Mental Illness and Neurological Disorders" provides an update on tobacco use by a particularly vulnerable portion of our population. Chapters examine shared genetic factors, fundamental cellular mechanisms, emerging therapeutic treatments, and state-of-the-art smoking cessation strategies.

Tobacco abuse is driven by nicotine addiction, a chronic relapsing disorder that is nearly twice as common among those with mental illness. The highest rates of tobacco use occur in those diagnosed with depression, bipolar disorder, or schizophrenia. Unfortunately, some health-care providers still consider treating tobacco dependence in subjects with chronic mental illness as too challenging. Because smoking is believed to provide a coping mechanism, some think that quitting may interfere with recovery from mental illness. This point of view contributes to significant disparities in tobacco-related morbidity and mortality. Smokers with serious mental illness have increased risk for cancer, lung disease, and cardiovascular disease, and on average, live shorter lives than the general population. Therefore, tobacco use in mental health patients produces substantial social and financial costs to individuals, their families, and society.

One of the goals of this publication is to highlight the involvement of the nicotinic cholinergic system in both the mechanisms of nicotine addiction and mental disease. People affected by mental disease may be particularly vulnerable to nicotine dependence because of the role of the cholinergic system in normal brain function. In the CNS, most acetylcholine (ACh) is synthesized and released by a relatively small number of cholinergic nuclei. However, through broad projections and terminal fields these sparse cholinergic nuclei deliver ACh to nearly every area of the brain. By binding to muscarinic and nicotinic receptors, ACh modifies individual synapses and cells and changes the activity of entire neuronal networks to influence cognition, emotions, and behavioral output. Learning and memory, attention, sleep and wakefulness, reward, and goal-directed behavior are all modulated by the cholinergic system. Therefore, it is not surprising that disruption or dysfunction of cholinergic transmission contributes to the pathophysiology of several neuropsychiatric disorders. The focus on the nicotinic cholinergic system and its role in mental illness reflects the fact that nicotine, the main addictive component found in tobacco products, binds to the nicotinic subtype of ACh receptors (nAChRs).

The chapter written by Dani provides an introduction to nAChR structure–function relationships and the cellular effects produced by the binding of nicotine to nAChRs. This introductory chapter is followed by the contribution of Featherstone and Siegel who review current research on the molecular mechanisms of schizophrenia. Among other mechanisms, they discuss the influence of nAChRs containing the α7 subunit. This topic is further reviewed by Wallace and Bertrand who, based on the unique biophysical and pharmacological properties of the receptor, discuss the development of α7 nAChR-based therapies. Turner and colleagues provide an example of how preclinical rodent models and human genetic data can be integrated to identify new potential therapeutic targets. In particular, they discuss the neuregulin signaling pathway as a potential contributor to both nicotine dependence and comorbid psychiatric disorders. Nestler highlights the complex relationship between mood disorders and the brain's reward circuitry, which is the target of addictive drugs such as nicotine. He discusses the role of transcriptional mechanisms in depression and argues that modeling key symptoms of depression in animals could enable the development of antidepressant medications with fundamentally new mechanisms of action. Those same models could then be used to study the interaction between nicotine dependence and mental disease. Gould and colleagues provide a summary of the research on nAChR-mediated mechanisms in depression as well as attention deficit hyperactivity disorder, anxiety disorders, and post-traumatic stress disorder. In Chapters 5 and 8, Evins, Audrain-McGovern, and their respective colleagues bring the focus to the clinical arena and discuss evidence-based smoking cessation strategies that can be used in patients with a diagnosis of either schizophrenia or depression. Their contributions provide evidence that tobacco cessation strategies can be successfully implemented in mental health patients. The final two chapters of the book are dedicated to the review of nAChR-mediated mechanisms in two neurological disorders, Parkinson's disease and Alzheimer's disease. Several epidemiological studies have shown an inverse association between smoking and Parkinson's disease, and Quik discusses the potential of nAChR ligands as neuroprotecting agents against nigrostriatal damage and the emergence of L-dopa-induced dyskinesias. Finally, Shen and Wu review the data on the involvement of α7-containing nAChRs in the mechanism of Alzheimer's disease.

It is clear that there is high demand for novel therapeutics that could effectively treat comorbid disorders such as tobacco smoking and mental disorders. Understanding the common cellular processes that link these disorders provides new therapeutic targets and informs clinical interventions. This book offers several examples of how such strategies are being implemented, and therefore, the content has appeal for both research scientists and clinicians.

Mariella De Biasi

PART 1

Introductory chapter

CHAPTER ONE

Neuronal Nicotinic Acetylcholine Receptor Structure and Function and Response to Nicotine

John A. Dani[1]
Department of Neuroscience, Mahoney Institute for Neurosciences, Perelman School of Medicine, University of Pennsylvania, Philadelphia, Pennsylvania, USA
[1]Corresponding author: e-mail address: johndani@upenn.edu

Contents

Abstract

Nicotinic acetylcholine receptors (nAChRs) belong to the "Cys-loop" superfamily of ligand-gated ion channels that includes $GABA_A$, glycine, and serotonin ($5\text{-}HT_3$) receptors. There are 16 homologous mammalian nAChR subunits encoded by a multigene family. These subunits combine to form many different nAChR subtypes with various expression patterns, diverse functional properties, and differing pharmacological characteristics. Because cholinergic innervation is pervasive and nAChR expression is extremely broad, practically every area of the brain is impinged upon by nicotinic mechanisms. This review briefly examines the structural and functional properties of the receptor/channel complex itself. The review also summarizes activation and desensitization of nAChRs by the low nicotine concentrations obtained from tobacco. Knowledge of the three-dimensional structure and the structural characteristics of channel gating has reached an advanced stage. Likewise, the basic functional properties of the channel also are reasonably well understood. It is these receptor/channel properties that underlie the participation of nAChRs in nearly every anatomical region of the mammalian brain.

International Review of Neurobiology, Volume 124
ISSN 0074-7742
http://dx.doi.org/10.1016/bs.irn.2015.07.001

1. INTRODUCTION

Mammalian nicotinic acetylcholine receptors (nAChRs) are composed on five subunits arranged around a water filled pore (Fig. 1). The neuronal subunits are divided into the alpha (α2–α7, α9, and α10) and beta (β2–β4) classifications based on the presence of adjacent cysteine groups in the extracellular domain of only the α subunits (Albuquerque, Pereira, Alkondon, & Rogers, 2009; Dani & Balfour, 2011; Dani & Bertrand, 2007; Fasoli & Gotti, 2015; Lewis & Picciotto, 2013; McGehee & Role, 1995; McKay et al., 2007; Papke, 2014; Unwin, 2013; Zoli, Pistillo, & Gotti, 2014). The α8 subunit has been found in avian tissue but not in mammalian tissue. Much of the structural and functional diversity of neuronal nAChRs arises from the many possible subunit combinations. The two most commonly found nAChR subtypes in the mammalian brain are the α4β2 heteromeric and the α7 homomeric subunit combinations, which are didactically represented in Fig. 1 showing their agonist-binding sites.

The α4β2 is a subtype with high affinity for nicotine, and the α7 subtype is the main contributor to the α-bungarotoxin-binding sites of the brain. Because each subunit has sidedness and is not completely symmetrical, the placement of the many different subunits within the pentameric complex can produce thousands of different nAChR subtypes. For example, the α5 subunit may combine as an "accessory" subunit that does not contribute to the agonist-binding site (Fig. 1, right), but its presence modifies the functional properties of the receptor/channel complex. Another complicating feature is illustrated by considering the α4β2 heteromeric receptor that can exist as a 2(α4)–3(β2) receptor (represented in Fig. 1, left). It also can

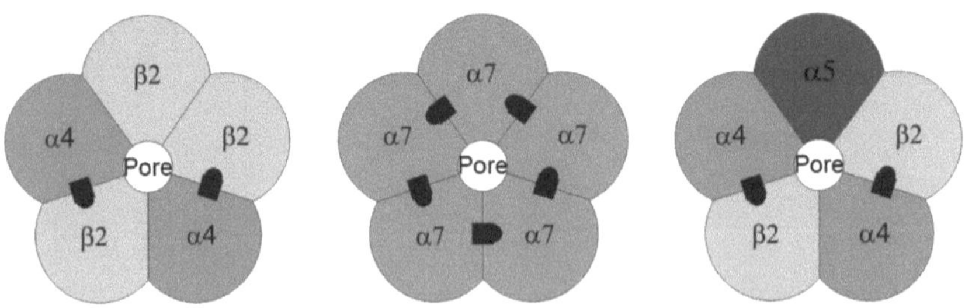

Figure 1 Didactic illustration of the nAChR subunits arranged as pentamers around the water-filled cation-permeable pore. The most common nAChRs in the brain are hetero-oligomeric α4β2 nAChRs and homo-oligomeric α7 nAChRs. The recognized ACh-binding sites are indicated by black asymmetric designs located between adjacent subunits. *Adapted from Fig. 1B of McKay, Placzek, and Dani (2007).*

exist as a 3(α4)–2(β2) receptor, which can potentially have another agonist-binding site arising from the sidedness of the α subunit (Fasoli & Gotti, 2015). Therefore, each different pentameric complex can, in principle, produce a nAChR receptor/channel with different functional characteristics: e.g., opening, closing, and desensitizing kinetics; ionic conductance; cationic selectivity; and pharmacological properties. In practice, however, these subtypes commonly share many structural and functional properties, leading to the grouping of nAChRs into a few main neuronal nAChR subtype classifications. For example, those that contain the α7 subunit (α7*) as a homomeric or heteromeric receptor most commonly also have accompanying characteristics. They bind α-bungarotoxin, have relatively low affinity for nicotine and have relatively fast kinetics. Those that contain β2 (β2*) commonly have high affinity for nicotine, desensitize to low agonist concentrations, have relatively slow kinetics, and do not bind α-bungarotoxin. Because these broad nAChR categories include such a diverse collection of subtypes, not all the members perfectly follow these broad functional characteristics.

Other nAChR subtypes have a much more restricted distribution in the brain, but in some cases they can constitute the most abundant receptor subtype in a restricted brain area where they are expressed. For example, α3β4* nAChRs, which are commonly found in the peripheral nervous system, are expressed at high levels only in the medial habenula, interpeduncular nucleus, and locus coeruleus. α3β4* nAChRs have low affinity for nicotine and have much slower desensitization kinetics than α4β2 nAChRs (Fenster, Rains, Noerager, Quick, & Lester, 1997).

All the mammalian neuronal nAChR subtypes do share the general functional property of being permeable to small monovalent and divalent cations. The main conducting species under biological conditions are Na^+, K^+, and Ca^{2+}. Agonists, such as endogenous acetylcholine (ACh) or exogenous nicotine (which can be obtained from smoking tobacco), stabilize the open conformation of the nAChR channel that transiently permeates small cations for several milliseconds before closing back to a resting state or closing to a desensitized state that is unresponsive to agonists. Brief exposure to high concentrations of the neurotransmitter, such as acetylcholine at a synaptic cleft, favors synchronous opening of the nAChRs' pores. However, prolonged expose to low concentrations of nicotine, as obtained from tobacco use, produces some activation but also significant desensitization of nAChRs to the unresponsive closed state (Dani, Radcliffe, & Pidoplichko, 2000; Giniatullin, Nistri, & Yakel, 2005; Quick & Lester, 2002; Wooltorton, Pidoplichko, Broide, & Dani, 2003).

2. NICOTINIC RECEPTOR STRUCTURE

The neuronal nAChR subunits share a similar linear structure and transmembrane topology with the muscle α1 subunit (Fig. 2A) (Papke, 2014). The relatively long extracellular N-terminal domain contributes to ligand binding, followed by the three hydrophobic transmembrane regions (M1–M3), a large intracellular loop, a fourth transmembrane region (M4), and ultimately a short extracellular C-terminus (Fig. 2A). The general hydrophobicity plot for the alpha subunits suggests the basic structural domains, including the four transmembrane domains (Fig. 2B). The Cys-loop, which is shared by the whole gene superfamily, is created by a disulfide bond that links a 15 amino acid sequence contained within the large N-terminal extracellular domain. The M2 transmembrane segment in each subunit provides the main lining of the ionic pore with some contribution from the M1 segment where the pore

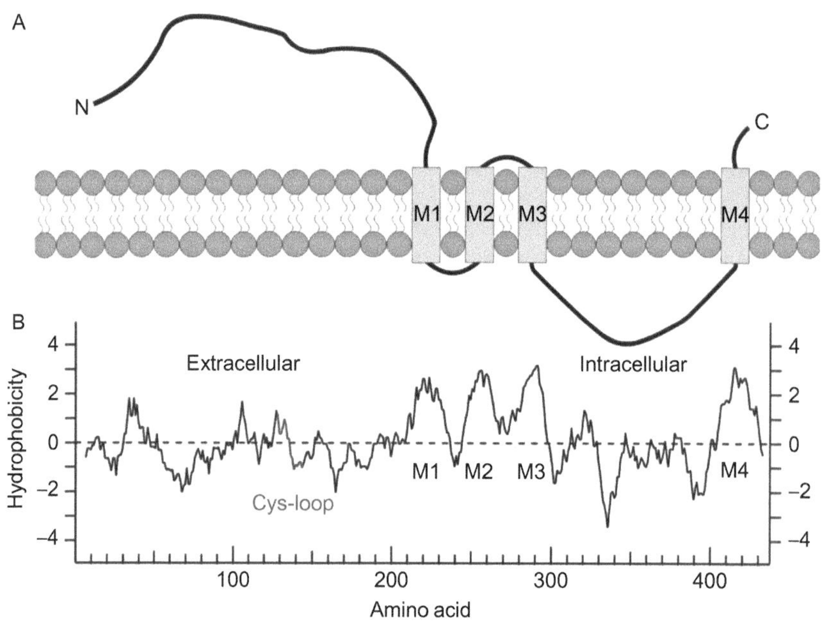

Figure 2 Transmembrane topology of a nAChR subunit. (A) A didactic illustration of the linear structure of the nAChR subunit with four transmembrane domains (M1–M4) passing through the lipid bilayer member. (B) A plot of the hydrophobicity profile of a human α1 subunit. The profile is aligned with the linear representation of the subunit just above. *Panel (A): Adapted from Fig. 1A of McKay et al. (2007). Panel (B): Adapted from Fig. 1A of Papke (2014).*

widens (Bertrand, Galzi, Devillers-Thiery, Bertrand, & Changeux, 1993a; Dani, 1989; Karlin, 2002; Unwin & Fujiyoshi, 2012). The M1, M3, and M4 segments separate the pore-lining region from the hydrophobic membrane (Papke, 2014). The intracellular domains are quite variable among the different subunits. This variability has functional consequences for intracellular modifications, such as phosphorylation, and for linking to intracellular cytoskeletal elements that control cellular trafficking and influence surface distribution and clustering (Kracun, Harkness, Gibb, & Millar, 2008; Pollock, Pastoor, Katnik, Cuevas, & Wecker, 2009).

The basic structure of neuronal nicotinic receptors is homologous to the muscle nAChR (Karlin, 2002; Papke, 2014). Negative staining of the pseudo crystalline form of the muscle-type nAChRs isolated from the *Torpedo Californica* electric organ revealed the structure at 3.6 Å resolution (Unwin, 2005; Unwin, Miyazawa, Li, & Fujiyoshi, 2002). The muscle subunits that compose the receptor/channel are two α1 combined with β1, γ, δ, and the γ subunit is replaced by an ε subunit in the adult animal. The arrangement around the central pore is illustrated in a ribbon diagram in Fig. 3A. A lateral cross-section of the nAChR (Fig. 3B) displays an

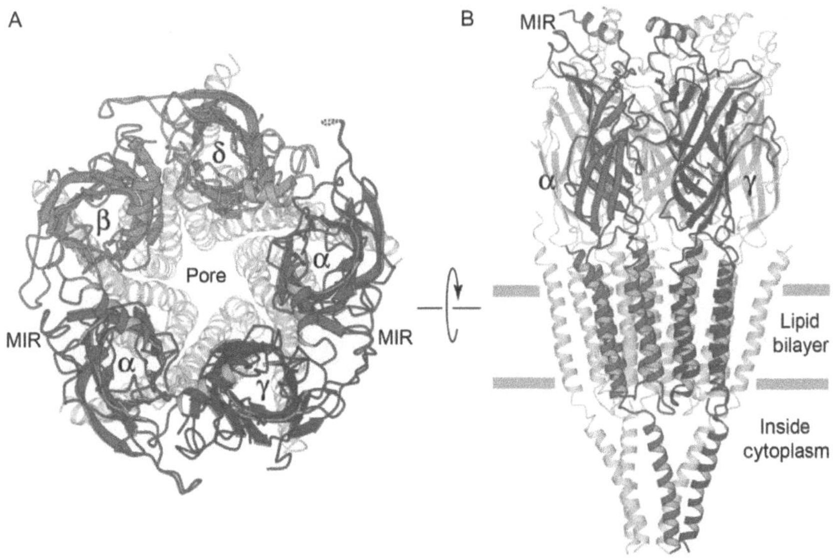

Figure 3 Illustrations of ribbon diagrams of the nAChR. (A) View from the top into the pore with only the upper most portion highlighted in colors (gray shades in the print version). (B) View from the side with only the front two subunits highlighted in colors (gray shades in the print version). The plane of the membrane is indicated by the two horizontal gray lines. *Adapted from Fig. 3 of Unwin (2005).*

extracellular water-filled vestibule that is about 20 Å in diameter and extends 60 Å from the membrane surface into the synaptic cleft. The pore narrows at the level of the surface membrane, and permeant ions pass along this ionic pore for about 40 Å (Unwin, 2005).

When viewed from the side (Fig. 3B), the three main domains of the nAChR are observable. (1) The large extracellular domain that contains the agonist-binding sites and also creates the entrance vestibule to the pore. (2) The transmembrane domain that creates the water-filled, hydrophilic ionic pathway through the lipid bilayer membrane when the pore is open. (3) The intracellular domain that is the most highly variable among the subunits and contains sites for modifications and interaction with cytoplasmic elements. The main immunogenic region (MIR, Fig. 3) is a short amino acid sequence of the α1 subunit where many antibodies bind (Tzartos, Kokla, Walgrave, & Conti-Tronconi, 1988), including the autoantibodies to muscle nAChRs in human myasthenia gravis (Luo et al., 2009). This region is very diverse among the nAChR subunit family, but in some cases antibodies produced to the muscle α1 subunit's MIR also bind to other neuronal nAChR subunits.

3. NICOTINIC RECEPTOR CHANNEL GATING

In most cases, there are two ACh-binding sites per muscle and heteromeric neuronal nAChR. Each binding site is formed by a pocket at the interface between adjacent subunits within the extracellular N-terminal domain (Albuquerque et al., 2009; Galzi et al., 1990; Karlin, 2002; Papke, 2014; Sine, 2002; Sine & Engel, 2006; Unwin, 2013). The situation is more complicated for the homomeric α7 nAChR subtype, where the sidedness of the interfaces between the alpha subunits provides five potential-binding sites (Fig. 1, middle). In the case of the muscle receptors, the two ACh molecules bind at the interface between the α and γ subunits (or α–ε in the adult form) and between the α and δ subunits (Fig. 4, where only the α and γ-binding site is indicated). The blue (light gray in the print version) shading in Fig. 4 indicates the protein structure pulling closer together when ACh binds. The red (dark gray in the print version) sphere indicates a bound ACh molecule in the binding pocket at the α–γ interface. The yellow (light gray in the print version) arrows indicate the general structural movement of the protein, including the C-loop (or loop C) closing over the ACh in the binding site. In the α4β2 receptor (Fig. 1, left), the ACh molecules bind between the α4

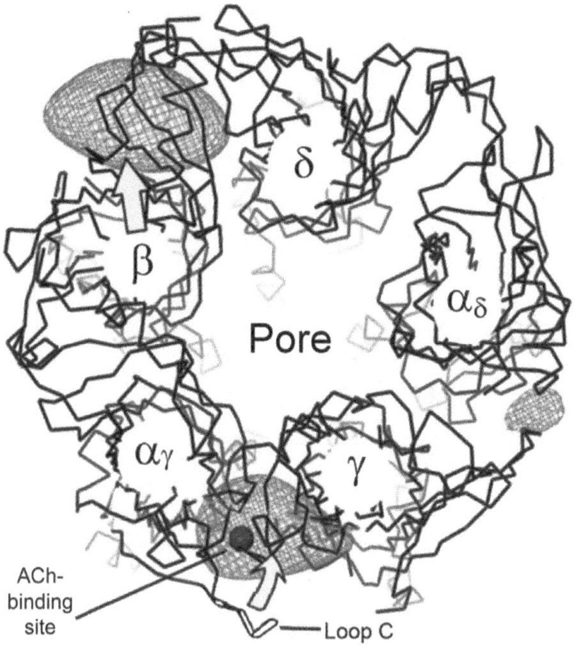

Figure 4 Illustration of the structural changes induced by ACh (red (dark gray in the print version) sphere) binding into the pocket formed by the closing of loop C, as viewed from the top into the pore. The blue (light gray in the print version) shaded regions represent the most significant increases in density of the open channel relative to the closed channel. The yellow (light gray in the print version) arrows indicate the general structural displacement caused by opening. *Adapted from Fig. 2 of Unwin and Fujiyoshi (2012).*

and β2 subunits as indicated. Therefore, both α and β subunits contribute to the pharmacology of the heteromeric-binding site.

The extracellular ligand-binding domain consists of six loops: three on the principal side of the α subunit and three on the adjacent subunit (Williams, Stokes, Horenstein, & Papke, 2011). Two important loops in the N-terminal extracellular domain are the Cys-loop (Fig. 2B) and the C-loop (Fig. 4). In the 3D crystal structure of the α1 subunit, the Cys-loop is a 13 amino acid sequence linked by a cysteine disulfide bond located at the bottom of a beta-barrel that lies in close proximity to the extracellular M2–M3 loop (Fig. 2A). When an agonist, such as nicotine or ACh enters the binding site, the C-loop moves and covers the ligand (Celie et al., 2004) (Fig. 4). The ligand-binding process also requires participation of a series of aromatic residues whose structural arrangement is shared by all members of the Cys-loop family of channels (Taly, Corringer, Guedin, Lestage, & Changeux, 2009).

The M2 segment lines the ion channel along the axis of symmetry so that it also provides the amino acids for the explicit gate that closes the pore. The closure gate is located near the middle of the membrane-spanning portion of the channel (near $Z=0$ in Fig. 5), where hydrophobic residues approach each other to narrow the closed structure of the pore (Fig. 5, black dashed contour). The hydrophobic environment is energetically unfavorable for ion permeation. Thus, this gate, which is composed of three rings of hydrophobic residues, prevents passage of permeant ions when the channel is in the closed conformation (Hilf & Dutzler, 2008; Unwin & Fujiyoshi, 2012). This area of the M2 region is allosterically coupled with the agonist-binding region (Taly et al., 2009). The analysis of bacterial proteins homologue to nAChRs has suggested that channel opening is produced by the concerted tilting of the M2 helices, the M2–M3 loop, and the M3

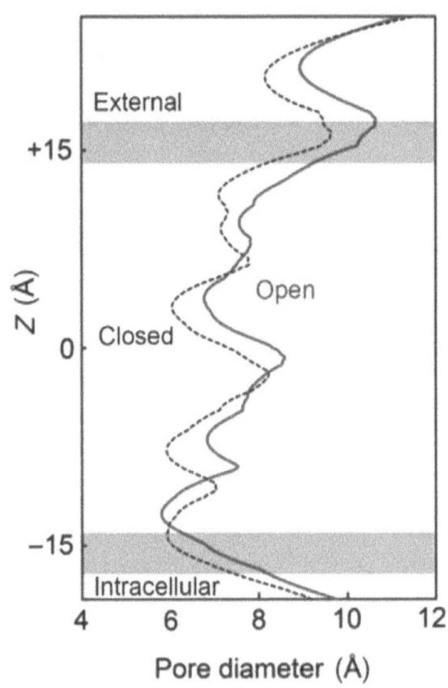

Figure 5 Representation of the structural change of the nAChR pore when transitioning from closed (dashed black) to open (red (dark gray in the print version)). The pore is represented to run vertically with the gray horizontal lines delimiting the position of the lipid bilayer membrane. Upon opening, the diameter of the pore increases in the constricting hydrophobic region near the middle of the membrane (at $Z=0$), and the narrowest region shifts near to the intracellular membrane surface where the pore is lined by polar residues (near $Z=-14$). *Adapted from Fig. 10 of Unwin and Fujiyoshi (2012).*

segment (Popot, Demel, Sobel, Van Deenen, & Changeux, 1978). A series of interacting residues participate to transmit the agonist-binding conformational changes to the channel gate (Lee & Sine, 2005; Sine & Engel, 2006). Unwin and Fujiyoshi (2012) and Unwin (2013) have presented evidence indicating that the M2 transmembrane domain also converts from a bent conformation to a more straightened conformation by flexing in a way that moves the hydrophobic gate residues in a radial direction away from the axis of the pore (Fig. 5, red (dark gray in the print version) contour).

Structural models in conjunction with single-channel current measurements of the muscle-like nAChR revealed invariant charged amino acids that electrostatically couple α subunit-binding domains, ultimately linking them to the channel-forming α-helix. Movement of these structures underlies nAChR channel gating. During channel opening, the narrowest region of the pore moves from near the middle of channel to near the intracellular membrane surface where the pore is lined by polar residues (Fig. 5, red (dark gray in the print version) contour). The width of the narrow region moves from a hydrophobic to a hydrophilic lining region of the pore. The pore's narrowest region does not become much wider, but it does become amenable to the permeation of cations by providing a polar, hydrophilic pathway. Permeation studies and structural data indicate that the narrowest cross-section near the inner surface of the membrane is short (3–6 Å) and about 6–7 Å in diameter (Dani, 1989; Karlin, 2002; Unwin, 2005).

4. CATIONIC PERMEABILITY OF THE NICOTINIC RECEPTOR PORE

Mammalian nAChRs are cation selective, being permeable to small monovalent and divalent cations that can fit through the narrowest hydrophilic region of the open pore (Albuquerque et al., 2009; Dani, 1989; Dani & Bertrand, 2007; Dani & Eisenman, 1987). When the linear sequences of homologous cationic nAChR and anionic channel domains are aligned, a proline residue in the anionic channel is found to be missing from the short intracellular segment between M1 and M2 of the nAChRs (as illustrated in Fig. 2A). Also near the inner mouth of the nAChR pore a negatively charged glutamate residue of the nAChR channel is missing from anionic channels, and a valine in M2 is replaced by a threonine in the channel lining of the nAChR (Galzi et al., 1992). When the amino acids of the anionic channel are inserted such that the absent proline is provided, the negatively charged glutamate is removed, and the polar threonine is replaced by valine, then the homomeric α7 nAChR is converted from cationic to anionic selectivity (Galzi et al., 1992).

Although sodium and potassium carry most of the nAChR current, calcium makes a significant contribution (Albuquerque et al., 2009; Dani, 2001; Dani & Mayer, 1995; Fucile, 2004; Vernino, Amador, Luetje, Patrick, & Dani, 1992; Vernino, Rogers, Radcliffe, & Dani, 1994). While nAChR activity causes depolarization, the divalent cation permeability plays an important physiological role by supplying ionic signals, including calcium (Bertrand, Galzi, Devillers-Thiery, Bertrand, & Changeux, 1993b; Dani & Bertrand, 2007; Decker & Dani, 1990; Gray, Rajan, Radcliffe, Yakehiro, & Dani, 1996; McGehee, Heath, Gelber, Devay, & Role, 1995; Vernino et al., 1992). The relative permeability of calcium to sodium estimated from permeability ratios is ~0.1 for muscle, ~2.0 for heteromeric neuronal, and ≥10 for homomeric α7 or the heteromeric α9/α10 nAChRs, which are expressed in cochlear hair cells (Bertrand et al., 1993b; Castro & Albuquerque, 1995; Fayuk & Yakel, 2005; Haghighi & Cooper, 2000; Lipovsek et al., 2014; Seguela, Wadiche, Dineley-Miller, Dani, & Patrick, 1993; Vernino et al., 1992). The higher calcium permeability of the α7 nAChRs arises from the arrangement of charged residues at the inner mouth of the ionic pore and polar residues in the outer part of the channel. These entrance vestibules form the transition from bulk solution to the narrow selectivity filter of the channel (Dani, 1986). For the nAChR, these vestibules have an overall net negative charge that enhances the cationic selectivity and contributes to the relatively high conductance of the nAChR channel. Substitution of the negatively charged glutamate residue found at the inner mouth of the α7 nAChRs by the neutral alanine residue suppresses calcium permeability (Bertrand et al., 1993b). Similarly, replacement of the α7 leucine at the extracellular entrance to the pore by threonine dramatically reduces the calcium permeability. However, substitution of the leucine by threonine at another polar ring of amino acids within the pore (position 247) did not alter divalent ionic selectivity, but altered agonist/antagonist relationships and aspects of desensitization (Bertrand et al., 1993a; Revah et al., 1991). These data illustrate the importance of particular conserved amino acids and the complex relationship between the structure of the pore and the resulting function.

The most basic conformational states of the nAChR are the closed state at rest, the open state, and the desensitized state. The kinetic rate at which the nicotinic receptor proceeds through the various conformational states and the selectivity with which it conducts cations in the open state depend on many factors, including the subunit composition. Therefore, the extensive nAChR diversity has the potential to produce many different responses to endogenous or exogenous agonists. The intensity of the membrane depolarization, the kinetics of gating activation, the rates of desensitization and recovery from desensitization, the size of the ionic signal, the pharmacology, and the regulatory controls of the ACh response all depend on the subunit composition of the nAChRs. In addition, the local environmental and regulatory factors influence the function of nAChRs. These influences include peptide transmitters, various protein kinases, the cytoskeleton, and calcium. Although calcium modulation can act intracellularly, nAChRs also are allosterically modulated by extracellular calcium, leading to dramatic changes in the channel opening probability (Amador & Dani, 1995; Mulle, Lena, & Changeux, 1992; Vernino et al., 1992). This modulation occurs over the physiological concentration range of external calcium. Therefore, high levels of neuronal activity that can diminish extracellular calcium (Wiest, Eagleman, King, & Montague, 2000) could cause a negative feedback that lowers the opening probability of nAChRs.

To add further complexity, the three basic conformational states (rest, open, and desensitized) do not account for the actual kinetic properties of nicotinic receptors. Rather, there are multiple conformations involved in the gating (Auerbach, 2014). Desensitization, in particular, encompasses many time constants (Steinbach & Sine, 1987). Thus, there may be short- and long-lived states of desensitization. Long exposures to low concentrations of agonist will favor deeper levels of desensitization, and this situation is often the case for smokers who maintain low concentrations of nicotine throughout the day (Dani & Heinemann, 1996; De Biasi & Dani, 2011).

5. NICOTINIC RECEPTOR RESPONSE TO NICOTINE FROM TOBACCO

Tobacco smoking activates and desensitizes nAChRs as 20–100 nM nicotine (Brody et al., 2006; Rose, Behm, Westman, & Coleman, 1999) reaches throughout the brain (Dani, Kosten, & Benowitz, 2014). Although many areas of the brain participate, nicotinic receptors of the midbrain dopamine (DA) area are particularly important during the initiation of the addiction process (Dani et al., 2014; De Biasi & Dani, 2011). On the midbrain DA and GABA neurons' cell bodies and postsynaptically, many of the nAChRs contain $\alpha4\beta2$ subunits that have a high affinity for nicotine. When nicotine first arrives in the midbrain DA area, it excites nAChRs, particularly the high-affinity $\alpha4\beta2^*$ nAChRs and related nAChR subtypes and, to a lesser degree, the lower-affinity α7* nAChRs. Activation of the presynaptic nAChRs (commonly but not exclusively α7* nAChRs) enhances the release of glutamate (Dani et al., 2000; Mansvelder & McGehee, 2000, 2002). Simultaneously, activity of postsynaptic (and somatic) α4β2* nAChRs depolarizes DA neurons leading to enhanced action potential firing (Zhang et al., 2009). This depolarization and firing of the DA neurons helps to relieve the divalent cation block of NMDA receptors and, thus, enables the NMDA receptors to participate in long-term synaptic potentiation of glutamatergic afferents onto midbrain dopamine neurons.

After the initial exposure to nicotine and potentiation of glutamatergic afferents onto the DA neurons, there is significant, but incomplete, desensitization of particularly the high-affinity α4β2* nAChR subtypes. Thus, α4β2 nAChRs that are predominantly expressed on GABA neurons are significantly desensitized, decreasing normal afferent cholinergic drives onto the local GABAergic circuitry. Consequently, there is decreased GABAergic inhibition onto the DA neurons. The DA neurons from the posterior ventral tegmental area that provide the main projection to the nucleus accumbens commonly express α6 and β3 subunits with the α4 and β2 subunits (Leslie, Mojica, & Reynaga, 2013; Zhao-Shea et al., 2011). At the low concentrations of nicotine achieved by smokers, the presence of the α6 subunit, particularly in α6α4β2* nAChRs, slows the rate and degree of desensitization seen with the higher affinity α4β2 nAChRs (Liu, Zhao-Shea, McIntosh, Gardner, & Tapper, 2012). Therefore, those α6-containing receptors are important to maintain the more prolonged activation of DA neurons caused by nicotine from tobacco (Leslie et al., 2013; Pidoplichko et al., 2004).

The desensitization of nAChRs arising from the relatively long-lived nicotine from tobacco has other immediate effects. The sensitive nAChR subtypes at cholinergic synapses are desensitized by prolonged nicotine. Thus, smoking will turn down the gain for activity arriving via nicotinic cholinergic synapses because fewer nAChRs will be able to respond to the synaptic ACh release. In summary, nicotine not only sends inappropriate information through the mesocorticolimbic DA system but also decreases the amplitude for normal nicotinic cholinergic information processing (Dani et al., 2014).

6. CONCLUSION

Nicotinic receptors of the brain share a basic fundamental property: they mediate a cationic conductance upon binding agonist. The tremendous diversity of nAChR subtypes provides the structural and functional flexibility necessary for them to play multiple, varied roles (Zoli et al., 2014). Broad, sparse cholinergic projects throughout the brain ensure that nicotinic mechanisms modulate the neuronal excitability of relatively wide circuits (Albuquerque et al., 2009; Dani & Bertrand, 2007). Presynaptic and preterminal nicotinic receptors regulate the release of many neurotransmitters. Postsynaptic nAChRs contribute a neuro-anatomically varied, but usually small, component of fast excitatory transmission. Nonsynaptic nAChRs modulate many neurotransmitter systems by influencing input impedance, neuronal set point, and neuronal excitability. While this review focused on nAChRs of the mammalian brain, nAChRs also are widely distributed and play even more diverse roles in the peripheral nervous system and in non-neuronal tissue. Thus, from a rather simple, basic underlying function (i.e., cationic permeability in response to agonist), nAChRs serve an extraordinary array of roles.

ACKNOWLEDGMENT

Research in the laboratory and effort for this review was supported by the following NIH grants: NIDA DA09411 and NINDS NS21229.

REFERENCES

Albuquerque, E. X., Pereira, E. F., Alkondon, M., & Rogers, S. W. (2009). Mammalian nicotinic acetylcholine receptors: From structure to function. *Physiological Reviews*, *89*, 73–120.

Amador, M., & Dani, J. A. (1995). Mechanism for modulation of nicotinic acetylcholine receptors that can influence synaptic transmission. *The Journal of Neuroscience: The Official Journal of the Society for Neuroscience*, *15*, 4525–4532.

Auerbach, A. (2014). Agonist activation of a nicotinic acetylcholine receptor. *Neuropharmacology*, *96*, 150–156.

Bertrand, D., Galzi, J. L., Devillers-Thiery, A., Bertrand, S., & Changeux, J. P. (1993b). Mutations at two distinct sites within the channel domain M2 alter calcium permeability of neuronal alpha 7 nicotinic receptor. *Proceedings of the National Academy of Sciences of the United States of America*, *90*, 6971–6975.

Bertrand, D., Galzi, J. L., Devillers-Thiery, A., Bertrand, S., & Changeux, J. P. (1993a). Stratification of the channel domain in neurotransmitter receptors. *Current Opinion in Cell Biology*, *5*, 688–693.

Brody, A. L., Mandelkern, M. A., London, E. D., Olmstead, R. E., Farahi, J., Scheibal, D., et al. (2006). Cigarette smoking saturates brain alpha 4 beta 2 nicotinic acetylcholine receptors. *Archives of General Psychiatry*, *63*, 907–915.

Castro, N. G., & Albuquerque, E. X. (1995). Alpha-Bungarotoxin-sensitive hippocampal nicotinic receptor channel has a high calcium permeability. *Biophysical Journal*, *68*, 516–524.

Celie, P. H., van Rossum-Fikkert, S. E., van Dijk, W. J., Brejc, K., Smit, A. B., & Sixma, T. K. (2004). Nicotine and carbamylcholine binding to nicotinic acetylcholine receptors as studied in AChBP crystal structures. *Neuron*, *41*, 907–914.

Dani, J. A. (1986). Ion-channel entrances influence permeation. Net charge, size, shape, and binding considerations. *Biophysical Journal, 49*, 607–618.
Dani, J. A. (1989). Open channel structure and ion binding sites of the nicotinic acetylcholine receptor channel. *The Journal of Neuroscience: The Official Journal of the Society for Neuroscience, 9*, 884–892.
Dani, J. A. (2001). Overview of nicotinic receptors and their roles in the central nervous system. *Biological Psychiatry, 49*, 166–174.
Dani, J. A., & Balfour, D. J. (2011). Historical and current perspective on tobacco use and nicotine addiction. *Trends in Neurosciences, 34*, 383–392.
Dani, J. A., & Bertrand, D. (2007). Nicotinic acetylcholine receptors and nicotinic cholinergic mechanisms of the central nervous system. *Annual Review of Pharmacology and Toxicology, 47*, 699–729.
Dani, J. A., & Eisenman, G. (1987). Monovalent and divalent cation permeation in acetylcholine receptor channels. Ion transport related to structure. *The Journal of General Physiology, 89*, 959–983.
Dani, J. A., & Heinemann, S. (1996). Molecular and cellular aspects of nicotine abuse. *Neuron, 16*, 905–908.
Dani, J. A., Kosten, T. R., & Benowitz, N. L. (2014). The pharmacology of nicotine and tobacco. In R. K. Ries, D. A. Fiellin, S. C. Miller, & R. Saitz (Eds.), *The ASAM principles of addiction medicine* (5th ed., pp. 201–216). Philadelphia, PA: Wolters Kluwer, (chapter 12).
Dani, J. A., & Mayer, M. L. (1995). Structure and function of glutamate and nicotinic acetylcholine receptors. *Current Opinion in Neurobiology, 5*, 310–317.
Dani, J. A., Radcliffe, K. A., & Pidoplichko, V. I. (2000). Variations in desensitization of nicotinic acetylcholine receptors from hippocampus and midbrain dopamine areas. *European Journal of Pharmacology, 393*, 31–38.
De Biasi, M., & Dani, J. A. (2011). Reward, addiction, withdrawal to nicotine. *Annual Review of Neuroscience, 34*, 105–130.
Decker, E. R., & Dani, J. A. (1990). Calcium permeability of the nicotinic acetylcholine receptor: The single-channel calcium influx is significant. *The Journal of Neuroscience: The Official Journal of the Society for Neuroscience, 10*, 3413–3420.
Fasoli, F., & Gotti, C. (2015). Structure of neuronal nicotinic receptors. *Current Topics in Behavioral Neurosciences, 23*, 1–17.
Fayuk, D., & Yakel, J. L. (2005). Ca2+ permeability of nicotinic acetylcholine receptors in rat hippocampal CA1 interneurones. *The Journal of Physiology, 566*, 759–768.
Fenster, C. P., Rains, M. F., Noerager, B., Quick, M. W., & Lester, R. A. (1997). Influence of subunit composition on desensitization of neuronal acetylcholine receptors at low concentrations of nicotine. *The Journal of Neuroscience: The Official Journal of the Society for Neuroscience, 17*, 5747–5759.
Fucile, S. (2004). Ca2+ permeability of nicotinic acetylcholine receptors. *Cell Calcium, 35*, 1–8.
Galzi, J. L., Devillers-Thiery, A., Hussy, N., Bertrand, S., Changeux, J. P., & Bertrand, D. (1992). Mutations in the channel domain of a neuronal nicotinic receptor convert ion selectivity from cationic to anionic. *Nature, 359*, 500–505.
Galzi, J. L., Revah, F., Black, D., Goeldner, M., Hirth, C., & Changeux, J. P. (1990). Identification of a novel amino acid alpha-tyrosine 93 within the cholinergic ligands-binding sites of the acetylcholine receptor by photoaffinity labeling. Additional evidence for a three-loop model of the cholinergic ligands-binding sites. *The Journal of Biological Chemistry, 265*, 10430–10437.
Giniatullin, R., Nistri, A., & Yakel, J. L. (2005). Desensitization of nicotinic ACh receptors: Shaping cholinergic signaling. *Trends in Neurosciences, 28*, 371–378.

Gray, R., Rajan, A. S., Radcliffe, K. A., Yakehiro, M., & Dani, J. A. (1996). Hippocampal synaptic transmission enhanced by low concentrations of nicotine. *Nature, 383*, 713–716.

Haghighi, A. P., & Cooper, E. (2000). A molecular link between inward rectification and calcium permeability of neuronal nicotinic acetylcholine alpha3beta4 and alpha4beta2 receptors. *The Journal of Neuroscience: The Official Journal of the Society for Neuroscience, 20*, 529–541.

Hilf, R. J., & Dutzler, R. (2008). X-ray structure of a prokaryotic pentameric ligand-gated ion channel. *Nature, 452*, 375–379.

Karlin, A. (2002). Emerging structure of the nicotinic acetylcholine receptors. *Nature Reviews Neuroscience, 3*, 102–114.

Kracun, S., Harkness, P. C., Gibb, A. J., & Millar, N. S. (2008). Influence of the M3–M4 intracellular domain upon nicotinic acetylcholine receptor assembly, targeting and function. *British Journal of Pharmacology, 153*, 1474–1484.

Lee, W. Y., & Sine, S. M. (2005). Principal pathway coupling agonist binding to channel gating in nicotinic receptors. *Nature, 438*, 243–247.

Leslie, F. M., Mojica, C. Y., & Reynaga, D. D. (2013). Nicotinic receptors in addiction pathways. *Molecular Pharmacology, 83*, 753–758.

Lewis, A. S., & Picciotto, M. R. (2013). High-affinity nicotinic acetylcholine receptor expression and trafficking abnormalities in psychiatric illness. *Psychopharmacology, 229*, 477–485.

Lipovsek, M., Fierro, A., Perez, E. G., Boffi, J. C., Millar, N. S., Fuchs, P. A., et al. (2014). Tracking the molecular evolution of calcium permeability in a nicotinic acetylcholine receptor. *Molecular Biology and Evolution, 31*, 3250–3265.

Liu, L., Zhao-Shea, R., McIntosh, J. M., Gardner, P. D., & Tapper, A. R. (2012). Nicotine persistently activates ventral tegmental area dopaminergic neurons via nicotinic acetylcholine receptors containing alpha4 and alpha6 subunits. *Molecular Pharmacology, 81*, 541–548.

Luo, J., Taylor, P., Losen, M., de Baets, M. H., Shelton, G. D., & Lindstrom, J. (2009). Main immunogenic region structure promotes binding of conformation-dependent myasthenia gravis autoantibodies, nicotinic acetylcholine receptor conformation maturation, and agonist sensitivity. *The Journal of Neuroscience: The Official Journal of the Society for Neuroscience, 29*, 13898–13908.

Mansvelder, H. D., & McGehee, D. S. (2000). Long-term potentiation of excitatory inputs to brain reward areas by nicotine. *Neuron, 27*, 349–357.

Mansvelder, H. D., & McGehee, D. S. (2002). Cellular and synaptic mechanisms of nicotine addiction. *Journal of Neurobiology, 53*, 606–617.

McGehee, D. S., Heath, M. J., Gelber, S., Devay, P., & Role, L. W. (1995). Nicotine enhancement of fast excitatory synaptic transmission in CNS by presynaptic receptors. *Science, 269*, 1692–1696.

McGehee, D. S., & Role, L. W. (1995). Physiological diversity of nicotinic acetylcholine receptors expressed by vertebrate neurons. *Annual Review of Physiology, 57*, 521–546.

McKay, B. E., Placzek, A. N., & Dani, J. A. (2007). Regulation of synaptic transmission and plasticity by neuronal nicotinic acetylcholine receptors. *Biochemical Pharmacology, 74*, 1120–1133.

Mulle, C., Lena, C., & Changeux, J. P. (1992). Potentiation of nicotinic receptor response by external calcium in rat central neurons. *Neuron, 8*, 937–945.

Papke, R. L. (2014). Merging old and new perspectives on nicotinic acetylcholine receptors. *Biochemical Pharmacology, 89*, 1–11.

Pidoplichko, V. I., Noguchi, J., Areola, O. O., Liang, Y., Peterson, J., Zhang, T., et al. (2004). Nicotinic cholinergic synaptic mechanisms in the ventral tegmental area contribute to nicotine addiction. *Learning & Memory, 11*, 60–69.

Pollock, V. V., Pastoor, T., Katnik, C., Cuevas, J., & Wecker, L. (2009). Cyclic AMP-dependent protein kinase A and protein kinase C phosphorylate alpha4beta2 nicotinic receptor subunits at distinct stages of receptor formation and maturation. *Neuroscience*, *158*, 1311–1325.

Popot, J. L., Demel, R. A., Sobel, A., Van Deenen, L. L., & Changeux, J. P. (1978). Interaction of the acetylcholine (nicotinic) receptor protein from Torpedo marmorata electric organ with monolayers of pure lipids. *European Journal of Biochemistry*, *85*, 27–42.

Quick, M. W., & Lester, R. A. (2002). Desensitization of neuronal nicotinic receptors. *Journal of Neurobiology*, *53*, 457–478.

Revah, F., Bertrand, D., Galzi, J. L., Devillers-Thiery, A., Mulle, C., Hussy, N., et al. (1991). Mutations in the channel domain alter desensitization of a neuronal nicotinic receptor. *Nature*, *353*, 846–849.

Rose, J. E., Behm, F. M., Westman, E. C., & Coleman, R. E. (1999). Arterial nicotine kinetics during cigarette smoking and intravenous nicotine administration: Implications for addiction. *Drug and Alcohol Dependence*, *56*, 99–107.

Seguela, P., Wadiche, J., Dineley-Miller, K., Dani, J. A., & Patrick, J. W. (1993). Molecular cloning, functional properties, and distribution of rat brain alpha 7: A nicotinic cation channel highly permeable to calcium. *The Journal of Neuroscience: The Official Journal of the Society for Neuroscience*, *13*, 596–604.

Sine, S. M. (2002). The nicotinic receptor ligand binding domain. *Journal of Neurobiology*, *53*, 431–446.

Sine, S. M., & Engel, A. G. (2006). Recent advances in Cys-loop receptor structure and function. *Nature*, *440*, 448–455.

Steinbach, J. H., & Sine, S. M. (1987). Function of nicotinic acetylcholine receptors. *Society of General Physiologists Series*, *41*, 19–42.

Taly, A., Corringer, P. J., Guedin, D., Lestage, P., & Changeux, J. P. (2009). Nicotinic receptors: Allosteric transitions and therapeutic targets in the nervous system. *Nature Reviews Drug Discovery*, *8*, 733–750.

Tzartos, S. J., Kokla, A., Walgrave, S. L., & Conti-Tronconi, B. M. (1988). Localization of the main immunogenic region of human muscle acetylcholine receptor to residues 67–76 of the alpha subunit. *Proceedings of the National Academy of Sciences of the United States of America*, *85*, 2899–2903.

Unwin, N. (2005). Refined structure of the nicotinic acetylcholine receptor at 4A resolution. *Journal of Molecular Biology*, *346*, 967–989.

Unwin, N. (2013). Nicotinic acetylcholine receptor and the structural basis of neuromuscular transmission: Insights from Torpedo postsynaptic membranes. *Quarterly Reviews of Biophysics*, *46*, 283–322.

Unwin, N., & Fujiyoshi, Y. (2012). Gating movement of acetylcholine receptor caught by plunge-freezing. *Journal of Molecular Biology*, *422*, 617–634.

Unwin, N., Miyazawa, A., Li, J., & Fujiyoshi, Y. (2002). Activation of the nicotinic acetylcholine receptor involves a switch in conformation of the alpha subunits. *Journal of Molecular Biology*, *319*, 1165–1176.

Vernino, S., Amador, M., Luetje, C. W., Patrick, J., & Dani, J. A. (1992). Calcium modulation and high calcium permeability of neuronal nicotinic acetylcholine receptors. *Neuron*, *8*, 127–134.

Vernino, S., Rogers, M., Radcliffe, K. A., & Dani, J. A. (1994). Quantitative measurement of calcium flux through muscle and neuronal nicotinic acetylcholine receptors. *The Journal of Neuroscience: The Official Journal of the Society for Neuroscience*, *14*, 5514–5524.

Wiest, M. C., Eagleman, D. M., King, R. D., & Montague, P. R. (2000). Dendritic spikes and their influence on extracellular calcium signaling. *Journal of Neurophysiology*, *83*, 1329–1337.

Williams, D. K., Stokes, C., Horenstein, N. A., & Papke, R. L. (2011). The effective opening of nicotinic acetylcholine receptors with single agonist binding sites. *The Journal of General Physiology*, *137*, 369–384.

Wooltorton, J. R., Pidoplichko, V. I., Broide, R. S., & Dani, J. A. (2003). Differential desensitization and distribution of nicotinic acetylcholine receptor subtypes in midbrain dopamine areas. *The Journal of Neuroscience: The Official Journal of the Society for Neuroscience*, *23*, 3176–3185.

Zhang, T., Zhang, L., Liang, Y., Siapas, A. G., Zhou, F. M., & Dani, J. A. (2009). Dopamine signaling differences in the nucleus accumbens and dorsal striatum exploited by nicotine. *The Journal of Neuroscience: The Official Journal of the Society for Neuroscience*, *29*, 4035–4043.

Zhao-Shea, R., Liu, L., Soll, L. G., Improgo, M. R., Meyers, E. E., McIntosh, J. M., et al. (2011). Nicotine-mediated activation of dopaminergic neurons in distinct regions of the ventral tegmental area. *Neuropsychopharmacology*, *36*, 1021–1032.

Zoli, M., Pistillo, F., & Gotti, C. (2014). Diversity of native nicotinic receptor subtypes in mammalian brain. *Neuropharmacology*, *96*, 302–311.

PART 2

Schizophrenia

CHAPTER TWO

The Role of Nicotine in Schizophrenia

Robert E. Featherstone[1], Steven J. Siegel
Translational Neuroscience Program, Department of Psychiatry, University of Pennsylvania, Philadelphia, Pennsylvania, USA
[1]Corresponding author: e-mail address: fear@upenn.edu

Contents

Abstract

Schizophrenia is associated with by severe disruptions in thought, cognition, emotion, and behavior. Patients show a marked increase in rates of smoking and nicotine dependence relative to nonaffected individuals, a finding commonly ascribed to the potential ameliorative effects of nicotine on symptom severity and cognitive impairment. Indeed, many studies have demonstrated improvement in patients following the administration of nicotine. Such findings have led to an increased emphasis on the development of therapeutic agents to target the nicotinic system as well as increasing the impetus to understand the genetic basis for nicotinic dysfunction in schizophrenia. The goal of this review article is to provide a critical summary of evidence for the role of the nicotinic system in schizophrenia. The first part will review the role of nicotine in normalization of primary dysfunctions and endophenotypical changes found in schizophrenia. The second part will provide a summary of genetic evidence linking polymorphisms in

International Review of Neurobiology, Volume 124
ISSN 0074-7742
http://dx.doi.org/10.1016/bs.irn.2015.07.002

nicotinic receptor genes to smoking and schizophrenia. The third part will summarize attempts to treat schizophrenia using agents specifically targeting nicotinic and nicotinic receptor subtypes. Although currently available antipsychotic treatments are generally able to manage some aspects of schizophrenia (e.g., positive symptoms) they fail to address several other critically effected aspects of the disease. As such, the search for novel mechanisms to treat this disease is necessary.

1. INTRODUCTION

1.1 Schizophrenia

Schizophrenia is a characterized by profound disruption in thought, cognition, emotion, and behavior. Symptoms typically emerge during late adolescence or early adulthood and severely disrupt the attainment of developmental milestones typical of people this age. Not surprisingly, the economic burden of this disease is quite substantial. For example, in 2002, the estimated cost of schizophrenia in the United States was estimated at $62.7 billion (Wu et al., 2005). More effective methods for treating schizophrenia would greatly reduce this burden.

Schizophrenia has been conceptualized as consisting of positive and negative symptoms (Crow, 1980). Positive symptoms reflect disease associated change in thinking and behavior that are seen in patients but not in healthy individuals, and include hallucinations and delusions (psychosis), as well as thought disorders. Negative symptoms reflect absence of behaviors and emotional states in patients that are part of normal everyday experience for healthy individuals, and include loss of pleasure (anhedonia), social impairment, and blunted affect. Current antipsychotic treatments, which primarily act upon dopamine D2 receptors (Kapur & Remington, 2001), are generally able to manage positive symptoms but much less able to address negative symptoms. This is a serious limitation since negative symptoms are highly correlated with reduced daily function in patients. Additionally, schizophrenia is associated with profound impairment across a range of cognitive domains, including long-term memory, working memory, and attention. Indeed, patients with schizophrenia often score at least two standard deviations below the mean of healthy controls (Saykin et al., 1991), suggesting severe dysfunction. Reduced cognitive function is related to poorer social and vocational outcome, poorer psychosocial rehabilitation, and difficulties maintaining independent living (Green, 1996). Despite increased awareness of the importance of cognition in schizophrenia, there are currently no therapeutic agents capable of producing meaningful improvement in this area. As such, the development of therapeutic interventions capable of addressing impaired cognition would constitute a major breakthrough in the treatment of schizophrenia and would help to ease the burden this disease places on individuals, families, and society.

1.2 Nicotine and Schizophrenia

Initial interest in the nicotinic system in schizophrenia stemmed from reports of higher rates of smoking in patients with schizophrenia relative to nonpatients (Goff, Henderson, & Amico, 1992; Hughes, Hatsukami, Mitchell, & Dahlgren, 1986; Kelly & McCreadie, 1999). Increased rates of smoking are relatively specific to schizophrenia and are not as frequently and as extensively observed in other psychiatric disorders. Several aspects of the relationship between smoking and schizophrenia point to a primary dysfunction in the nicotinic acetylcholine receptor system and highlight the likelihood that this dysfunction could play a central role in the development of the disease. Higher rates of smoking are observed in unaffected relatives of patients, suggesting that the physiological factors underlying smoking are inherited as part of an underlying genetic vulnerability to schizophrenia (Lyons et al., 2002). Most (90%) patients who smoke typically begin smoking prior to the onset of schizophrenia (Kelly & McCreadie, 1999) and higher rates of smoking are observed in adolescents who later develop schizophrenia compared to those that do not (de Leon, 1996). This evidence argues against the notion that increased rates of smoking stem from a desire to alleviate the effects of schizophrenia symptoms. Increased rates of smoking are also associated with earlier disease onset (Goff et al., 1992), increased disease severity (Herran et al., 2000), poorer prognosis (Aguilar, Gurpegui, Diaz, & de Leon, 2005), poorer quality of life (Dixon et al., 2007), and greater number of hospital admissions (Aguilar et al., 2005; Goff et al., 1992; Kelly & McCreadie, 1999; Patkar et al., 2002), implying that the increased need to smoke may be closely associated with key regulators of disease onset and progression. Several key early studies showed that smoking was associated with symptom remediation, demonstrating that normalization of a presumed nicotinic imbalance could have a significant positive impact on disease prognosis. Of particular interest in this regard has been findings suggesting a positive impact of nicotine on various aspects of cognitive performance. Such findings have led to an increased emphasis on the development of therapeutic agents to target the nicotinic system as well as increasing the impetus to understand the genetic basis for nicotinic dysfunction in schizophrenia.

2. EFFECTS OF NICOTINE ON SCHIZOPHRENIA ENDOPHENOTYPES AND SYMPTOM DOMAINS

When asked why they smoke, patients often claim that smoking helps alleviate the effects of symptoms and helps them to concentrate (Glynn & Sussman, 1990). This has led to several studies that have assessed the effects of smoking or nicotine administration on symptom severity and other aspects of schizophrenia pathology. Understanding how smoking can benefit symptoms in patients can provide insight into the potential mechanisms that underlie symptom manifestation. Moreover, such studies can help to understand what could be expected of a therapeutic agent capable of mimicking the effects of nicotine.

2.1 Electrophysiological Measures

2.1.1 P50 Sensory Gating

Some of the first experimental evidence for a possible role of nicotine in managing schizophrenia symptomology came from studies showing restoration of sensory gating in patients following administration of nicotine. Averaging EEG responses to sensory stimuli across repeated trials reveals a stereotypical pattern of time-locked voltage change, referred to as the "event-related potential" or ERP. When two stimuli are presented within close temporal proximity to one another the ERP response to the second stimuli (S2) will show attenuation in amplitude relative to the first (S1), a phenomenon termed sensory gating or suppression. A ratio measure consisting of the amplitude of S2 divided by the amplitude of S1 is used to quantify sensory gating, with 0 representing complete gating and 1 representing no gating (S2 = S1). Although it is now understood that gating is common across all ERP components, including both the later negative response (N1) and positive response (P300), early studies on sensory gating in schizophrenia focused almost exclusively on the P50 response. Reductions in P50 gating and/or P50 amplitude have been widely reported in schizophrenia (Adler et al., 1982; Freedman, Adler, Waldo, Pachtman, & Franks, 1983; Judd, McAdams, Budnick, & Braff, 1992; Patterson et al., 2008; Siegel, Waldo, Mizner, Adler, & Freedman, 1984). The overall effect size for schizophrenia across studies as shown in meta-analyses appears to be large (0.85) (Bramon, Rabe-Hesketh, Sham, Murray, & Frangou, 2004), suggesting that patients are markedly different on this measure than controls. It should be noted, however, that several studies have failed to show P50 gating deficits in patients (Arnfred, Chen, Glenthoj, & Hemmingsen, 2003; Bachmann et al., 2010; Turetsky, Bilker, Siegel, Kohler, & Gur, 2009).

2.1.1.1 Effects of Nicotine on P50 Gating

Several lines of evidence have linked disease-related deficits in P50 gating and amplitude to dysfunction in the nicotinic system. Cigarette smoking or acute nicotine administration significantly improves P50 gating in patients with schizophrenia (Adler, Hoffer, Wiser, & Freedman, 1993; Adler et al., 1982; Griffith et al., 1998). Likewise, administration of nicotinic receptor agonists and/or antagonists has been shown to alter sensory gating in both rodents and humans (Amann, Phillips, Halene, & Siegel, 2008; Metzger, Maxwell, Liang, & Siegel, 2007; Olincy et al., 2006; Rudnick et al., 2010; Simosky, Stevens, Kem, & Freedman, 2001; Stevens, Kem, Mahnir, & Freedman, 1998). Both P50 amplitude and sensory gating of the P50 have been localized to the hippocampus (Bak, Glenthoj, Rostrup, Larsson, & Oranje, 2011; Bak, Rostrup, Larsson, Glenthoj, & Oranje, 2014; Korzyukov et al., 2007; Tregellas et al., 2007; Williams, Nuechterlein, Subotnik, & Yee, 2011), a brain region that receives widespread cholinergic input from the medial septum (Dani & Bertrand, 2007; Dutar, Bassant, Senut, & Lamour, 1995). In rats, removal of cholinergic input to the hippocampus disrupts normal sensory gating of the P20 component (equivalent to P50 in humans), a loss that can be restored by administration of nicotine (Bickford & Wear, 1995; Stevens & Wear, 1997). Mice with reduced nicotinic receptors, especially α7, show reduced P20/N40 gating (Stevens et al., 1996; Stevens & Wear, 1997). Given the prominence that sensory gating has played in guiding the study of the nicotinic system in schizophrenia and in generating theories about schizophrenia, it is worth discussing the extent to which sensory gating deficits relate to the constellation of deficits observed in the disease.

2.1.1.2 P50 Gating as a Predictor of Disease-Related Symptoms

The finding that gating deficits can be observed in unaffected family members of patients demonstrates that, for the most part, reduced sensory gating likely does not strongly predict symptom presence or severity (Turetsky, Dent, Jaeger, & Zukin, 2012). The relationship between P50 gating deficits and symptoms has been formally assessed in several studies. In regards to negative symptoms, no differences in sensory gating were observed between patients showing primarily negative symptoms versus those showing non-negative symptoms (Adler et al., 1990; Santos et al., 2010). Likewise, dividing patients based on high or low levels of P50 sensory gating did not predict severity of negative symptoms (Erwin, Turetsky, Moberg, Gur, & Gur, 1998). Several studies have also failed to discover significant correlations between P50 gating and symptom severity (Light, Geyer, Clementz, Cadenhead, & Braff, 2000; Thoma et al., 2005; Yee, Nuechterlein, Morris, & White, 1998). In contrast, a few studies have shown a relationship between P50 gating and negative symptoms (Louchart-de la Chapelle et al., 2005; Micoulaud-Franchi et al., 2014; Ringel et al., 2004). The discrepancies between findings may relate to subtle differences in patient populations or variance in underlying neurological changes. For example, a significant relationship has been reported between gating and negative symptoms using magnetoencephalography (MEG) centered in the frontal temporal cortex despite finding no such relationship using standard nonlocalized EEG (Thoma et al., 2005). Likewise, a number of studies have failed to show a relationship between sensory gating and level of positive symptoms (Arnfred & Chen, 2004; Erwin et al., 1998; Light et al., 2000; Ringel et al., 2004). In contrast, some studies have found alterations in sensory gating in subjects with ongoing auditory hallucinations or perceptual abnormalities (Micoulaud-Franchi et al., 2014; Smith et al., 2013; Smucny, Olincy, Eichman, Lyons, & Tregellas, 2013). Smith showed that the relationship was most predictive for trait rather than measures of hallucination. Increased depression and anxiety scores have been found in patients with low P50 gating (Micoulaud-Franchi et al., 2014; Yee et al., 1998). In regards to cognition, patients with reduced sensory gating show greater attentional dysfunction on the continuous performance task (CPT) than patients with normal gating (Erwin et al., 1998), while patients identified based on impaired attention on the scale for assessment of negative symptoms show reduced gating (Yee et al., 1998). Likewise gating ratio shows significant correlations with attentional performance (Cullum et al., 1993; Smith et al., 2010; Smucny et al., 2013). In contrast, some studies have reported impaired cognition (attention) in patients that do not show P50 deficits (Adler et al., 1990; Guterman & Josiassen, 1994) and one study reported no association between P50 gating and several cognitive domains (Sanchez-Morla et al., 2013). There appears to be little association between P50 gating and long-term memory (Cullum et al., 1993; Erwin et al., 1998; Hsieh et al., 2004; Sanchez-Morla et al., 2013; Smith et al., 2010). More significant associations have been reported for N100 amplitude and gating and memory (Hsieh et al., 2004; Smith et al., 2010).

At present, there is limited evidence to suggest that sensory gating in patients will be an effective screen for identifying agents capable of addressing either positive or negative symptoms in patients. There is close overlap between the sensory gating and attentional impairments, but not long-term memory. Differences between patients in regards to antipsychotic status, smoking status, or symptom remission could contribute to some of the discrepancies across studies. The P50 component is a low amplitude response that is difficult to detect relative to background noise. As such, several factors can influence size of S1 response and gating ratio, including filter setting, method of quantifying amplitude of deflection (increase over baseline vs. increase over preceding trough), differences in criteria used for trial inclusion and artifact rejection, size of selection window, intensity of auditory stimulus, and location where study was conducted (de Wilde, Bour, Dingemans, Koelman, & Linszen, 2007; Patterson et al., 2008). Although effect sizes for P50 gating in schizophrenia tend to be large, there is some discrepancy across research groups (de Wilde et al., 2007). Similarly, P50 gating has been criticized for having poor test–retest reliability (Lamberti, Schwarzkopf, Boutros, Crilly, & Martin, 1993). Despite these caveats, P50 sensory gating has provided a measure of dysfunction in sensory processing in schizophrenia that is highly translatable across species and which has played an important role in highlighting a potential role for nicotine in the treatment of some aspects of schizophrenia.

2.1.2 Mismatch Negativity

More recently, research focus has shifted to later occurring ERP phenomena elicited during stimulus preparations requiring more elaborate stimulus processing. These measures are interesting in that they better predict disease correlates and may offer a more direct assessment of cognitive processing than can be obtained with sensory gating. When a rare occurring stimulus is presented within a series of a more frequent stimulus the resulting waveform is much larger than that produced by the standard stimulus, a phenomenon referred to as mismatch negativity (MMN). MMN was originally shown using stimuli that differed according to either pitch or intensity (Naatanen, Gaillard, & Mantysalo, 1978), but it can also be seen using stimuli that differ based on duration and latency, and even in response to the omission of a stimulus (Naatanen, Jacobsen, & Winkler, 2005). Much evidence suggests that the difference in response is not simply due to gating or suppression of the N1 response to the standard stimulus. For example, the N1 and MMN have different latencies and durations (Naatanen et al., 2005), and are derived from different generators in the brain (Javitt, Steinschneider, Schroeder, & Arezzo, 1996; Naatanen et al., 2005). Early explanations suggested that MMN occurred due to the detection of a mismatch between the memory trace formed by the frequent stimulus and current sensory input from the deviant stimulus (Naatanen et al., 1978). As such, MMN is thought to reflect primitive sensory memory.

2.1.2.1 Effects of Nicotine on MMN

Studies assessing the effects of nicotine on MMN amplitude have produced mixed results. While several studies have demonstrated increased MMN amplitude following acute administration of nicotine (gum or patch) to healthy subjects (Baldeweg, Wong, & Stephan, 2006; Fisher et al., 2010; Harkrider & Hedrick, 2005; Knott, Impey, et al., 2014; Martin, Davalos, & Kisley, 2009), other studies have shown no change following nicotine (Inami, Kirino, Inoue, & Arai, 2005; Inami, Kirino, Inoue, Suzuki, & Arai, 2007; Knott, Blais, et al., 2006; Knott, Heenan, et al., 2011; Mathalon et al., 2014). Studies that have specifically addressed the effects of nicotine on amelioration of disease-related reductions in MMN amplitude in patients have likewise shown inconsistent results (Dulude, Labelle, & Knott, 2010; Fisher et al., 2012). Dulude and colleagues (2010) found that acute administration of nicotine normalized disease-related decrements in duration but not frequency MMN. In contrast, Fisher et al. (2012) found no effect of nicotine on duration, intensity, or frequency MMN in patients with persistent auditory verbal hallucinations. Several differences between the procedures used in each study could account for these discrepancies. For example, Dulude used a two stimulus (standard and deviant) preparation with infrequent presentation of the deviant (5% of trials) that may have been more optimal for producing MMN, and used a higher dose of nicotine. Deviant probability was higher in the Fisher study (17% of trials) with frequency, duration, and intensity deviants all being presented within the same session. Consistent with the Dulude study, nicotine appears to be able to block ketamine-induced attenuation of MMN (Knott et al., 2012).

In contrast to MMN amplitude, several studies have shown shortened MMN latency following acute nicotine administration in healthy controls (Engeland, Mahoney, Mohr, Ilivitsky, & Knott, 2002a,2002b; Inami et al., 2005, 2007; Knott, Blais, et al., 2006). Increased MMN latency presumably reflects quicker processing of differences between the standard and novel stimulus. Studies in patients have reported shorter (Domjan, Csifcsak, Drotos, Janka, & Szendi, 2012; Horton, Millar, Labelle, & Knott, 2011), longer (Kathmann, Wagner, Rendtorff, & Engel, 1995; Shutara et al., 1996) or no change (Murphy et al., 2013) in MMN latency. Two studies have shown that acute nicotine does not alter MMN latency for frequency or duration deviants in patients (Fisher et al., 2012; Inami et al., 2007), while nicotine shortened MMN latency for intensity deviants (Fisher et al., 2012). The significance of a change in MMN latency is not well understood. Increased MMN latency has been linked to poorer working memory (Toyomaki et al., 2008), but not degree of symptom severity (Horton et al., 2011; Kathmann et al., 1995), although see (Grzella et al., 2001).

2.1.2.2 MMN as a Predictor of Disease-Related Symptoms

Numerous studies have shown reduced MMN in patients with schizophrenia (Javitt, Doneshka, Zylberman, Ritter, & Vaughan, 1993; Shelley et al., 1991; Umbricht & Krljes, 2005). The mean effect size across studies is large (0.99) suggesting that impaired MMN is a reliable feature of schizophrenia (Umbricht & Krljes, 2005). MMN deficits are strongly associated with global functioning and degree of independent living (Jahshan et al., 2012; Light & Braff, 2005a,2005b; Rissling et al., 2014; Wynn, Sugar, Horan, Kern, & Green, 2010). Additionally, MMN appears to be associated with social function (Bar-Haim, Marshall, Fox, Schorr, & Gordon-Salant, 2003; Wynn et al., 2010), language ability (Kawakubo & Kasai, 2006; Revheim et al., 2014; Turetsky et al., 2009), and cognition (Baldeweg, Klugman, Gruzelier, & Hirsch, 2004; Rissling et al., 2013, 2014). In contrast, the majority of studies have not shown strong associations between degree of negative or positive symptomology and MMN (Light & Braff, 2005a,2005b; Umbricht & Krljes, 2005), although a handful of reports have linked MMN deficits to severity of negative symptoms (Catts et al., 1995; Javitt, Shelley, & Ritter, 2000; Turetsky et al., 2009). Dysfunction in the glutamate system is now widely regarded as one of the most likely etiological factors for the development of schizophrenia (Goff & Coyle, 2001; Jentsch & Roth, 1999; Olney & Farber, 1995; Walsh et al., 2008). Numerous studies have shown that administration of NMDA receptor antagonists can disrupt MMN, both in humans and laboratory animals (Ehrlichman, Maxwell, Majumdar, & Siegel, 2008; Gunduz-Bruce et al., 2012; Javitt et al., 1996; Kreitschmann-Andermahr et al., 2001; Umbricht et al., 2000). This differs sharply from P50 sensory gating which seems relatively immune from the effects of NMDA antagonists such as ketamine (Oranje, Gispen-de Wied, Verbaten, & Kahn, 2002; van Berckel, Oranje, van Ree, Verbaten, & Kahn, 1998). Thus, to the extent that schizophrenia is due to NMDA dysfunction, MMN appears to be better able to adequately detect these dysfunctions than measures based on change in P50.

2.1.3 P300

MMN reflects a relatively passive response to change in stimuli that can occur even when the subject is not required to attend to those differences. In contrast, the ERP contains a late occurring positive deflection occurring around 300 ms (P300) that is elicited by active processing of sensory information. In the original demonstration, subjects were presented with stimuli pairs consisting of a cue and test stimulus and were told to guess which test stimulus would follow the cue (Sutton, Braren, Zubin, & John, 1965). One cue was always followed with the same stimulus (certain condition) while a second cue was followed with one of two test stimuli (tone or light) presented randomly (uncertain condition). Subjects showed a larger positive deflection in the ERP during uncertain trials than during certain trials, and this occurred roughly 300 ms following the test stimulus. Simply detecting a difference was not sufficient to produce the P300 response since the increased response to the uncertain stimuli was not seen when the subject simply had to count the occurrence of clicks and flashes. Modern studies use three stimulus paradigms in which the subject is presented with a standard, target, and distracter stimulus. The subject is told to respond to the target while ignoring the distracter (Polich, 2007). In this arrangement, both the distracter and target produce a response, referred to as P3a and P3b, respectively. The P3a reflects detection of stimulus difference while P3b reflects active processing of the target stimulus. The P300 (P3b) is thought to be an electrophysiological marker of context updating that occurs when incoming sensory information does not match representations of previous stimuli held in working memory and must be updated (Polich, 2007).

2.1.3.1 Effects of Nicotine on P300

In healthy subjects, chronic smokers show reduced P300 amplitude relative to nonsmokers (Anokhin et al., 2000; Evans et al., 2013; Guney, Genc, Kutlu, & Ilhan, 2009; Mobascher et al., 2010; Neuhaus et al., 2006; Polich & Ochoa, 2004). It does not appear that reduced P300 amplitude is due to ongoing brain changes associated with chronic smoking, since length of smoking history does not correlate with degree of amplitude reduction (Neuhaus et al., 2006). Similarly, decreases in P300 amplitude are seen in abstinent former smokers (Neuhaus et al., 2006). P300 amplitude reduction has been proposed as a potential endophenotype of smoking dependence (Perlman, Markin, & Iacono, 2013; Yoon, Iacono, Malone, & McGue, 2006) and has been linked to increased hereditary vulnerability to other forms of substance abuse (Euser et al., 2012). Indeed, 14 year olds who showed reduced P300 amplitude showed higher rates of early adolescent nicotine use (Perlman et al., 2013). As such, it is likely that reduced P300 amplitude represents an underlying trait linked to increased likelihood for smoking.

Studies assessing the effects of acute nicotine on P300 amplitude in healthy controls have provided less consistent results. In general, nicotine does not appear to enhance P300 amplitude elicited during oddball or choice reaction time tasks (Evans, Jentink, Sutton, Van Rensburg, & Drobes, 2014; Knott et al., 2009; Knott, Choueiry, et al., 2014; Knott, Scherling, et al., 2006; Lindgren, Molander, Verbaan, Lunell, & Rosen, 1999; but see Knott, Bosman, Mahoney, Ilivitsky, & Quirt, 1999). Knott, Choueiry, et al. (2014) showed enhancing effects of nicotine on P300a and P300b but these were only seen in subjects with low initial baseline amplitude (Knott, Choueiry, et al., 2014). Two studies have demonstrated nicotine-induced enhancements in P300 amplitude during a recognition memory task where subjects were required to evaluate whether a probe stimulus had been seen during a previous sample phase (Houlihan, Pritchard, & Robinson, 2001; Pineda, Herrera, Kang, & Sandler, 1998) (although, see Ilan & Polich, 1999). Similarly, mixed results have been reported for P300 latency with some studies showing increased latency following nicotine (Houlihan, Pritchard, & Robinson, 1996; Knott, Millar, et al., 2011) and others showing no effect (Houlihan et al., 2001; Ilan & Polich, 1999; Lindgren et al., 1999).

Only one study has evaluated the effects of nicotine on P300 in patients with schizophrenia (Mobascher et al., 2012). Nicotine did not enhance P300 amplitude in patients or controls in this study, but nicotine did increase P300 amplitude-correlated fMRI BOLD signal.

2.1.3.2 P300 as a Predictor of Disease-Related Symptoms

The P300 response is reduced in schizophrenia (Jeon & Polich, 2003; Roth & Cannon, 1972). In meta-analyses, auditory P300 has shown mean effect size of 0.89 (Jeon & Polich, 2003). Similarly, a recent cross site study found a mean effect size of 0.62 for P300 deficits in schizophrenia (Turetsky et al., 2014). Decreased P300 amplitude has been suggested to be one of the most reliable biomarkers of schizophrenia due to the robustness and universality of this finding (Ford, 1999; Turetsky et al., 2007). P300 reductions in schizophrenia have been correlated with performance on tasks assessing long-term memory (Fabiani, Karis, & Donchin, 1986; Kim et al., 2003; Nieman et al., 2002; Shajahan, O'Carroll, Glabus, Ebmeier, & Blackwood, 1997; Souza et al., 1995; Vianin et al., 2002) and attention (Ford, 1999; Nieman et al., 2002; Turetsky et al., 2009). Although not as numerous, increases in P300 latency have also been reported in schizophrenia (O'Donnell et al., 1995). Several studies have linked decreases in P300 amplitude and/or increased latency to severity of positive and negative symptoms (Kim, Shim, Kim, Im, & Lee, 2014; Liu, Tam, Xue, Yao, & Wu, 2004; Pfefferbaum, Ford, White, & Roth, 1989; Rissling et al., 2014; Strik, Dierks, & Maurer, 1993), although P300 deficits remain even after such symptoms have undergone remission (Turetsky et al., 2007).

2.2 Cognition

Several studies suggest nicotine use is linked to deficits in attention and working memory in people with schizophrenia. Patients who smoke show impaired attention relative to nonsmoking patients (Harris et al., 2004). Likewise, both degree of dependency and amount of cigarettes smoked per day are associated with reduced attention in patients who smoke (Patkar et al., 2002; Taiminen et al., 1998). The reasons for this relationship are not clear. It is possible that patients with reduced attentional capacity smoke in order to improve attentional performance or that both smoking and reduced attention emerge as part of a common genetic dysfunction. In contrast to the effects on attentional performance, no differences were observed between smoking and nonsmoking patients on long-delay memory (Harris et al., 2004).

Much evidence suggests that acute nicotine can facilitate some aspects of cognition in patients. For example, acute nicotine has been shown to improve sustained attention, recognition memory, and working memory in patients (Depatie et al., 2002; Jacobsen et al., 2004; Myers et al., 2004; Quisenaerts et al., 2014; Smith, Singh, Infante, Khandat, & Kloos, 2002; Smith et al., 2006), but not long-term memory (Quisenaerts et al., 2014; Smith et al., 2006). Short-term abstinence increases the severity of impairment in attention and visual–spatial working memory in patients (AhnAllen,

Nestor, Shenton, McCarley, & Niznikiewicz, 2008; Sacco et al., 2005), raising the possibility that acute nicotine administration is simply counteracting the negative effects of withdrawal. However, beneficial effects of nicotine on attentional tasks have also been observed in nonsmokers and, thus, may not be solely due to withdrawal (Barr et al., 2008). Interestingly, withdrawal-induced decrements in spatial working memory performance are observed in patients even after longer-term withdrawal (George et al., 2002).

Although the evidence is generally favorable for procognitive effects of nicotine, it should be noted that in many of these studies the effects of nicotine have been relatively modest. For example, while Depatie and colleagues (2002) showed an increased number of hits on the CPT following nicotine in patients when compared to placebo, the hit rate remained far below that of healthy controls, suggesting that nicotine did not normalize performance. Nicotine also did not alter sensitivity (d′) in patients. Also, some studies have failed to show positive effects of nicotine on either attention or working memory (Quisenaerts et al., 2014; Sherr et al., 2002).

2.3 Disease-Related Symptoms

Increased rates of smoking in patients may stem from an attempt to alleviate symptoms associated with schizophrenia and/or limit the exacerbation of symptoms during nicotine withdrawal. Self-report and case studies have suggested an increase in severity of symptoms during withdrawal from nicotine (Dalack & Meador-Woodruff, 1996; Glynn & Sussman, 1990; Hamera, Schneider, & Deviney, 1995). Negative symptoms: Higher rates of negative symptoms have been reported during smoking abstinence in experimental studies (Smith et al., 2002), although the effect of abstinence may not be lasting (Dalack, Becks, Hill, Pomerleau, & Meador-Woodruff, 1999). Others have not found evidence for this relationship (Tidey, Colby, & Xavier, 2014). Several studies have addressed differences in negative symptom severity between smoking and nonsmoking patients. Early studies reported greater severity in smoking patients compared to nonsmoking patients (BPRS) (Goff et al., 1992), as well as greater PANSS score on some cognitive domains (Taiminen et al., 1998). Consequently, several studies have reported no difference in negative symptom severity between smoking and nonsmoking patients (Aguilar et al., 2005; Herran et al., 2000; Kelly & McCreadie, 1999; Tang, George, Mao, Cai, & Chen, 2007; Zhang et al., 2013). Additionally, several studies have assessed the relationship between degree of dependence/amount smoked and negative symptoms. Patkar et al. (2002) found a positive association between degree of nicotine dependence and blunted affect and social withdrawal (PANSS) (Patkar et al., 2002). Krishnadas, Jauhar, Telfer, Shivashankar, and McCreadie (2012) found mildly dependent patients had higher rates of negative symptoms than nonsmokers, but not heavily dependent smokers (Krishnadas et al., 2012). In contrast, several reports have reported no relationship between degree of dependence/amount smoked and severity of negative symptoms (Barnes et al., 2006; Kotov, Guey, Bromet, & Schwartz, 2010; Meszaros et al., 2011; Ucok, Polat, Bozkurt, & Meteris, 2004), while other studies have shown lower rates of negative symptoms in patients who are heavy smokers (Zhang et al., 2007, 2014; Ziedonis, Kosten, Glazer, & Frances, 1994). Mood and anxiety: In patients, the number of cigarettes and/or degree of nicotine dependence is related to increased anxiety (Herran et al., 2000; Liao, Yang, Lee, Chen, & Tsai, 2002; Patkar et al., 2002) and blunted affect/mood (Kotov et al., 2010; Liao et al., 2002; Patkar et al., 2002; Tidey et al., 2014). Positive Symptoms: Several studies have reported no difference in positive symptom severity as a function of smoking status or amount smoked (Aguilar et al., 2005; Barnes et al., 2006; Herran et al., 2000; Kelly & McCreadie, 1999; Kotov et al., 2010; Patkar et al., 2002; Smith et al., 2002; Taiminen et al., 1998; Zhang et al., 2014). In contrast, some studies have shown positive correlations between smoking and positive symptoms (Fisher et al., 2012; Goff et al., 1992; Krishnadas et al., 2012; Ucok et al., 2004; Zhang et al., 2013; Ziedonis et al., 1994). Finally, lower rates of positive symptoms have also been demonstrated in smokers compared to nonsmokers (Zhang et al., 2007).

2.4 Conclusions

The effects of nicotine on most aspects schizophrenia symptomology are not compelling. Positive effects of nicotine have reliably been observed on P50 sensory gating, attention, and working memory, but, with the exception of sensory gating, most of these effects have been relatively modest. Little evidence suggests that nicotine can improve deficits in long-term memory, which can be severe in patients. As well, nicotine does not appear to strongly alter severity of negative or positive symptoms, but may be associated with changes in depression or anxiety present in patients. Positive effects of nicotine on ERP measures such as MMN and P300 are also relatively modest.

As a caveat, most studies have used healthy subjects, and it is unknown how well the results may extend to patient populations. Ceiling effects may also limit the ability for nicotine to alter these measures in healthy controls in the absence of baseline reductions in response. As well, the majority of studies assessing the effects of acute nicotine have used the gum or patch to administer nicotine, which may not activate nicotinic receptors as strongly as smoking. Studies of MMN have typically focused on the acute effects of nicotine and have not explicitly addressed the effects of chronic nicotine, as has been done in studies of sensory gating and P300. Despite the close link between smoking and schizophrenia and the numerous reports of reduced MMN andP300 in patients, few studies have addressed the relationship between smoking and change in these measures. The relationship between P300 amplitude reduction and substance abuse is intriguing, but has only been demonstrated in health controls and not patients. Like MMN and P300, smoking appears to be highly related to measures of overall function and well-being, and these are highly correlated with MMN and P300.

3. NICOTINIC RECEPTOR GENES IN SCHIZOPHRENIA

Although schizophrenia is largely inherited, few specific genes have been linked to development of the disease. There is a close relationship between familial risk for schizophrenia and familial risk for becoming a smoker (Lyons et al., 2002), raising the possibility that the genetic changes underlying increased smoking risk may be inherited as part of a shared underlying genetic vulnerability to schizophrenia. Nicotine receptors are ligand-gated oligomers comprised of five subunits. Subunits belong to one of two classes, α or β. In the brain, there are three β subunits (β2, β3, and β4) and six α subtypes (α2, 3, 4, 5, 6, and 7). Expression of α9 and 10 in the brain is very limited and will not be discussed further (Gotti & Clementi, 2004; Gotti et al., 2009; Gotti, Zoli, & Clementi, 2006). Aside from α7, which only forms receptors by combining with itself, all nicotinic receptors consist of at least one α subunit and one β subunit, although it is common for a nicotine receptor to be made from multiple subtypes from one class (e.g., α6α4β2). Examination of the genes which code for these subunits is necessary to better understand the contribution of these genes to disease etiology and treatment. Additionally, with the exception of agents for α4β2 and α7, there are few readily available compounds which selectively target individual subunits of nicotinic receptors, making pharmaceutical study of subunit function difficult.

3.1 Canonical Nicotinic-Binding Sites

3.1.1 CHRNA7

Two types of nicotine receptor were initially identified based on response to alpha-bungarotoxin and affinity for nicotine, with one receptor type characterized by high alpha-bungarotoxin binding/low nicotine affinity and the other characterized by lack of bungarotoxin binding/high affinity for nicotine (Cordero-Erausquin, Marubio, Klink, & Changeux, 2000; Lindstrom, 1997; Zoli, Lena, Picciotto, & Changeux, 1998). A role for the low nicotine affinity bungarotoxin-binding receptor in sensory gating was first established in rats using intraventricular administration of alpha-bungarotoxin, which was shown to reduce sensory gating (Luntz-Leybman, Bickford, & Freedman, 1992). Cross-strain variation in alpha-bungarotoxin-binding level has been shown to predict sensory gating in mice (Stevens et al., 1996). Specifically, DBA mice have the lowest level of alpha-bungarotoxin binding and, correspondingly, the poorest gating (Stevens et al., 1996). Reductions in sensory gating in DBA mice are reversible though administration of nicotine (Stevens & Wear, 1997). Contemporaneous research established that the alpha-bungarotoxin-binding site was located exclusively on α7 receptors in the brain (Couturier et al., 1990; Orr-Urtreger et al., 1997), suggesting that α7 may be the primary site upon which nicotine acts to regulate sensory gating. Validation of the role of α7 in schizophrenia in sensory gating in humans was later obtained through linkage analysis (Freedman et al., 1997, 2001). Subsequent studies have provided further validation of the association between variation in CHRNA7 and schizophrenia (Leonard et al., 1998; Stephens et al., 2012). Additionally, polymorphisms in CHRNA7 have been linked to increased risk for smoking in patients (De Luca et al., 2004), and attention (Rigbi et al., 2008).

Likewise, support for a role of α7 in schizophrenia has come from studies in transgenic mice lacking the gene for α7 (Chrna7). Two types of α7 KO mice have been developed, differing based on background strain [C57BL/6 (Orr-Urtreger et al., 1997) and C3H (Adams, Yonchek, Zheng, Collins, & Stevens, 2008)]. α7 KO mice show poorer sensory gating albeit with greatly increased amplitude of S1 and S2 responses (Adams et al., 2008). In contrast, sensorimotor gating appears to be intact (Paylor et al., 1998; Young, Meves, Tarantino, Caldwell, & Geyer, 2011) or relatively mildly impaired (Azzopardi, Typlt, Jenkins, & Schmid, 2013). α7 KO mice show intact hippocampal dependent learning and memory in both the standard Morris water maze and Barnes maze (Azzopardi et al., 2013; Paylor et al., 1998), and are unimpaired on novel object recognition (Young et al., 2011).

Attentional performance is disrupted by removal of α7 receptors. α7 KO mice acquire the 5-choice serial reaction time task (5CSRT) more slowly than controls and show higher rates of omission, after acquisition (Young et al., 2004). In a separate study, KO mice showed modest deficits in accuracy on 5-choice performance, but these could be reversed by changing task parameters to minimize the effects of impulsivity (variable ITI, punishment of premature responses) (Hoyle, Genn, Fernandes, & Stolerman, 2006). Interestingly, increased impulsivity has been found on other tasks in these mice (Keller, Keller, Bowers, & Wehner, 2005). Young et al. (2007) showed increased rates of omission in α7 KO mice on the 5CSRT task during increased attentional load (increased session length), with accuracy being comparable to wild-type mice during both standard and high attention load (Young et al., 2007). More profound disruptions have been shown in α7 KO mice during performance of a working memory task (odor span) (Young et al., 2007). One feature of this task is the ability to manipulate memory load by steadily increasing the number of items to remember, thus providing a measure of maximal memory capacity. More substantial impairments on the odor span task could suggest a preferential involvement of α7 in working memory or may indicate that the tasks used to assess attention and memory have not been sufficiently taxing to capture deficits in KO mice. Notably, KO mice showed impairments in spatial working memory in the Morris water maze (Fernandes, Hoyle, Dempster, Schalkwyk, & Collier, 2006). Deficits have also been found in appetitive learning. For example, KO mice showed greater trials to criterion during acquisition of the simple discrimination component of the attentional set-shifting task, but no effects on reversal learning or extradimensional shift learning, as well as deficits acquiring a radial arm maze task (Young et al., 2011). Removal of α7 receptors does not produce a strong anxiety phenotype on elevated plus maze (Azzopardi et al., 2013; Fernandes et al., 2006), while reductions in anxiety are seen on the light–dark test (Paylor et al., 1998). α7 KO mice show normal behavioral response to nicotine on a number of measures, including enhancement of learning and memory (Davis & Gould, 2007), nicotine self-administration (Pons et al., 2008), nicotine conditioned place preference (CPP; Harenza, Muldoon, De Biasi, Damaj, & Miles, 2014; Walters, Brown, Changeux, Martin, & Damaj, 2006), and show normal response (somatic signs and anxiety) to withdrawal from nicotine (Grabus, Martin, & Imad, 2005; Jackson, Martin, Changeux, & Damaj, 2008), although fail to they fail to show increased pain sensitivity during withdrawal (Grabus et al., 2005; Jackson et al., 2008). Similarly, KO mice showed normal enhancement of intracranial self-stimulation following withdrawal from nicotine (Stoker, Olivier, & Markou, 2012). $GABA_A$ receptor expression is reduced in mice with partial deletion of α7 (Adams et al., 2012), as are several markers of GABAergic interneuron integrity (Lin, Hsu, Baumann, Coulter, et al., 2014), suggesting a possible basis for the behavioral changes observed in KO mice. Additionally, α7 KO mice show reduced levels of NMDA receptor levels (Lin, Hsu, Baumann, Coulter, & Lynch, 2014).

A second line of research implicating the α7 receptor in schizophrenia has come from studies of individuals with copy number variations (CNVs) in the chromosomal region containing CHRNA7. CNVs are deletions or duplications of large, multigene sections of a chromosome that occur due to errors in DNA replication. Since CNVs typically occur in one chromosome, most result in a heterozygous knockout for all genes in the deleted region. Carriers of a deletion on the long (q) arm of chromosome 15 in a region designated 13.3 (15q13.3) have an increased incidence of schizophrenia (Ben-Shachar et al., 2009; Stefansson et al., 2008). Behavioral changes observed in deletion carriers include cognitive disability, developmental delay, reduced attention, and increased aggression (Ben-Shachar et al., 2009). The 15q13.3 region includes six genes that code for known proteins, including CHRNA7. However, individuals with CNV deletions restricted to CHRNA7 show the full range of disabilities seen in individuals with the full deletion, suggesting that intellectual impairments seen in 15q13.3 microdeletion syndrome are primarily due to loss of CHRNA7 (Hoppman-Chaney, Wain, Seger, Superneau, & Hodge, 2013). It should be noted that the behavioral phenotype of deletion carriers is much more severe than that seen in patients with schizophrenia, possibly reflecting greater loss of α7 function in deletion carriers. Contrarily, some carriers of the deletion do not show the deletion syndrome, suggesting that reductions of α7 can occur with little obvious dysfunction. Two groups have been developed mouse models of 15q13.3 microdeletion characterized by a heterozygous deletion of the mouse analog (7c) of the human 15q13.3 region. Like the full knockout mouse behavioral deficits in these mice are typically mild and, interestingly, do not include deficits in sensory gating (Fejgin et al., 2014). There is close overlap between the effects of α7 removal and the effects of nicotine. Both alter attention and working memory (in KO mice, 5-choice, radial arm maze, odor span) and sensory gating, but have little impact on long-term memory (water maze, novel object).

3.1.2 CHRNA4-CHRNB2

3.1.2.1 CHRNB2

In addition to the low nicotine affinity alpha-bungarotoxin-binding receptor, there is a high affinity nicotine receptor that shows little response to alpha-bungarotoxin (Cordero-Erausquin et al., 2000; Lindstrom, 1997; Zoli et al., 1998). The high affinity receptor is the most widely distributed nicotinic receptor in the brain (Flores, Rogers, Pabreza, Wolfe, & Kellar, 1992). Removal of β2 receptors in mice leads to loss of high affinity nicotine binding, suggesting that the high affinity nicotine receptor was at least partly comprised of β2 subunits (Picciotto et al., 1995). β2 Subunits show high levels of coexpression with the α4 subunit (Flores et al., 1992), demonstrating that the high affinity receptor is primarily an α4β2 receptor. These receptors are strongly upregulated in chronic smokers, a physiological change thought to underlie the development of nicotine dependence (De Biasi & Dani, 2011; Flores, Davila-Garcia, Ulrich, & Kellar, 1997).

Transgenic mice lacking the β2 subunit do not self-administer nicotine, do not develop place preferences with nicotine and do not show increased striatal dopamine efflux following nicotine administration (Epping-Jordan, Picciotto, Changeux, & Pich, 1999; Orejarena et al., 2012; Picciotto et al., 1998; Pons et al., 2008; Walters et al., 2006). Although these data strongly suggest that the β2 subunit may play a critical role in regulating nicotine reward, several studies in humans have failed to demonstrate an association between CHRNB2 polymorphisms and nicotine dependence (Etter et al., 2009; Li et al., 2005; Lueders et al., 2002; Saccone et al., 2007; Silverman et al., 2000). It is possible that the role of β2 subunit may relate to smoking cessation rather than dependence, per se. The negative effects of nicotine withdrawal appear to be reduced in β2 subunit KO mice (Portugal, Kenney, & Gould, 2008; Stoker, Marks, & Markou, 2014), although see (Besson et al., 2006). Likewise, polymorphisms in CHRNB2 have been related to smoking cessation and severity of withdrawal symptoms in humans (Conti et al., 2008; Perkins, Lerman, Mercincavage, Fonte, & Briski, 2009). Overall, it is not clear how these polymorphisms relate to actual β2 receptor function.

One study has assessed ERPs following transgenic removal of Chrnb2. In intact mice, administration of nicotine increases amplitude of the P20 response and decreases N40 amplitude (equivalent to the human N100) (Amann et al., 2008; Featherstone et al., 2012; Metzger et al., 2007; Phillips, Ehrlichman, & Siegel, 2007). Removal of the β2 receptor did not block the ability for nicotine to increase P20 amplitude or P20 sensory gating. In contrast, nicotine-induced change in the N40 component was blocked in β2 KO mice (Rudnick, Koehler, Picciotto, & Siegel, 2009). No baseline differences were observed between wild-type and KO mice in the absence of nicotine. This data suggests that some of the effects of nicotine on ERP measures may be influenced by the β2 subunit.

β2 KO mice showed normal performance of Morris water maze as adults (6 months), although aged KO mice did not (Zoli, Picciotto, Ferrari, Cocchi, & Changeux, 1999). Likewise, removal of β2 did not produce impairments in novel object recognition (Granon, Faure, & Changeux, 2003; Wiklund et al., 2009), although several aspects exploratory behavior in β2 KO mice suggest abnormally rigid behavior and poor adaptation to environmental change (Granon et al., 2003). In contrast, these mice show normal set shifting (Cole, Poole, Guzman, Gould, & Parikh, 2014), suggesting that the lack of flexibility seen during exploratory behavior does not translate across all behavioral contexts. β2 KO mice have been reported to show improved spatial learning (plus maze) (Granon et al., 2003) and passive avoidance (Picciotto et al., 1995), possibly due to an alteration in ancillary behaviors rather than a change in learning per se (Granon et al., 2003). Such exploratory behaviors are restored following targeted reexpression of β2 receptors in the VTA (Maskos et al., 2005). β2 KO mice do not show altered responding on DRL reinforcement schedules (Horst et al., 2012), or delay discounting (Serreau, Chabout, Suarez, Naude, & Granon, 2011), suggesting that they are not abnormally impulsive.

While cognitive performance is only modestly impaired at baseline, the ability for nicotine to influence cognitive performance appears to be lost following removal of β2 receptors (Davis & Gould, 2007; Picciotto et al., 1995; Wehner et al., 2004). This suggests that β2 receptors may not be directly involved in cognition but might play a key role in mediating procognitive effects of nicotine. Detrimental effects of nicotine on cognition also appear to be lost in KO mice (Cole et al., 2014). These mice do not show increased anxiety associated with nicotine withdrawal, but show normal hyperalgesia and normal or increased somatic signs of withdrawal (Jackson et al., 2008; Salas, Pieri, & De Biasi, 2004).

3.1.2.2 CHRNA4

Several studies have linked polymorphisms in CHRNA4 to smoking in humans (Breitling et al., 2009; Conti et al., 2008; Feng et al., 2004; Hutchison et al., 2007; Kishi et al., 2013; Li et al., 2005; Saccone, Saccone, et al., 2009; Saccone et al., 2010; Voineskos et al., 2007; Wessel et al., 2010), but negative data have also been found (Saccone et al., 2007). A single nucleotide polymorphism (SNP) in CHRNA4 has been linked to reduced fMRI BOLD signal in anterior cingulate and parietal cortex during a visual oddball task (similar to that used in P300 studies) (Winterer et al., 2007). The same SNP was subsequently found to be associated with reduced ERP response (N1 and P1 amplitude) in an auditory and visual oddball stimulus preparation (Espeseth, Endestad, Rootwelt, & Reinvang, 2007), as well as enhanced visual attention (Greenwood et al., 2009; Parasuraman, Greenwood, Kumar, & Fossella, 2005).

Targeted deletion of α4 receptors also leads to loss of high affinity-binding nicotine receptors (Marubio et al., 1999). α4 Null mutant mice show reduced nicotine self-administration (Pons et al., 2008) and reduced nicotine evoked dopamine release (Marubio et al., 2003). Mice with hypersensitive α4 receptors show increased sensitivity to the rewarding effects of nicotine (Tapper et al., 2004). The behavior of α4 mice has not been very extensively characterized in regards to psychiatry or cognition, but some evidence suggests they show increased anxiety on the elevated plus maze (Labarca et al., 2001; Ross et al., 2000).

Only a handful of studies have addressed the issue of CHRNB2 or CHRNA4 changes in schizophrenia, or the role of polymorphisms in these genes and smoking in patient populations. Linkage analysis has identified the 1q21–22 region, which contains the CHRNB2 gene, in relationship to familial schizophrenia (Brzustowicz, Hodgkinson, Chow, Honer, & Bassett, 2000). It should be noted, however, the 1q21.1 region, which does not contain CHRNB2, has also been linked to schizophrenia and could explain this link (Stefansson et al., 2008). A recent case report described a teenager with duplication in the region of 1q21.3 containing CHRNB2 who showed absence seizures and learning impairment, but no apparent psychiatric problems (Muhle et al., 2010). De Luca et al. found no association between CHRNA4 or CHRNB2 polymorphisms and schizophrenia when assessed alone, but did see evidence for an association between some combinations of CHRNA4/CHRNB2 SNPs and schizophrenia (De Luca, Voineskos, Wong, & Kennedy, 2006), although a later study failed to replicate this result (Kishi et al., 2008).

3.2 Noncanonical Nicotinic-Binding Sites

Much initial interest was centered upon α7 and α4/β2 containing receptors since these appeared to characterize most of the responses observed to nicotine or alpha-bungarotoxin and are the most dominant nicotinic receptors in the brain. However, it has now become clear that the subunit constitution of nicotinic receptors is much more complex than this simple dichotomy. There are a wide variety of subunit combinations that can exist which have never been extensively studied in the context of schizophrenia and, consequently, have poorly understood function (for example, α6α4β2β3). It is clear that the addition of some subunits (α5 or α6) to a α4/β2 receptor can greatly alter several properties of the α4/β2 receptor, but little is understood how this might regulate function at the organismal level, especially in terms of smoking dependence and schizophrenia symptomology. Importantly, many of these subunit combinations do not show wide distribution in the brain but are highly expressed in brain areas known to be abnormal in schizophrenia and cognitive dysfunction. This more limited expression pattern may confer greater selectivity of agents targeting these receptors, making these subunits highly attractive targets for drug development relative to compounds that act on more widely distributed subunit combinations.

3.2.1 CHRNA5–CHRNA3–CHRNB4

Recent studies have identified a number of SNPs within a region of chromosome 15 (15q25.1) containing nicotinic receptor genes CHRNA5, CHRNA3, and CHRNB4 in association with risk for smoking and nicotine dependence (Berrettini et al., 2008; Bierut et al., 2008; Saccone et al., 2007; Saccone, Wang, et al., 2009; Thorgeirsson et al., 2008; Weiss et al., 2008). Most notably, several studies have identified a SNP in CHRNA5 (rs16969968) in relation to increase in nicotine dependence (Lips et al., 2010; Saccone et al., 2007). Polymorphisms in this region have also been linked to increased risk for lung cancer in smokers as well (Amos et al., 2008, 2010; Spitz, Amos, Dong, Lin, & Wu, 2008; Thorgeirsson et al., 2008). Although less prominent, SNPs in both CHRNA3 and CHRNB4 have also been associated with smoking and nicotine dependence.

3.2.1.1 CHRNA5

In rats, expression of α5 is highest in hippocampus, where a little over a third of nicotinic receptors express this subunit, followed by striatum and cortex, whereas expression in the cerebellum is minimal (Mao, Perry, Yasuda, Wolfe, & Kellar, 2008). Primates show similar patterns of expression, although expression may be less intensive in hippocampus (Han et al., 2000). Current changes in response to nicotine are not observed in cells expressing α5 subunits when coexpressed with a single α4 or β2 subunit (Ramirez-Latorre et al., 1996). However, the inclusion of α5 along with α4/β2 results in a much greater response to nicotine than that seen with α4/β2 alone (Ramirez-Latorre et al., 1996). A similar enhancing effect of α5 has been shown when this subunit is coexpressed with α3/β2, but not α3/β4 (Gerzanich, Wang, Kuryatov, & Lindstrom, 1998; Wang et al., 1996). Deletion or loss of function of the α5 subunit would be expected to lead to reduced responsively in α4/β2 and α3/β2 receptors. Additionally, and somewhat paradoxically, inclusion of the α5 subunit has been shown to limit the upregulation of α4/β2 receptors normally seen in response to chronic nicotine (Mao et al., 2008). Long-term receptor upregulation in chronic smokers is thought to be a key physiological factor underlying the development of nicotine addiction and enhanced response to nicotine seen in smokers (De Biasi & Dani, 2011). As such, individuals with loss of function in α5 would be expected to show increased receptor upregulation compared to intact individuals, possibly leading to increased susceptibility for nicotine dependence. Importantly, mutations in rs16969968 have been shown to produce an animo acid change resulting in reduced response of α5 containing receptors to nicotinic agonists (Bierut et al., 2008). Other SNPs in this region alter α5 expression in the brain (Jackson, Fanous, Chen, Kendler, & Chen, 2013; Wang et al., 2009).

Two human studies have linked specific SNPs in 15q25.1 with schizophrenia (Hong, Yang, et al., 2011; Jackson et al., 2013). Both showed associations between the α5 receptor regulating variant rs6969968 and schizophrenia, amongst others. rs6969968 was also shown to be associated with smoking severity in both patients and controls (Hong, Yang, et al., 2011). It is not clear how these polymorphisms relate to the etiology of schizophrenia. rs6969968 was not associated with negative symptoms in patients, but other SNPs involved in the regulation of α5 expression levels were (Jackson et al., 2013). SNPs in 15q25, including rs6969968, have been associated with performance on attention and working memory tasks (Winterer et al., 2010), as well as cognitive flexibility (Wisconsin card sort) (Zhang, Kranzler, Poling, & Gelernter, 2010). Additionally, one study has assessed P300 response as a function of 15q25 status using a three stimulus oddball task (Evans, MacQueen, et al., 2014). Consistent with lower P300 amplitude in chronic smokers, nicotine deprivation reduced P3a amplitude and several SNPs enhanced the effects of nicotine deprivation on this reduction. Interestingly, no effect was observed for the minor allele of rs6969968. These results may suggest that polymorphisms in this region may regulate the reductions seen in P300 response in chronic smokers and that P3 amplitude may serve as a biomarker for individuals with loss of function polymorphisms in the CHRNA5–CHRNA3–CHRNB4 region and, possibly, enhanced susceptibility to development of nicotine dependence.

Transgenic mice have been generated that lack α5 receptors (Salas, Orr-Urtreger, et al., 2003). These mice do not show altered startle response, prepulse inhibition, passive avoidance, or locomotor activity (Salas, Orr-Urtreger, et al., 2003). Abnormal responses to nicotine are seen, including loss of nicotine-induced antinociception (Jackson et al., 2010), decreased nicotine-induced dopamine release *in vitro* (Salminen et al., 2004), loss of nicotine-induced hypolocomotion and seizures (Salas, Orr-Urtreger, et al., 2003), and reduction of some symptoms of nicotine withdrawal (Jackson et al., 2008; Salas, Sturm, Boulter, & De Biasi, 2009). In contrast, KO mice show normal CPP in response to nicotine (Jackson et al., 2010). Female, but not male, α5 KO mice show reduced anxiety on the elevated plus maze (Gangitano, Salas, Teng, Perez, & De Biasi, 2009). To date, there is a lack of studies that have assessed these mice on the types of cognitive tasks used to characterize α7 or β2 mice, or on behavioral measures more closely related to schizophrenia.

3.2.1.2 CHRNA3

The α3 subunit is expressed at high levels in the thalamus and habenula (Wada et al., 1989). In contrast, expression in canonical schizophrenia sites such as the medial temporal lobe and cortex is much more limited (Wada et al., 1989). Two SNPs within CHRNA3 have been linked to reduced prepulse inhibition in both patients with schizophrenia and nonpatients, as well as increased rates of positive and negative symptoms in patients (Jackson et al., 2013; Petrovsky et al., 2010). Loss of function variants of rs1051730 have been linked to reduced working memory and attention in healthy subjects (Winterer et al., 2010).

Study of the function of α3 *in vivo* is limited by the fact that homozygous deletion of CHRNA3 is lethal (Xu et al., 1999), although behavior has been assessed in mice with heterozygous deletion. Partial loss of α3 results in decreased susceptibility to nicotine-induced seizures, but normal hypolocomotor response to nicotine (Salas, Cook, Bassetto, & De Biasi, 2004).

3.2.1.3 CHRNB4

Expression of β4 is limited to the medial habenula, olfactory bulbs, and a handful other areas (Duvoisin, Deneris, Patrick, & Heinemann, 1989;

Salas, Pieri, Fung, Dani, & De Biasi, 2003). While studies of the CHRNA5–CHRNA3–CHRNB4 region have mostly focused on CHRNA5 or CHRNA3, associations have been shown between polymorphisms in CHRNB4 and some aspects of smoking (Schlaepfer et al., 2008). A SNP within CHRNB4 has been linked to increased risk for schizophrenia and regulation of α5 expression (Jackson et al., 2013).

Mice with homozygous deletion of β4 show reduced anxiety on elevated plus maze and light–dark box, but increased anxiety during social isolation (Salas, Pieri, et al., 2003; Semenova, Contet, Roberts, & Markou, 2012). Deletion of β4 does not impair spatial learning or contextual fear conditioning, suggesting that this receptor is not highly involved in cognitive function (Semenova et al., 2012; Wehner et al., 2004). Physiological and behavioral changes to nicotine are drastically altered in these mice, however, characterized by reduced seizure and hypolocomotor response to nicotine (Kedmi, Beaudet, & Orr-Urtreger, 2004; Salas, Cook, et al., 2004), reduced behaviors associated with nicotine withdrawal (Salas, Pieri, et al., 2004; Stoker et al., 2012) and reduced nicotine-induced analgesia (Semenova et al., 2012). KO mice showed normal enhancement of intercranial self-stimulation following withdrawal from nicotine (Stoker et al., 2012). In contrast, these mice showed normal nicotine facilitation of contextual fear conditioning (Wehner et al., 2004).

3.2.2 CHRNA6–CHRNB3

Both CHRNA6 and CHRNB3 are located close together on chromosome 8 and are coexpressed in the brain. The α6 receptor is primarily located on midbrain dopamine cells where it is colocalized with the β2 subunit (Han et al., 2000; Le Novere, Zoli, & Changeux, 1996). α6* (α6α4β2 and α6β2) receptors are the major α subunit found on dopaminergic cells in the midbrain (Gotti et al., 2006; Le Novere et al., 1996) where they play an important role in the regulation of nicotine reward. Specifically, null mutation of α6 abolishes nicotine self-administration, which can be reinstated by selective reexpression of α6 receptors in the VTA of KO mice (Pons et al., 2008). The role of α6 in regulating dopamine release is likely dependent upon interaction with α4/β2 subunits (Exley et al., 2011). Nicotine-induced dopamine release is not decreased in α6 KO mice (Champtiaux et al., 2003).

The β3 subunit is mainly expressed in dopamine cells in the midbrain and colocalizes with α6. β3 KO mice show reduced prepulse inhibition, increased locomotor activity, and reduced α-Conotoxin MII-sensitive nicotine-induced dopamine release (Cui et al., 2003). β3 KO mice do not show impairments in contextual fear conditioning and show normal nicotine facilitation of conditioning (Wehner et al., 2004).

Several genes located on chromosome 8p have been linked to schizophrenia, but the precise role of either CHRNA6 or CHRNB3 in mediating this relationship is not presently clear (Tabares-Seisdedos & Rubenstein, 2009).

3.2.3 CHRNA2

Relative to α7, α4, or β2, α2 is not highly expressed in brain areas known to be affected in schizophrenia (Han et al., 2000; Wada et al., 1989). In rats, expression is highest in the interpeduncular nucleus, with moderate expression in the medial septum and some parts of the hippocampus and amygdala (Wada et al., 1989). In the hippocampus, α2 receptors are located on interneurons (Sudweeks & Yakel, 2000). Several human studies have identified modest associations between CHRNA2 and smoking/ND (Heitjan et al., 2008; Sullivan et al., 2004; Wang et al., 2014; Yang et al., 2014). Polymorphisms in CHRNA2 were associated reduced response inhibition during cognitive performance, regardless of smoking status (Rigbi et al., 2008). CHRNA2 does not appear to be associated with risk for schizophrenia (Blaveri et al., 2001), but might be associated with increased rates of smoking in patients (Faraone et al., 2004).

Removal of these receptors abolishes the ability for nicotine to influence the induction of hippocampal LTP (Nakauchi, Brennan, Boulter, & Sumikawa, 2007). α2 KO mice do not show changes in anxiety, locomotor activity, or baseline cognitive performance, but show enhanced nicotine-induced analgesia, self-administration, and increased nicotine enhancement of cued fear conditioning (Lotfipour et al., 2013). The effects of nicotine withdrawal are attenuated by removal of α2 receptors (Salas et al., 2009).

3.2.4 Conclusions

There is a lot of evidence linking changes in canonical genes (CHRNA7 and CHRNA4) to schizophrenia. Many other candidate genes appear promising but have only received a limited amount of study in relationship to schizophrenia. Overall, little work has been done examining the links between changes in these latter genes and electrophysiological endophenotpes known to be disrupted in schizophrenia.

4. DEVELOPMENT OF NICOTINIC AGENTS FOR THE TREATMENT OF SCHIZOPHRENIA

Much evidence suggests that nicotine may be useful as a treatment for several aspects of schizophrenia symptoms, including cognition. There is now a large amount of data on the effects of agents selective for activating receptor subtypes associated with canonical receptor subunits (α7 and α4/β2). This strategy is relatively mature and there are now compounds in late stage clinical trial. Much less research has been conducted with other receptor subunits.

4.1 α7

Several lines of evidence have implicated CHRNA7 and the α7 receptor in schizophrenia (see above) which has led to the development of several compounds with selective agonist properties at this receptor subtype. These include full and partial agonists, as well as positive allosteric modulators. The effect of these agents on preclinical and clinical measures in patients has been reasonably well established.

4.1.1 Sensory Gating

A number of early studies demonstrated normalization of P20/N40 sensory gating reductions in DBA mice (Simosky et al., 2001; Stevens et al., 1998) and fimbria-fornix-lesioned rats (Adams, Stevens, Kem, & Freedman, 2000) following administration of the partial α7 agonist DMXB-A (GTS-21). Subsequent studies have replicated these results using other α7 agents (Feuerbach et al., 2009; Hajos et al., 2005; Radek, Robb, Stevens, Gopalakrishnan, & Bitner, 2012; Tietje et al., 2008). Positive effects on P50 gating have been found in patient populations for DMXB-A (Olincy et al., 2006) and the α7 agonist/5-HT_3 antagonist tropisetron (Koike et al., 2005; Shiina et al., 2010), although other α7 agonists have not been as successful (Preskorn et al., 2014; Winterer et al., 2013).

4.1.2 MMN/P300

Neither MMN or P300 amplitude were altered in patients following administration of the α7 positive allosteric modulator JNJ-39393406 (Winterer et al., 2013). In contrast, both were increased following administration of the nicotinic activator EVP-6124 (Preskorn et al., 2014).

4.1.3 Disease-Related Symptoms

Several rodent studies have assessed the effects of α7 agents on surrogate measures of positive symptoms. α7 Agonists restore PPI following disruption by apomorphine or MK-801 (Callahan, Terry, & Tehim, 2014; Hauser et al., 2009; Roncarati et al., 2009; Wallace et al., 2011), enhance the ability for antipsychotics to restore PPI in DBA mice (Kohlhaas, Bitner, Gopalakrishnan, & Rueter, 2012) and reverse the detrimental effects of isolation-rearing-induced deficits in PPI in rats (Cilia et al., 2005). Likewise, disruptive effects of amphetamine on latent inhibition (LI) are blocked by administration of an α7 partial agonist (Barak et al., 2009). In regards to negative symptoms, increased social behavior has been reported following administration of α7 agonists (Hauser et al., 2009). To date, there is a lack of studies that have assessed the effects of α7 agents on rodent models of anhedonia.

Some evidence suggests that the anxiolytic effects of nicotine may be due in part to activation of α7 receptors. Anxiolytic effects of α7 activators have been reported on social exploration in rodents (Feuerbach et al., 2009), although not on the elevated plus maze (Hauser et al., 2009).

In patients, improvements in negative symptoms have been reported following administration of α7 activators (Freedman et al., 2008; Lieberman et al., 2013; Umbricht et al., 2014). These effects were particularly strong on the anhedonia subdomain, suggesting a potential role for α7 in treating this aspect of the disease (Freedman et al., 2008; Lieberman et al., 2013). Modest or null effects of α7 activators have been reported for positive symptoms (Freedman et al., 2008; Olincy et al., 2006; Umbricht et al., 2014). The lack of effect of α7 agents on positive symptoms in humans contrasts sharply with positive findings in preclinical rodent models. This could be due to the use of surrogate measures in rodents, which may not closely overlap with the human measures. To date, there is a lack of studies that have assessed the effects of α7 agents on PPI or LI in patients, which would be helpful in resolving this discrepancy.

4.1.4 Cognition

4.1.4.1 Attention

Nicotine produces robust enhancements in performance on rodent measures of attention, such as the 5-choice serial reaction time task (5CSRT) and the sustained attention task (Grottick & Higgins, 2000). Given the prominence of attentional deficits in schizophrenia, as well as the close overlap between attention and sensory gating, α7 has been considered as the likely mechanism that mediates these effects of nicotine. α7 Activators do not improve performance on the 5-choice continuous performance task (5CPT) or 5CSRT (although in most of these studies α4/β2 agonist did) (Grottick, Haman, Wyler, & Higgins, 2003; Grottick & Higgins, 2000; Hahn, Sharples, Wonnacott, Shoaib, & Stolerman, 2003; Young, Meves, & Geyer, 2013). The lack of effect of α7 agonists is consistent with studies showing a failure of the α7 antagonist methyllycaconitine (MLA) to block the enhancing effects of nicotine on sustained attention tasks (Blondel, Sanger, & Moser, 2000). It should be noted that many of these studies used the full agonist AR-R17779, which has not been assessed in sensory gating or in human trials. The Young et al. (2013) study is notable since it used PNU 282987, which is a full α7 agonist shown to be effective in sensory gating (Hajos et al., 2005), although it has not been tested in human studies. The α7 agonist/5-HT_3 antagonist RG3487/MEM3454 has been shown to be effective on a sustained attention task (Rezvani et al., 2009), but this could be due to actions at 5-HT_3 rather than α7. This study used an operant signal detection task while the former studies used the 5CSRT or 5CPT task, raising the possibility that this latter task is more sensitive to detecting potential effects of α7 activators. This seems unlikely, however, since other studies have failed to show a strong role for α7 in mediating the performance enhancing effects of nicotine on this task (Howe et al., 2010). The limited effectiveness of α7 agents on sustained attention tasks are generally consistent with the relatively mild changes in accuracy observed in α7 KO mice. In contrast, both the positive allosteric modulator (PAM) PNU-120596 and partial agonist SSR 180711 have been shown to be effective in reversing deficits in attentional set shifting in rodent models of schizophrenia (Brooks et al., 2012; McLean et al., 2012).

Studies in humans have failed to show significant effects of α7 activators on attention. The partial agonist DMXB-A improved overall performance on RBANs test (Repeatable Battery for the Assessment of Neuropsychological Status), although no significance was found for any of the individual subtests (Olincy et al., 2006). A follow up study failed to show an overall effect of chronic DMXB-A treatment on attention in patients, although some positive effects were observed during the first treatment arm of the study (Freedman et al., 2008). Likewise, no effects were seen on attention following treatment with the partial agonist EVP-6124 (Preskorn et al., 2014).

4.1.4.2 Working Memory

The partial agonist AZD0328 improved performance on a spatial working memory task in monkeys (Castner et al., 2011) while the partial agonist

A-582941 increased performance during a delayed matching to sample task (Buccafusco, Terry, Decker, & Gopalakrishnan, 2007). DMXB-A treatment did not improve performance on the MATRICS^{TM} cognitive battery in patients, although some positive effects were observed for working memory during the first treatment arm of the study (Freedman et al., 2008). Positive effects on working memory were found for the partial agonist TC-5619 (Lieberman et al., 2013).

4.1.4.3 Long-Term Memory

Several studies have shown positive effects of α7 activators on object recognition in rodents (Callahan et al., 2014; Marquis et al., 2011; McLean et al., 2011; Pichat et al., 2007; Prickaerts et al., 2012; Roncarati et al., 2009; Werkheiser et al., 2011). These results are surprising given the lack of effectiveness of nicotine in improving long-term memory in humans as well as the lack of memory deficits in α7 KO mice. Object recognition has several drawbacks as a preclinical model of human memory deficits and may not provide a good predictor of clinical efficacy (Young, Powell, Risbrough, Marston, & Geyer, 2009). However, positive effects of α7 activators have also been observed on other rodent measures of long-term memory, including contextual fear conditioning (Yildirim, Connor, & Gould, 2014) and Morris water maze (Meyer et al., 1997; Vicens, Ribes, Torrente, & Domingo, 2011).

In humans, DMXB-A increased overall performance as assessed by the RBANs test, but no significance was found for any of the individual subtests (Olincy et al., 2006). No overall change on the MATRICS^{TM} cognitive battery, which includes measures of visual–spatial memory, was observed following administration of either DMXB-A or RG3487 (Freedman et al., 2008; Umbricht et al., 2014). In contrast, positive effects on memory have been observed for EVP-6124 (Preskorn et al., 2014).

4.1.5 Conclusions

There is only limited support for the effectiveness of α7 activators in reversing cognitive deficits in patients with schizophrenia. Whereas nicotine appears to have beneficial effects on both rodent and human measures of attention, it would appear that this may not be strongly influenced by α7 receptors. Much preclinical data would suggest a role of α7 agents in reversing memory deficits in patients, but this does not appear to translate to human subjects.

4.2 α4β2

Like α7, several selective α4β2 agents have been developed. Most notable and widely used is the partial agonist varenciline, which has been approved for smoking cessation and assessed in several studies. Additionally, several α4β2 selective agonists have been developed by pharmaceutical companies, with some having undergone limited clinical evaluation.

4.2.1 Sensory Gating

Some data supports a role for α4β2 in sensory gating and P20/50 amplitude. Pretreatment with the α4β2 agonist DhβE blocked the effects of nicotine on P20 amplitude in mice, but did not affect gating (Featherstone et al., 2012). Varenicline, a partial agonist at α4β2 receptors, has been shown to improve sensory gating in DBA mice (Wildeboer-Andrud & Stevens, 2011) and increase P20 amplitude in normal mice (Rudnick et al., 2010). In contrast, the selective α4β2 agonist AZD3480/TC-1734 mimicked the effects of nicotine on the N40 component but did not alter P20 response (Featherstone et al., 2012), consistent with results showing a lack of change in P20 response to nicotine in β2 KO mice (Rudnick et al., 2009). Short-term varenicline treatment blocked the detrimental effects of nicotine abstinence on P50 amplitude in humans (Rudnick et al., 2010). Similarly, long-term (8 week) treatment with varenicline improved sensory gating in patients (Hong, Thaker, et al., 2011).

4.2.2 MMN/P300

The α4β2 positive allosteric modulator NS9283 enhanced P3-like response to a target stimulus during a rodent operant auditory discrimination task designed to model human P300 stimulus preparations (Grupe, Grunnet, Laursen, & Bastlund, 2014). In contrast, the α4β2 partial agonist AZD3480 enhanced MMN amplitude in healthy humans, but did not alter P300 amplitude (Dunbar et al., 2007)

4.2.3 Disease-Related Symptoms

In rodents, administration of the α4β2 PAM NS9283 reversed reductions in PPI induced by the NMDA antagonist PCP (Timmermann et al., 2012). An early case report suggested that the α4β2 partial agonist varenicline could play a role in symptom relapse or exacerbation of symptom severity (Freedman, 2007). However, the majority of published studies have not born this out (Cerimele & Durango, 2012). Several reports have failed to demonstrate significant change in symptom severity in patients receiving chronic varenicline treatment (Hong, Thaker, et al., 2011; Shim et al., 2012; Smith et al., 2009; Williams et al., 2012), although in some of these studies strong trends were seen for decreased positive symptoms following treatment (Hong, Thaker, et al., 2011; Smith et al., 2009). Chronic varenicline does not alter PPI in patients (Hong, Thaker, et al., 2011).

Anxiolytic effects were observed for ABT-418 on elevated plus maze (Decker et al., 1994). Varenicline was anxiolytic in marble burying and novelty induced hypohagia (Turner, Castellano, & Blendy, 2010), but not elevated plus maze.

4.2.4 Cognition

4.2.4.1 Attention

There have been several published reports showing a positive effect of α4β2 receptor agonists and PAMs on rodent measures of sustained attention (Grottick et al., 2003; Grottick & Higgins, 2000; Hahn et al., 2003; Howe et al., 2010; Mohler et al., 2010; Rezvani, Cauley, Johnson, Gatto, & Levin, 2012; Timmermann et al., 2012). In humans, acute and short-term AZD3480 treatment has been shown to improve attentional performance in healthy controls (Dunbar et al., 2007), although a subsequent study failed to demonstrate significant improvements on the CPT in patients with schizophrenia following chronic treatment (Velligan et al., 2012). Likewise, varenicline treatment does not appear to improve CPT performance in patients (Hong, Thaker, et al., 2011; Roh et al., 2014; Shim et al., 2012).

4.2.4.2 Working Memory

Some studies suggest a positive role for α4/β2 agonists on working memory. In rodents, TC-1734 (AZD3480) improved working memory performance on a radial maze (Gatto et al., 2004). The partial agonist ABT-594 improved working memory as assessed by a delayed matching to sample task in monkeys (Buccafusco et al., 2007). In humans, modest effects of AZD3480 have been reported on working memory in healthy controls (Dunbar et al., 2007) but not in patients (Velligan et al., 2012). In nonschizophrenic subjects, varenicline has been shown to improve working memory as assessed by the n-back test (Mocking et al., 2013) but in patients did not improve spatial working memory (Hong, Thaker, et al., 2011) or working memory as assessed by RBANs (Smith et al., 2009). In contrast, varenicline treatment was effective in blocking the detrimental effects of smoking abstinence in patients (Wing, Wass, Bacher, Rabin, & George, 2013).

4.2.4.3 Long-Term Memory

In rodents, administration of ABT 418 prior to passive avoidance training was found to enhance retention 24 h later (Decker et al., 1994), while TC-1734 (AZD-3480) was able to reverse passive avoidance deficits induced by scopolamine (Gatto et al., 2004). The partial agonist, TC-1734 (AZD-3480) improved novel object recognition and reference memory in mice and rats (Gatto et al., 2004). Likewise, NS9283 reversed the antimnemonic effects of PCP on water maze performance and improved social recognition memory (Timmermann et al., 2012). In humans, no effect of AZD3480 was found on visual or verbal learning in patients (Velligan et al., 2012), but positive effects have been seen on episodic memory (Dunbar et al., 2007). Varenicline produced positive effects on delayed recall in healthy controls (Mocking et al., 2013). In patients, varenicline produced modest effects on performance of a human water maze task and improved list recall (Smith et al., 2009).

4.3 Other Receptor Subtypes

Very little work has been done to generate novel agents with selective capabilities for non-α7 or α4β2 receptor subtypes. Agents have been developed which target α3β4 receptors but the behavioral effects of these compounds have not been extensively characterized in relation to schizophrenia (Chatterjee et al., 2011). Given the extensive literature suggesting a role for CHRNA3/CHRNβ4/CHRNA5, as well as CHRNA6, in both schizophrenia and smoking it is surprising that there are no compounds capable of targeting disease-related reduction of receptor function in the relevant gene products.

5. CONCLUSIONS

The bulk of evidence suggests that nicotine has a beneficial effect on some aspects of schizophrenia symptomology. Overall, however, attempts to recreate these effects using subtype selective nicotinic activators have been relatively unsuccessful. Preclinical studies have been focused largely on sensory gating, which appears to have little predictive capacity for schizophrenia symptomology. In contrast, much less work has been done on the role of nicotine or nicotinic subtype selective agonists on electrophysiological measures more closely related to core aspects of schizophrenia, such as MMN or P300. Additionally, most work has focused on the development of agents capable of selectively targeting either α7 or α4β2 subtype containing receptors. There is no reason, however, to believe that the effects observed following nicotine are necessarily reducible to selective effects on only one receptor subtype. Perhaps drug development should focus on broad spectrum agents that activate multiple receptor subtypes. Additionally, increased understanding of the role of noncanonical nicotinic subtypes could lead to the development of novel methods to alter cholinergic function.

REFERENCES

Adams, C. E., Stevens, K. E., Kem, W. R., & Freedman, R. (2000). Inhibition of nitric oxide synthase prevents alpha 7 nicotinic receptor-mediated restoration of inhibitory auditory gating in rat hippocampus. *Brain Research, 877*, 235–244.

Adams, C. E., Yonchek, J. C., Schulz, K. M., Graw, S. L., Stitzel, J., Teschke, P. U., et al. (2012). Reduced Chrna7 expression in mice is associated with decreases in hippocampal markers of inhibitory function: Implications for neuropsychiatric diseases. *Neuroscience, 207*, 274–282.

Adams, C. E., Yonchek, J. C., Zheng, L., Collins, A. C., & Stevens, K. E. (2008). Altered hippocampal circuit function in C3H alpha7 null mutant heterozygous mice. *Brain Research, 1194*, 138–145.

Adler, L. E., Hoffer, L. D., Wiser, A., & Freedman, R. (1993). Normalization of auditory physiology by cigarette smoking in schizophrenic patients. *The American Journal of Psychiatry, 150*, 1856–1861.

Adler, L. E., Pachtman, E., Franks, R. D., Pecevich, M., Waldo, M. C., & Freedman, R. (1982). Neurophysiological evidence for a defect in neuronal mechanisms involved in sensory gating in schizophrenia. *Biological Psychiatry, 17*, 639–654.

Adler, L. E., Waldo, M. C., Tatcher, A., Cawthra, E., Baker, N., & Freedman, R. (1990). Lack of relationship of auditory gating defects to negative symptoms in schizophrenia. *Schizophrenia Research, 3*, 131–138.

Aguilar, M. C., Gurpegui, M., Diaz, F. J., & de Leon, J. (2005). Nicotine dependence and symptoms in schizophrenia: Naturalistic study of complex interactions. *The British Journal of Psychiatry, 186*, 215–221.

AhnAllen, C. G., Nestor, P. G., Shenton, M. E., McCarley, R. W., & Niznikiewicz, M. A. (2008). Early nicotine withdrawal and transdermal nicotine effects on neurocognitive performance in schizophrenia. *Schizophrenia Research, 100*, 261–269.

Amann, L. C., Phillips, J. M., Halene, T. B., & Siegel, S. J. (2008). Male and female mice differ for baseline and nicotine-induced event related potentials. *Behavioral Neuroscience, 122*, 982–990.

Amos, C. I., Gorlov, I. P., Dong, Q., Wu, X., Zhang, H., Lu, E. Y., et al. (2010). Nicotinic acetylcholine receptor region on chromosome 15q25 and lung cancer risk among African Americans: A case–control study. *Journal of the National Cancer Institute, 102*, 1199–1205.

Amos, C. I., Wu, X., Broderick, P., Gorlov, I. P., Gu, J., Eisen, T., et al. (2008). Genome-wide association scan of tag SNPs identifies a susceptibility locus for lung cancer at 15q25.1. *Nature Genetics, 40*, 616–622.

Anokhin, A. P., Vedeniapin, A. B., Sirevaag, E. J., Bauer, L. O., O'Connor, S. J., Kuperman, S., et al. (2000). The P300 brain potential is reduced in smokers. *Psychopharmacology, 149*, 409–413.

Arnfred, S. M., & Chen, A. C. (2004). Exploration of somatosensory P50 gating in schizophrenia spectrum patients: Reduced P50 amplitude correlates to social anhedonia. *Psychiatry Research, 125*, 147–160.

Arnfred, S. M., Chen, A. C., Glenthoj, B. Y., & Hemmingsen, R. P. (2003). Normal p50 gating in unmedicated schizophrenia outpatients. *The American Journal of Psychiatry*, *160*, 2236–2238.

Azzopardi, E., Typlt, M., Jenkins, B., & Schmid, S. (2013). Sensorimotor gating and spatial learning in alpha7-nicotinic receptor knockout mice. *Genes, Brain, and Behavior*, *12*, 414–423.

Bachmann, S., Weisbrod, M., Rohrig, M., Schroder, J., Thomas, C., Scherg, M., et al. (2010). MEG does not reveal impaired sensory gating in first-episode schizophrenia. *Schizophrenia Research*, *121*, 131–138.

Bak, N., Glenthoj, B. Y., Rostrup, E., Larsson, H. B., & Oranje, B. (2011). Source localization of sensory gating: A combined EEG and fMRI study in healthy volunteers. *NeuroImage*, *54*, 2711–2718.

Bak, N., Rostrup, E., Larsson, H. B., Glenthoj, B. Y., & Oranje, B. (2014). Concurrent functional magnetic resonance imaging and electroencephalography assessment of sensory gating in schizophrenia. *Human Brain Mapping*, *35*, 3578–3587.

Baldeweg, T., Klugman, A., Gruzelier, J., & Hirsch, S. R. (2004). Mismatch negativity potentials and cognitive impairment in schizophrenia. *Schizophrenia Research*, *69*, 203–217.

Baldeweg, T., Wong, D., & Stephan, K. E. (2006). Nicotinic modulation of human auditory sensory memory: Evidence from mismatch negativity potentials. *International Journal of Psychophysiology*, *59*, 49–58.

Barak, S., Arad, M., De Levie, A., Black, M. D., Griebel, G., & Weiner, I. (2009). Pro-cognitive and antipsychotic efficacy of the alpha7 nicotinic partial agonist SSR180711 in pharmacological and neurodevelopmental latent inhibition models of schizophrenia. *Neuropsychopharmacology*, *34*, 1753–1763.

Bar-Haim, Y., Marshall, P. J., Fox, N. A., Schorr, E. A., & Gordon-Salant, S. (2003). Mismatch negativity in socially withdrawn children. *Biological Psychiatry*, *54*, 17–24.

Barnes, M., Lawford, B. R., Burton, S. C., Heslop, K. R., Noble, E. P., Hausdorf, K., et al. (2006). Smoking and schizophrenia: Is symptom profile related to smoking and which antipsychotic medication is of benefit in reducing cigarette use? *The Australian and New Zealand Journal of Psychiatry*, *40*, 575–580.

Barr, R. S., Culhane, M. A., Jubelt, L. E., Mufti, R. S., Dyer, M. A., Weiss, A. P., et al. (2008). The effects of transdermal nicotine on cognition in nonsmokers with schizophrenia and nonpsychiatric controls. *Neuropsychopharmacology*, *33*, 480–490.

Ben-Shachar, S., Lanpher, B., German, J. R., Qasaymeh, M., Potocki, L., Nagamani, S. C., et al. (2009). Microdeletion 15q13.3: A locus with incomplete penetrance for autism, mental retardation, and psychiatric disorders. *Journal of Medical Genetics*, *46*, 382–388.

Berrettini, W., Yuan, X., Tozzi, F., Song, K., Francks, C., Chilcoat, H., et al. (2008). Alpha-5/alpha-3 nicotinic receptor subunit alleles increase risk for heavy smoking. *Molecular Psychiatry*, *13*, 368–373.

Besson, M., David, V., Suarez, S., Cormier, A., Cazala, P., Changeux, J. P., et al. (2006). Genetic dissociation of two behaviors associated with nicotine addiction: Beta-2 containing nicotinic receptors are involved in nicotine reinforcement but not in withdrawal syndrome. *Psychopharmacology*, *187*, 189–199.

Bickford, P. C., & Wear, K. D. (1995). Restoration of sensory gating of auditory evoked response by nicotine in fimbria-fornix lesioned rats. *Brain Research*, *705*, 235–240.

Bierut, L. J., Stitzel, J. A., Wang, J. C., Hinrichs, A. L., Grucza, R. A., Xuei, X., et al. (2008). Variants in nicotinic receptors and risk for nicotine dependence. *The American Journal of Psychiatry*, *165*, 1163–1171.

Blaveri, E., Kalsi, G., Lawrence, J., Quested, D., Moorey, H., Lamb, G., et al. (2001). Genetic association studies of schizophrenia using the 8p21-22 genes: Prepronociceptin (PNOC), neuronal nicotinic cholinergic receptor alpha polypeptide 2 (CHRNA2) and arylamine N-acetyltransferase 1 (NAT1). *European Journal of Human Genetics*, *9*, 469–472.

Blondel, A., Sanger, D. J., & Moser, P. C. (2000). Characterisation of the effects of nicotine in the five-choice serial reaction time task in rats: Antagonist studies. *Psychopharmacology, 149*, 293–305.
Bramon, E., Rabe-Hesketh, S., Sham, P., Murray, R. M., & Frangou, S. (2004). Meta-analysis of the P300 and P50 waveforms in schizophrenia. *Schizophrenia Research, 70*, 315–329.
Breitling, L. P., Dahmen, N., Mittelstrass, K., Rujescu, D., Gallinat, J., Fehr, C., et al. (2009). Association of nicotinic acetylcholine receptor subunit alpha 4 polymorphisms with nicotine dependence in 5500 Germans. *The Pharmacogenomics Journal, 9*, 219–224.
Brooks, J. M., Pershing, M. L., Thomsen, M. S., Mikkelsen, J. D., Sarter, M., & Bruno, J. P. (2012). Transient inactivation of the neonatal ventral hippocampus impairs attentional set-shifting behavior: Reversal with an alpha7 nicotinic agonist. *Neuropsychopharmacology, 37*, 2476–2486.
Brzustowicz, L. M., Hodgkinson, K. A., Chow, E. W., Honer, W. G., & Bassett, A. S. (2000). Location of a major susceptibility locus for familial schizophrenia on chromosome 1q21-q22. *Science, 288*, 678–682.
Buccafusco, J. J., Terry, A. V., Jr., Decker, M. W., & Gopalakrishnan, M. (2007). Profile of nicotinic acetylcholine receptor agonists ABT-594 and A-582941, with differential subtype selectivity, on delayed matching accuracy by young monkeys. *Biochemical Pharmacology, 74*, 1202–1211.
Callahan, P. M., Terry, A. V., Jr., & Tehim, A. (2014). Effects of the nicotinic alpha7 receptor partial agonist GTS-21 on NMDA-glutamatergic receptor related deficits in sensorimotor gating and recognition memory in rats. *Psychopharmacology, 231*, 3695–3706.
Castner, S. A., Smagin, G. N., Piser, T. M., Wang, Y., Smith, J. S., Christian, E. P., et al. (2011). Immediate and sustained improvements in working memory after selective stimulation of alpha7 nicotinic acetylcholine receptors. *Biological Psychiatry, 69*, 12–18.
Catts, S. V., Shelley, A. M., Ward, P. B., Liebert, B., McConaghy, N., Andrews, S., et al. (1995). Brain potential evidence for an auditory sensory memory deficit in schizophrenia. *The American Journal of Psychiatry, 152*, 213–219.
Cerimele, J. M., & Durango, A. (2012). Does varenicline worsen psychiatric symptoms in patients with schizophrenia or schizoaffective disorder? A review of published studies. *The Journal of Clinical Psychiatry, 73*, e1039–e1047.
Champtiaux, N., Gotti, C., Cordero-Erausquin, M., David, D. J., Przybylski, C., Lena, C., et al. (2003). Subunit composition of functional nicotinic receptors in dopaminergic neurons investigated with knock-out mice. *The Journal of Neuroscience, 23*, 7820–7829.
Chatterjee, S., Steensland, P., Simms, J. A., Holgate, J., Coe, J. W., Hurst, R. S., et al. (2011). Partial agonists of the alpha3beta4* neuronal nicotinic acetylcholine receptor reduce ethanol consumption and seeking in rats. *Neuropsychopharmacology, 36*, 603–615.
Cilia, J., Cluderay, J. E., Robbins, M. J., Reavill, C., Southam, E., Kew, J. N., et al. (2005). Reversal of isolation-rearing-induced PPI deficits by an alpha7 nicotinic receptor agonist. *Psychopharmacology, 182*, 214–219.
Cole, R. D., Poole, R. L., Guzman, D. M., Gould, T. J., & Parikh, V. (2014). Contributions of beta2 subunit-containing nAChRs to chronic nicotine-induced alterations in cognitive flexibility in mice. *Psychopharmacology, 232*(7), 1207–1217.
Conti, D. V., Lee, W., Li, D., Liu, J., Van Den Berg, D., Thomas, P. D., et al. (2008). Nicotinic acetylcholine receptor beta2 subunit gene implicated in a systems-based candidate gene study of smoking cessation. *Human Molecular Genetics, 17*, 2834–2848.
Cordero-Erausquin, M., Marubio, L. M., Klink, R., & Changeux, J. P. (2000). Nicotinic receptor function: New perspectives from knockout mice. *Trends in Pharmacological Sciences, 21*, 211–217.
Couturier, S., Bertrand, D., Matter, J. M., Hernandez, M. C., Bertrand, S., Millar, N., et al. (1990). A neuronal nicotinic acetylcholine receptor subunit (alpha 7) is

developmentally regulated and forms a homo-oligomeric channel blocked by alpha-BTX. *Neuron, 5*, 847–856.

Crow, T. J. (1980). Positive and negative schizophrenic symptoms and the role of dopamine. *The British Journal of Psychiatry, 137*, 383–386.

Cui, C., Booker, T. K., Allen, R. S., Grady, S. R., Whiteaker, P., Marks, M. J., et al. (2003). The beta3 nicotinic receptor subunit: A component of alpha-conotoxin MII-binding nicotinic acetylcholine receptors that modulate dopamine release and related behaviors. *The Journal of Neuroscience, 23*, 11045–11053.

Cullum, C. M., Harris, J. G., Waldo, M. C., Smernoff, E., Madison, A., Nagamoto, H. T., et al. (1993). Neurophysiological and neuropsychological evidence for attentional dysfunction in schizophrenia. *Schizophrenia Research, 10*, 131–141.

Dalack, G. W., Becks, L., Hill, E., Pomerleau, O. F., & Meador-Woodruff, J. H. (1999). Nicotine withdrawal and psychiatric symptoms in cigarette smokers with schizophrenia. *Neuropsychopharmacology, 21*, 195–202.

Dalack, G. W., & Meador-Woodruff, J. H. (1996). Smoking, smoking withdrawal and schizophrenia: Case reports and a review of the literature. *Schizophrenia Research, 22*, 133–141.

Dani, J. A., & Bertrand, D. (2007). Nicotinic acetylcholine receptors and nicotinic cholinergic mechanisms of the central nervous system. *Annual Review of Pharmacology and Toxicology, 47*, 699–729.

Davis, J. A., & Gould, T. J. (2007). beta2 subunit-containing nicotinic receptors mediate the enhancing effect of nicotine on trace cued fear conditioning in C57BL/6 mice. *Psychopharmacology, 190*, 343–352.

De Biasi, M., & Dani, J. A. (2011). Reward, addiction, withdrawal to nicotine. *Annual Review of Neuroscience, 34*, 105–130.

Decker, M. W., Brioni, J. D., Sullivan, J. P., Buckley, M. J., Radek, R. J., Raszkiewicz, J. L., et al. (1994). (S)-3-methyl-5-(1-methyl-2-pyrrolidinyl)isoxazole (ABT 418): A novel cholinergic ligand with cognition-enhancing and anxiolytic activities: II. In vivo characterization. *The Journal of Pharmacology and Experimental Therapeutics, 270*, 319–328.

de Leon, J. (1996). Smoking and vulnerability for schizophrenia. *Schizophrenia Bulletin, 22*, 405–409.

De Luca, V., Voineskos, S., Wong, G., & Kennedy, J. L. (2006). Genetic interaction between alpha4 and beta2 subunits of high affinity nicotinic receptor: Analysis in schizophrenia. *Experimental Brain Research, 174*, 292–296.

De Luca, V., Wong, A. H., Muller, D. J., Wong, G. W., Tyndale, R. F., & Kennedy, J. L. (2004). Evidence of association between smoking and alpha7 nicotinic receptor subunit gene in schizophrenia patients. *Neuropsychopharmacology, 29*, 1522–1526.

Depatie, L., O'Driscoll, G. A., Holahan, A. L., Atkinson, V., Thavundayil, J. X., Kin, N. N., et al. (2002). Nicotine and behavioral markers of risk for schizophrenia: A double-blind, placebo-controlled, cross-over study. *Neuropsychopharmacology, 27*, 1056–1070.

de Wilde, O. M., Bour, L. J., Dingemans, P. M., Koelman, J. H., & Linszen, D. H. (2007). A meta-analysis of P50 studies in patients with schizophrenia and relatives: Differences in methodology between research groups. *Schizophrenia Research, 97*, 137–151.

Dixon, L., Medoff, D. R., Wohlheiter, K., DiClemente, C., Goldberg, R., Kreyenbuhl, J., et al. (2007). Correlates of severity of smoking among persons with severe mental illness. *The American Journal on Addictions, 16*, 101–110.

Domjan, N., Csifcsak, G., Drotos, G., Janka, Z., & Szendi, I. (2012). Different patterns of auditory information processing deficits in chronic schizophrenia and bipolar disorder with psychotic features. *Schizophrenia Research, 139*, 253–259.

Dulude, L., Labelle, A., & Knott, V. J. (2010). Acute nicotine alteration of sensory memory impairment in smokers with schizophrenia. *Journal of Clinical Psychopharmacology, 30*, 541–548.

Dunbar, G., Boeijinga, P. H., Demazieres, A., Cisterni, C., Kuchibhatla, R., Wesnes, K., et al. (2007). Effects of TC-1734 (AZD3480), a selective neuronal nicotinic receptor agonist, on cognitive performance and the EEG of young healthy male volunteers. *Psychopharmacology, 191*, 919–929.

Dutar, P., Bassant, M. H., Senut, M. C., & Lamour, Y. (1995). The septohippocampal pathway: Structure and function of a central cholinergic system. *Physiological Reviews, 75*, 393–427.

Duvoisin, R. M., Deneris, E. S., Patrick, J., & Heinemann, S. (1989). The functional diversity of the neuronal nicotinic acetylcholine receptors is increased by a novel subunit: Beta 4. *Neuron, 3*, 487–496.

Ehrlichman, R. S., Maxwell, C. R., Majumdar, S., & Siegel, S. J. (2008). Deviance-elicited changes in event-related potentials are attenuated by ketamine in mice. *Journal of Cognitive Neuroscience, 20*, 1403–1414.

Engeland, C., Mahoney, C., Mohr, E., Ilivitsky, V., & Knott, V. (2002a). Nicotine and sensory memory in Alzheimer's disease: An event-related potential study. *Brain and Cognition, 49*, 232–234.

Engeland, C., Mahoney, C., Mohr, E., Ilivitsky, V., & Knott, V. J. (2002b). Acute nicotine effects on auditory sensory memory in tacrine-treated and nontreated patients with Alzheimer's disease: An event-related potential study. *Pharmacology, Biochemistry, and Behavior, 72*, 457–464.

Epping-Jordan, M. P., Picciotto, M. R., Changeux, J. P., & Pich, E. M. (1999). Assessment of nicotinic acetylcholine receptor subunit contributions to nicotine self-administration in mutant mice. *Psychopharmacology, 147*, 25–26.

Erwin, R. J., Turetsky, B. I., Moberg, P., Gur, R. C., & Gur, R. E. (1998). P50 abnormalities in schizophrenia: Relationship to clinical and neuropsychological indices of attention. *Schizophrenia Research, 33*, 157–167.

Espeseth, T., Endestad, T., Rootwelt, H., & Reinvang, I. (2007). Nicotine receptor gene CHRNA4 modulates early event-related potentials in auditory and visual oddball target detection tasks. *Neuroscience, 147*, 974–985.

Etter, J. F., Hoda, J. C., Perroud, N., Munafo, M., Buresi, C., Duret, C., et al. (2009). Association of genes coding for the alpha-4, alpha-5, beta-2 and beta-3 subunits of nicotinic receptors with cigarette smoking and nicotine dependence. *Addictive Behaviors, 34*, 772–775.

Euser, A. S., Arends, L. R., Evans, B. E., Greaves-Lord, K., Huizink, A. C., & Franken, I. H. (2012). The P300 event-related brain potential as a neurobiological endophenotype for substance use disorders: A meta-analytic investigation. *Neuroscience and Biobehavioral Reviews, 36*, 572–603.

Evans, D. E., Jentink, K. G., Sutton, S. K., Van Rensburg, K. J., & Drobes, D. J. (2014). 7mg nicotine patch fails to enhance P300 neural indices of cognitive control among nonsmokers. *Pharmacology, Biochemistry, and Behavior, 126C*, 77–82.

Evans, D. E., MacQueen, D. A., Jentink, K. G., Park, J. Y., Lin, H. Y., & Drobes, D. J. (2014). CHRNA5 variants moderate the effect of nicotine deprivation on a neural index of cognitive control. *Genes, Brain, and Behavior, 13*, 626–632.

Evans, D. E., Maxfield, N. D., Van Rensburg, K. J., Oliver, J. A., Jentink, K. G., & Drobes, D. J. (2013). Nicotine deprivation influences P300 markers of cognitive control. *Neuropsychopharmacology, 38*, 2525–2531.

Exley, R., Maubourguet, N., David, V., Eddine, R., Evrard, A., Pons, S., et al. (2011). Distinct contributions of nicotinic acetylcholine receptor subunit alpha4 and subunit alpha6 to the reinforcing effects of nicotine. *Proceedings of the National Academy of Sciences of the United States of America, 108*, 7577–7582.

Fabiani, M., Karis, D., & Donchin, E. (1986). P300 and recall in an incidental memory paradigm. *Psychophysiology, 23*, 298–308.

Faraone, S. V., Su, J., Taylor, L., Wilcox, M., Van Eerdewegh, P., & Tsuang, M. T. (2004). A novel permutation testing method implicates sixteen nicotinic acetylcholine receptor genes as risk factors for smoking in schizophrenia families. *Human Heredity, 57*, 59–68.

Featherstone, R. E., Phillips, J. M., Thieu, T., Ehrlichman, R. S., Halene, T. B., Leiser, S. C., et al. (2012). Nicotine receptor subtype-specific effects on auditory evoked oscillations and potentials. *PLoS One*, 7, e39775.

Fejgin, K., Nielsen, J., Birknow, M. R., Bastlund, J. F., Nielsen, V., Lauridsen, J. B., et al. (2014). A mouse model that recapitulates cardinal features of the 15q13.3 microdeletion syndrome including schizophrenia- and epilepsy-related alterations. *Biological Psychiatry, 76*, 128–137.

Feng, Y., Niu, T., Xing, H., Xu, X., Chen, C., Peng, S., et al. (2004). A common haplotype of the nicotine acetylcholine receptor alpha 4 subunit gene is associated with vulnerability to nicotine addiction in men. *American Journal of Human Genetics, 75*, 112–121.

Fernandes, C., Hoyle, E., Dempster, E., Schalkwyk, L. C., & Collier, D. A. (2006). Performance deficit of alpha7 nicotinic receptor knockout mice in a delayed matching-to-place task suggests a mild impairment of working/episodic-like memory. *Genes, Brain, and Behavior, 5*, 433–440.

Feuerbach, D., Lingenhoehl, K., Olpe, H. R., Vassout, A., Gentsch, C., Chaperon, F., et al. (2009). The selective nicotinic acetylcholine receptor alpha7 agonist JN403 is active in animal models of cognition, sensory gating, epilepsy and pain. *Neuropharmacology, 56*, 254–263.

Fisher, D. J., Grant, B., Smith, D. M., Borracci, G., Labelle, A., & Knott, V. J. (2012). Nicotine and the hallucinating brain: Effects on mismatch negativity (MMN) in schizophrenia. *Psychiatry Research, 196*, 181–187.

Fisher, D. J., Scott, T. L., Shah, D. K., Prise, S., Thompson, M., & Knott, V. J. (2010). Light up and see: Enhancement of the visual mismatch negativity (vMMN) by nicotine. *Brain Research, 1313*, 162–171.

Flores, C. M., Davila-Garcia, M. I., Ulrich, Y. M., & Kellar, K. J. (1997). Differential regulation of neuronal nicotinic receptor binding sites following chronic nicotine administration. *Journal of Neurochemistry, 69*, 2216–2219.

Flores, C. M., Rogers, S. W., Pabreza, L. A., Wolfe, B. B., & Kellar, K. J. (1992). A subtype of nicotinic cholinergic receptor in rat brain is composed of alpha 4 and beta 2 subunits and is up-regulated by chronic nicotine treatment. *Molecular Pharmacology, 41*, 31–37.

Ford, J. M. (1999). Schizophrenia: The broken P300 and beyond. *Psychophysiology, 36*, 667–682.

Freedman, R. (2007). Exacerbation of schizophrenia by varenicline. *The American Journal of Psychiatry, 164*, 1269.

Freedman, R., Adler, L. E., Waldo, M. C., Pachtman, E., & Franks, R. D. (1983). Neurophysiological evidence for a defect in inhibitory pathways in schizophrenia: Comparison of medicated and drug-free patients. *Biological Psychiatry, 18*, 537–551.

Freedman, R., Coon, H., Myles-Worsley, M., Orr-Urtreger, A., Olincy, A., Davis, A., et al. (1997). Linkage of a neurophysiological deficit in schizophrenia to a chromosome 15 locus. *Proceedings of the National Academy of Sciences of the United States of America, 94*, 587–592.

Freedman, R., Leonard, S., Gault, J. M., Hopkins, J., Cloninger, C. R., Kaufmann, C. A., et al. (2001). Linkage disequilibrium for schizophrenia at the chromosome 15q13-14 locus of the alpha7-nicotinic acetylcholine receptor subunit gene (CHRNA7). *American Journal of Medical Genetics, 105*, 20–22.

Freedman, R., Olincy, A., Buchanan, R. W., Harris, J. G., Gold, J. M., Johnson, L., et al. (2008). Initial phase 2 trial of a nicotinic agonist in schizophrenia. *The American Journal of Psychiatry, 165*, 1040–1047.

Gangitano, D., Salas, R., Teng, Y., Perez, E., & De Biasi, M. (2009). Progesterone modulation of alpha5 nAChR subunits influences anxiety-related behavior during estrus cycle. *Genes, Brain, and Behavior, 8*, 398–406.
Gatto, G. J., Bohme, G. A., Caldwell, W. S., Letchworth, S. R., Traina, V. M., Obinu, M. C., et al. (2004). TC-1734: An orally active neuronal nicotinic acetylcholine receptor modulator with antidepressant, neuroprotective and long-lasting cognitive effects. *CNS Drug Reviews, 10*, 147–166.
George, T. P., Vessicchio, J. C., Termine, A., Sahady, D. M., Head, C. A., Pepper, W. T., et al. (2002). Effects of smoking abstinence on visuospatial working memory function in schizophrenia. *Neuropsychopharmacology, 26*, 75–85.
Gerzanich, V., Wang, F., Kuryatov, A., & Lindstrom, J. (1998). Alpha 5 subunit alters desensitization, pharmacology, Ca++ permeability and Ca++ modulation of human neuronal alpha 3 nicotinic receptors. *The Journal of Pharmacology and Experimental Therapeutics, 286*, 311–320.
Glynn, S. M., & Sussman, S. (1990). Why patients smoke. *Hospital & Community Psychiatry, 41*, 1027–1028.
Goff, D. C., & Coyle, J. T. (2001). The emerging role of glutamate in the pathophysiology and treatment of schizophrenia. *The American Journal of Psychiatry, 158*, 1367–1377.
Goff, D. C., Henderson, D. C., & Amico, E. (1992). Cigarette smoking in schizophrenia: Relationship to psychopathology and medication side effects. *The American Journal of Psychiatry, 149*, 1189–1194.
Gotti, C., & Clementi, F. (2004). Neuronal nicotinic receptors: From structure to pathology. *Progress in Neurobiology, 74*, 363–396.
Gotti, C., Clementi, F., Fornari, A., Gaimarri, A., Guiducci, S., Manfredi, I., et al. (2009). Structural and functional diversity of native brain neuronal nicotinic receptors. *Biochemical Pharmacology, 78*, 703–711.
Gotti, C., Zoli, M., & Clementi, F. (2006). Brain nicotinic acetylcholine receptors: Native subtypes and their relevance. *Trends in Pharmacological Sciences, 27*, 482–491.
Grabus, S. D., Martin, B. R., & Imad, Damaj M. (2005). Nicotine physical dependence in the mouse: Involvement of the alpha7 nicotinic receptor subtype. *European Journal of Pharmacology, 515*, 90–93.
Granon, S., Faure, P., & Changeux, J. P. (2003). Executive and social behaviors under nicotinic receptor regulation. *Proceedings of the National Academy of Sciences of the United States of America, 100*, 9596–9601.
Green, M. F. (1996). What are the functional consequences of neurocognitive deficits in schizophrenia? *The American Journal of Psychiatry, 153*, 321–330.
Greenwood, P. M., Sundararajan, R., Lin, M. K., Kumar, R., Fryxell, K. J., & Parasuraman, R. (2009). Both a nicotinic single nucleotide polymorphism (SNP) and a noradrenergic SNP modulate working memory performance when attention is manipulated. *Journal of Cognitive Neuroscience, 21*, 2139–2153.
Griffith, J. M., O'Neill, J. E., Petty, F., Garver, D., Young, D., & Freedman, R. (1998). Nicotinic receptor desensitization and sensory gating deficits in schizophrenia. *Biological Psychiatry, 44*, 98–106.
Grottick, A. J., Haman, M., Wyler, R., & Higgins, G. A. (2003). Reversal of a vigilance decrement in the aged rat by subtype-selective nicotinic ligands. *Neuropsychopharmacology, 28*, 880–887.
Grottick, A. J., & Higgins, G. A. (2000). Effect of subtype selective nicotinic compounds on attention as assessed by the five-choice serial reaction time task. *Behavioural Brain Research, 117*, 197–208.
Grupe, M., Grunnet, M., Laursen, B., & Bastlund, J. F. (2014). Neuropharmacological modulation of the P3-like event-related potential in a rat two-tone auditory discrimination

task with modafinil and NS9283, a positive allosteric modulator of alpha4beta2 nAChRs. *Neuropharmacology*, *79*, 444–455.

Grzella, I., Muller, B. W., Oades, R. D., Bender, S., Schall, U., Zerbin, D., et al. (2001). Novelty-elicited mismatch negativity in patients with schizophrenia on admission and discharge. *Journal of Psychiatry & Neuroscience*, *26*, 235–246.

Gunduz-Bruce, H., Reinhart, R. M., Roach, B. J., Gueorguieva, R., Oliver, S., D'Souza, D. C., et al. (2012). Glutamatergic modulation of auditory information processing in the human brain. *Biological Psychiatry*, *71*, 969–977.

Guney, F., Genc, B. O., Kutlu, R., & Ilhan, B. C. (2009). Auditory P300 event-related potential in tobacco smokers. *Journal of Clinical Neuroscience*, *16*, 1311–1315.

Guterman, Y., & Josiassen, R. C. (1994). Sensory gating deviance in schizophrenia in the context of task related effects. *International Journal of Psychophysiology*, *18*, 1–12.

Hahn, B., Sharples, C. G., Wonnacott, S., Shoaib, M., & Stolerman, I. P. (2003). Attentional effects of nicotinic agonists in rats. *Neuropharmacology*, *44*, 1054–1067.

Hajos, M., Hurst, R. S., Hoffmann, W. E., Krause, M., Wall, T. M., Higdon, N. R., et al. (2005). The selective alpha7 nicotinic acetylcholine receptor agonist PNU-282987 [N-[(3R)-1-Azabicyclo[2.2.2]oct-3-yl]-4-chlorobenzamide hydrochloride] enhances GABAergic synaptic activity in brain slices and restores auditory gating deficits in anesthetized rats. *The Journal of Pharmacology and Experimental Therapeutics*, *312*, 1213–1222.

Hamera, E., Schneider, J. K., & Deviney, S. (1995). Alcohol, cannabis, nicotine, and caffeine use and symptom distress in schizophrenia. *The Journal of Nervous and Mental Disease*, *183*, 559–565.

Han, Z. Y., Le Novere, N., Zoli, M., Hill, J. A., Jr., Champtiaux, N., & Changeux, J. P. (2000). Localization of nAChR subunit mRNAs in the brain of Macaca mulatta. *The European Journal of Neuroscience*, *12*, 3664–3674.

Harenza, J. L., Muldoon, P. P., De Biasi, M., Damaj, M. I., & Miles, M. F. (2014). Genetic variation within the Chrna7 gene modulates nicotine reward-like phenotypes in mice. *Genes, Brain, and Behavior*, *13*, 213–225.

Harkrider, A. W., & Hedrick, M. S. (2005). Acute effect of nicotine on auditory gating in smokers and non-smokers. *Hearing Research*, *202*, 114–128.

Harris, J. G., Kongs, S., Allensworth, D., Martin, L., Tregellas, J., Sullivan, B., et al. (2004). Effects of nicotine on cognitive deficits in schizophrenia. *Neuropsychopharmacology*, *29*, 1378–1385.

Hauser, T. A., Kucinski, A., Jordan, K. G., Gatto, G. J., Wersinger, S. R., Hesse, R. A., et al. (2009). TC-5619: An alpha7 neuronal nicotinic receptor-selective agonist that demonstrates efficacy in animal models of the positive and negative symptoms and cognitive dysfunction of schizophrenia. *Biochemical Pharmacology*, *78*, 803–812.

Heitjan, D. F., Guo, M., Ray, R., Wileyto, E. P., Epstein, L. H., & Lerman, C. (2008). Identification of pharmacogenetic markers in smoking cessation therapy. *American Journal of Medical Genetics. Part B, Neuropsychiatric Genetics*, *147B*, 712–719.

Herran, A., de Santiago, A., Sandoya, M., Fernandez, M. J., Diez-Manrique, J. F., & Vazquez-Barquero, J. L. (2000). Determinants of smoking behaviour in outpatients with schizophrenia. *Schizophrenia Research*, *41*, 373–381.

Hong, L. E., Thaker, G. K., McMahon, R. P., Summerfelt, A., Rachbeisel, J., Fuller, R. L., et al. (2011). Effects of moderate-dose treatment with varenicline on neurobiological and cognitive biomarkers in smokers and nonsmokers with schizophrenia or schizoaffective disorder. *Archives of General Psychiatry*, *68*, 1195–1206.

Hong, L. E., Yang, X., Wonodi, I., Hodgkinson, C. A., Goldman, D., Stine, O. C., et al. (2011). A CHRNA5 allele related to nicotine addiction and schizophrenia. *Genes, Brain, and Behavior*, *10*, 530–535.

Hoppman-Chaney, N., Wain, K., Seger, P. R., Superneau, D. W., & Hodge, J. C. (2013). Identification of single gene deletions at 15q13.3: Further evidence that CHRNA7 causes the 15q13.3 microdeletion syndrome phenotype. *Clinical Genetics, 83*, 345–351.

Horst, N. K., Heath, C. J., Neugebauer, N. M., Kimchi, E. Y., Laubach, M., & Picciotto, M. R. (2012). Impaired auditory discrimination learning following perinatal nicotine exposure or beta2 nicotinic acetylcholine receptor subunit deletion. *Behavioural Brain Research, 231*, 170–180.

Horton, J., Millar, A., Labelle, A., & Knott, V. J. (2011). MMN responsivity to manipulations of frequency and duration deviants in chronic, clozapine-treated schizophrenia patients. *Schizophrenia Research, 126*, 202–211.

Houlihan, M. E., Pritchard, W. S., & Robinson, J. H. (1996). Faster P300 latency after smoking in visual but not auditory oddball tasks. *Psychopharmacology, 123*, 231–238.

Houlihan, M. E., Pritchard, W. S., & Robinson, J. H. (2001). Effects of smoking/nicotine on performance and event-related potentials during a short-term memory scanning task. *Psychopharmacology, 156*, 388–396.

Howe, W. M., Ji, J., Parikh, V., Williams, S., Mocaer, E., Trocme-Thibierge, C., et al. (2010). Enhancement of attentional performance by selective stimulation of alpha4beta2(*) nAChRs: Underlying cholinergic mechanisms. *Neuropsychopharmacology, 35*, 1391–1401.

Hoyle, E., Genn, R. F., Fernandes, C., & Stolerman, I. P. (2006). Impaired performance of alpha7 nicotinic receptor knockout mice in the five-choice serial reaction time task. *Psychopharmacology, 189*, 211–223.

Hsieh, M. H., Liu, K., Liu, S. K., Chiu, M. J., Hwu, H. G., & Chen, A. C. (2004). Memory impairment and auditory evoked potential gating deficit in schizophrenia. *Psychiatry Research, 130*, 161–169.

Hughes, J. R., Hatsukami, D. K., Mitchell, J. E., & Dahlgren, L. A. (1986). Prevalence of smoking among psychiatric outpatients. *The American Journal of Psychiatry, 143*, 993–997.

Hutchison, K. E., Allen, D. L., Filbey, F. M., Jepson, C., Lerman, C., Benowitz, N. L., et al. (2007). CHRNA4 and tobacco dependence: From gene regulation to treatment outcome. *Archives of General Psychiatry, 64*, 1078–1086.

Ilan, A. B., & Polich, J. (1999). Tobacco smoking and memory scanning: Behavioral and event-related potential effects. *Nicotine & Tobacco Research, 1*, 233–240.

Inami, R., Kirino, E., Inoue, R., & Arai, H. (2005). Transdermal nicotine administration enhances automatic auditory processing reflected by mismatch negativity. *Pharmacology, Biochemistry, and Behavior, 80*, 453–461.

Inami, R., Kirino, E., Inoue, R., Suzuki, T., & Arai, H. (2007). Nicotine effects on mismatch negativity in nonsmoking schizophrenic patients. *Neuropsychobiology, 56*, 64–72.

Jackson, K. J., Fanous, A. H., Chen, J., Kendler, K. S., & Chen, X. (2013). Variants in the 15q25 gene cluster are associated with risk for schizophrenia and bipolar disorder. *Psychiatric Genetics, 23*, 20–28.

Jackson, K. J., Marks, M. J., Vann, R. E., Chen, X., Gamage, T. F., Warner, J. A., et al. (2010). Role of alpha5 nicotinic acetylcholine receptors in pharmacological and behavioral effects of nicotine in mice. *The Journal of Pharmacology and Experimental Therapeutics, 334*, 137–146.

Jackson, K. J., Martin, B. R., Changeux, J. P., & Damaj, M. I. (2008). Differential role of nicotinic acetylcholine receptor subunits in physical and affective nicotine withdrawal signs. *The Journal of Pharmacology and Experimental Therapeutics, 325*, 302–312.

Jacobsen, L. K., D'Souza, D. C., Mencl, W. E., Pugh, K. R., Skudlarski, P., & Krystal, J. H. (2004). Nicotine effects on brain function and functional connectivity in schizophrenia. *Biological Psychiatry, 55*, 850–858.

Jahshan, C., Cadenhead, K. S., Rissling, A. J., Kirihara, K., Braff, D. L., & Light, G. A. (2012). Automatic sensory information processing abnormalities across the illness course of schizophrenia. *Psychological Medicine, 42*, 85–97.

Javitt, D. C., Doneshka, P., Zylberman, I., Ritter, W., & Vaughan, H. G., Jr. (1993). Impairment of early cortical processing in schizophrenia: An event-related potential confirmation study. *Biological Psychiatry, 33*, 513–519.

Javitt, D. C., Shelley, A., & Ritter, W. (2000). Associated deficits in mismatch negativity generation and tone matching in schizophrenia. *Clinical Neurophysiology, 111*, 1733–1737.

Javitt, D. C., Steinschneider, M., Schroeder, C. E., & Arezzo, J. C. (1996). Role of cortical N-methyl-D-aspartate receptors in auditory sensory memory and mismatch negativity generation: Implications for schizophrenia. *Proceedings of the National Academy of Sciences of the United States of America, 93*, 11962–11967.

Jentsch, J. D., & Roth, R. H. (1999). The neuropsychopharmacology of phencyclidine: From NMDA receptor hypofunction to the dopamine hypothesis of schizophrenia. *Neuropsychopharmacology, 20*, 201–225.

Jeon, Y. W., & Polich, J. (2003). Meta-analysis of P300 and schizophrenia: Patients, paradigms, and practical implications. *Psychophysiology, 40*, 684–701.

Judd, L. L., McAdams, L., Budnick, B., & Braff, D. L. (1992). Sensory gating deficits in schizophrenia: New results. *The American Journal of Psychiatry, 149*, 488–493.

Kapur, S., & Remington, G. (2001). Dopamine D(2) receptors and their role in atypical antipsychotic action: Still necessary and may even be sufficient. *Biological Psychiatry, 50*, 873–883.

Kathmann, N., Wagner, M., Rendtorff, N., & Engel, R. R. (1995). Delayed peak latency of the mismatch negativity in schizophrenics and alcoholics. *Biological Psychiatry, 37*, 754–757.

Kawakubo, Y., & Kasai, K. (2006). Support for an association between mismatch negativity and social functioning in schizophrenia. *Progress in Neuro-Psychopharmacology & Biological Psychiatry, 30*, 1367–1368.

Kedmi, M., Beaudet, A. L., & Orr-Urtreger, A. (2004). Mice lacking neuronal nicotinic acetylcholine receptor beta4-subunit and mice lacking both alpha5- and beta4-subunits are highly resistant to nicotine-induced seizures. *Physiological Genomics, 17*, 221–229.

Keller, J. J., Keller, A. B., Bowers, B. J., & Wehner, J. M. (2005). Performance of alpha7 nicotinic receptor null mutants is impaired in appetitive learning measured in a signaled nose poke task. *Behavioural Brain Research, 162*, 143–152.

Kelly, C., & McCreadie, R. G. (1999). Smoking habits, current symptoms, and premorbid characteristics of schizophrenic patients in Nithsdale, Scotland. *The American Journal of Psychiatry, 156*, 1751–1757.

Kim, M. S., Kang, S. S., Youn, T., Kang, D. H., Kim, J. J., & Kwon, J. S. (2003). Neuropsychological correlates of P300 abnormalities in patients with schizophrenia and obsessive-compulsive disorder. *Psychiatry Research, 123*, 109–123.

Kim, D. W., Shim, M., Kim, J. I., Im, C. H., & Lee, S. H. (2014). Source activation of P300 correlates with negative symptom severity in patients with schizophrenia. *Brain Topography, 27*, 307–317.

Kishi, T., Fukuo, Y., Okochi, T., Kawashima, K., Moriwaki, M., Furukawa, O., et al. (2013). Relationship between nicotine dependence and the endophenotype-related trait of cognitive function but not acoustic startle reponses in Japanese patients with schizophrenia. *Human Psychopharmacology, 28*, 220–229.

Kishi, T., Ikeda, M., Kitajima, T., Yamanouchi, Y., Kinoshita, Y., Kawashima, K., et al. (2008). Genetic association analysis of tagging SNPs in alpha4 and beta2 subunits of neuronal nicotinic acetylcholine receptor genes (CHRNA4 and CHRNB2) with schizophrenia in the Japanese population. *Journal of Neural Transmission, 115*, 1457–1461.

Knott, V., Blais, C., Scherling, C., Camarda, J., Millar, A., Fisher, D., et al. (2006). Neural effects of nicotine during auditory selective attention in smokers: An event-related potential study. *Neuropsychobiology, 53*, 115–126.

Knott, V. J., Bolton, K., Heenan, A., Shah, D., Fisher, D. J., & Villeneuve, C. (2009). Effects of acute nicotine on event-related potential and performance indices of auditory distraction in nonsmokers. *Nicotine & Tobacco Research, 11*, 519–530.

Knott, V., Bosman, M., Mahoney, C., Ilivitsky, V., & Quirt, K. (1999). Transdermal nicotine: Single dose effects on mood, EEG, performance, and event-related potentials. *Pharmacology, Biochemistry, and Behavior, 63*, 253–261.

Knott, V., Choueiry, J., Dort, H., Smith, D., Impey, D., de la Salle, S., et al. (2014). Baseline-dependent modulating effects of nicotine on voluntary and involuntary attention measured with brain event-related P3 potentials. *Pharmacology, Biochemistry, and Behavior, 122*, 107–117.

Knott, V., Heenan, A., Shah, D., Bolton, K., Fisher, D., & Villeneuve, C. (2011). Electrophysiological evidence of nicotine's distracter-filtering properties in non-smokers. *Journal of Psychopharmacology, 25*, 239–248.

Knott, V., Impey, D., Philippe, T., Smith, D., Choueiry, J., de la Salle, S., et al. (2014). Modulation of auditory deviance detection by acute nicotine is baseline and deviant dependent in healthy nonsmokers: A mismatch negativity study. *Human Psychopharmacology, 29*, 446–458.

Knott, V. J., Millar, A. M., McIntosh, J. F., Shah, D. K., Fisher, D. J., Blais, C. M., et al. (2011). Separate and combined effects of low dose ketamine and nicotine on behavioural and neural correlates of sustained attention. *Biological Psychology, 88*, 83–93.

Knott, V. J., Scherling, C. S., Blais, C. M., Camarda, J., Fisher, D. J., Millar, A., et al. (2006). Acute nicotine fails to alter event-related potential or behavioral performance indices of auditory distraction in cigarette smokers. *Nicotine & Tobacco Research, 8*, 263–273.

Knott, V., Shah, D., Millar, A., McIntosh, J., Fisher, D., Blais, C., et al. (2012). Nicotine, auditory sensory memory, and sustained attention in a human ketamine model of schizophrenia: Moderating influence of a hallucinatory trait. *Frontiers in Pharmacology, 3*, 172.

Kohlhaas, K. L., Bitner, R. S., Gopalakrishnan, M., & Rueter, L. E. (2012). Effects of alpha7 nicotinic acetylcholine receptor agonists on antipsychotic efficacy in a preclinical mouse model of psychosis. *Psychopharmacology, 220*, 823–833.

Koike, K., Hashimoto, K., Takai, N., Shimizu, E., Komatsu, N., Watanabe, H., et al. (2005). Tropisetron improves deficits in auditory P50 suppression in schizophrenia. *Schizophrenia Research, 76*, 67–72.

Korzyukov, O., Pflieger, M. E., Wagner, M., Bowyer, S. M., Rosburg, T., Sundaresan, K., et al. (2007). Generators of the intracranial P50 response in auditory sensory gating. *NeuroImage, 35*, 814–826.

Kotov, R., Guey, L. T., Bromet, E. J., & Schwartz, J. E. (2010). Smoking in schizophrenia: Diagnostic specificity, symptom correlates, and illness severity. *Schizophrenia Bulletin, 36*, 173–181.

Kreitschmann-Andermahr, I., Rosburg, T., Demme, U., Gaser, E., Nowak, H., & Sauer, H. (2001). Effect of ketamine on the neuromagnetic mismatch field in healthy humans. *Brain Research. Cognitive Brain Research, 12*, 109–116.

Krishnadas, R., Jauhar, S., Telfer, S., Shivashankar, S., & McCreadie, R. G. (2012). Nicotine dependence and illness severity in schizophrenia. *The British Journal of Psychiatry, 201*, 306–312.

Labarca, C., Schwarz, J., Deshpande, P., Schwarz, S., Nowak, M. W., Fonck, C., et al. (2001). Point mutant mice with hypersensitive alpha 4 nicotinic receptors show dopaminergic deficits and increased anxiety. *Proceedings of the National Academy of Sciences of the United States of America, 98*, 2786–2791.

Lamberti, J. S., Schwarzkopf, S. B., Boutros, N., Crilly, J. F., & Martin, R. (1993). Within-session changes in sensory gating assessed by P50 evoked potentials in normal subjects. *Progress in Neuro-Psychopharmacology & Biological Psychiatry, 17*, 781–791.

Le Novere, N., Zoli, M., & Changeux, J. P. (1996). Neuronal nicotinic receptor alpha 6 subunit mRNA is selectively concentrated in catecholaminergic nuclei of the rat brain. *The European Journal of Neuroscience*, *8*, 2428–2439.

Leonard, S., Gault, J., Moore, T., Hopkins, J., Robinson, M., Olincy, A., et al. (1998). Further investigation of a chromosome 15 locus in schizophrenia: Analysis of affected sibpairs from the NIMH genetics initiative. *American Journal of Medical Genetics*, *81*, 308–312.

Li, M. D., Beuten, J., Ma, J. Z., Payne, T. J., Lou, X. Y., Garcia, V., et al. (2005). Ethnic- and gender-specific association of the nicotinic acetylcholine receptor alpha4 subunit gene (CHRNA4) with nicotine dependence. *Human Molecular Genetics*, *14*, 1211–1219.

Liao, D. L., Yang, J. Y., Lee, S. M., Chen, H., & Tsai, S. J. (2002). Smoking in chronic schizophrenic inpatients in taiwan. *Neuropsychobiology*, *45*, 172–175.

Lieberman, J. A., Dunbar, G., Segreti, A. C., Girgis, R. R., Seoane, F., Beaver, J. S., et al. (2013). A randomized exploratory trial of an alpha-7 nicotinic receptor agonist (TC-5619) for cognitive enhancement in schizophrenia. *Neuropsychopharmacology*, *38*, 968–975.

Light, G. A., & Braff, D. L. (2005a). Mismatch negativity deficits are associated with poor functioning in schizophrenia patients. *Archives of General Psychiatry*, *62*, 127–136.

Light, G. A., & Braff, D. L. (2005b). Stability of mismatch negativity deficits and their relationship to functional impairments in chronic schizophrenia. *The American Journal of Psychiatry*, *162*, 1741–1743.

Light, G. A., Geyer, M. A., Clementz, B. A., Cadenhead, K. S., & Braff, D. L. (2000). Normal P50 suppression in schizophrenia patients treated with atypical antipsychotic medications. *The American Journal of Psychiatry*, *157*, 767–771.

Lin, H., Hsu, F. C., Baumann, B. H., Coulter, D. A., Anderson, S. A., & Lynch, D. R. (2014). Cortical parvalbumin GABAergic deficits with alpha7 nicotinic acetylcholine receptor deletion: Implications for schizophrenia. *Molecular and Cellular Neurosciences*, *61*, 163–175.

Lin, H., Hsu, F. C., Baumann, B. H., Coulter, D. A., & Lynch, D. R. (2014). Cortical synaptic NMDA receptor deficits in alpha7 nicotinic acetylcholine receptor gene deletion models: Implications for neuropsychiatric diseases. *Neurobiology of Disease*, *63*, 129–140.

Lindgren, M., Molander, L., Verbaan, C., Lunell, E., & Rosen, I. (1999). Electroencephalographic effects of intravenous nicotine—A dose–response study. *Psychopharmacology*, *145*, 342–350.

Lindstrom, J. (1997). Nicotinic acetylcholine receptors in health and disease. *Molecular Neurobiology*, *15*, 193–222.

Lips, E. H., Gaborieau, V., McKay, J. D., Chabrier, A., Hung, R. J., Boffetta, P., et al. (2010). Association between a 15q25 gene variant, smoking quantity and tobacco-related cancers among 17 000 individuals. *International Journal of Epidemiology*, *39*, 563–577.

Liu, Z., Tam, W. C., Xue, Z., Yao, S., & Wu, D. (2004). Positive and negative symptom profile schizophrenia and abnormalities in the P300 component of the event-related potential: A longitudinal controlled study. *Psychiatry Research*, *132*, 131–139.

Lotfipour, S., Byun, J. S., Leach, P., Fowler, C. D., Murphy, N. P., Kenny, P. J., et al. (2013). Targeted deletion of the mouse alpha2 nicotinic acetylcholine receptor subunit gene (Chrna2) potentiates nicotine-modulated behaviors. *The Journal of Neuroscience*, *33*, 7728–7741.

Louchart-de la Chapelle, S., Levillain, D., Menard, J. F., Van der Elst, A., Allio, G., Haouzir, S., et al. (2005). P50 inhibitory gating deficit is correlated with the negative symptomatology of schizophrenia. *Psychiatry Research*, *136*, 27–34.

Lueders, K. K., Hu, S., McHugh, L., Myakishev, M. V., Sirota, L. A., & Hamer, D. H. (2002). Genetic and functional analysis of single nucleotide polymorphisms in the

beta2-neuronal nicotinic acetylcholine receptor gene (CHRNB2). *Nicotine & Tobacco Research*, *4*, 115–125.
Luntz-Leybman, V., Bickford, P. C., & Freedman, R. (1992). Cholinergic gating of response to auditory stimuli in rat hippocampus. *Brain Research*, *587*, 130–136.
Lyons, M. J., Bar, J. L., Kremen, W. S., Toomey, R., Eisen, S. A., Goldberg, J., et al. (2002). Nicotine and familial vulnerability to schizophrenia: A discordant twin study. *Journal of Abnormal Psychology*, *111*, 687–693.
Mao, D., Perry, D. C., Yasuda, R. P., Wolfe, B. B., & Kellar, K. J. (2008). The alpha4beta2alpha5 nicotinic cholinergic receptor in rat brain is resistant to up-regulation by nicotine in vivo. *Journal of Neurochemistry*, *104*, 446–456.
Marquis, K. L., Comery, T. A., Jow, F., Navarra, R. L., Grauer, S. M., Pulicicchio, C., et al. (2011). Preclinical assessment of an adjunctive treatment approach for cognitive impairment associated with schizophrenia using the alpha7 nicotinic acetylcholine receptor agonist WYE-103914/SEN34625. *Psychopharmacology*, *218*, 635–647.
Martin, L. F., Davalos, D. B., & Kisley, M. A. (2009). Nicotine enhances automatic temporal processing as measured by the mismatch negativity waveform. *Nicotine & Tobacco Research*, *11*, 698–706.
Marubio, L. M., del Mar Arroyo-Jimenez, M., Cordero-Erausquin, M., Lena, C., Le Novere, N., de Kerchove d'Exaerde, A., et al. (1999). Reduced antinociception in mice lacking neuronal nicotinic receptor subunits. *Nature*, *398*, 805–810.
Marubio, L. M., Gardier, A. M., Durier, S., David, D., Klink, R., Arroyo-Jimenez, M. M., et al. (2003). Effects of nicotine in the dopaminergic system of mice lacking the alpha4 subunit of neuronal nicotinic acetylcholine receptors. *The European Journal of Neuroscience*, *17*, 1329–1337.
Maskos, U., Molles, B. E., Pons, S., Besson, M., Guiard, B. P., Guilloux, J. P., et al. (2005). Nicotine reinforcement and cognition restored by targeted expression of nicotinic receptors. *Nature*, *436*, 103–107.
Mathalon, D. H., Ahn, K. H., Perry, E. B., Jr., Cho, H. S., Roach, B. J., Blais, R. K., et al. (2014). Effects of nicotine on the neurophysiological and behavioral effects of ketamine in humans. *Frontiers in Psychiatry*, *5*, 3.
McLean, S. L., Grayson, B., Idris, N. F., Lesage, A. S., Pemberton, D. J., Mackie, C., et al. (2011). Activation of alpha7 nicotinic receptors improves phencyclidine-induced deficits in cognitive tasks in rats: Implications for therapy of cognitive dysfunction in schizophrenia. *European Neuropsychopharmacology*, *21*, 333–343.
McLean, S. L., Idris, N. F., Grayson, B., Gendle, D. F., Mackie, C., Lesage, A. S., et al. (2012). PNU-120596, a positive allosteric modulator of alpha7 nicotinic acetylcholine receptors, reverses a sub-chronic phencyclidine-induced cognitive deficit in the attentional set-shifting task in female rats. *Journal of Psychopharmacology*, *26*, 1265–1270.
Meszaros, Z. S., Dimmock, J. A., Ploutz-Snyder, R. J., Abdul-Malak, Y., Leontieva, L., Canfield, K., et al. (2011). Predictors of smoking severity in patients with schizophrenia and alcohol use disorders. *The American Journal on Addictions*, *20*, 462–467.
Metzger, K. L., Maxwell, C. R., Liang, Y., & Siegel, S. J. (2007). Effects of nicotine vary across two auditory evoked potentials in the mouse. *Biological Psychiatry*, *61*, 23–30.
Meyer, E. M., Tay, E. T., Papke, R. L., Meyers, C., Huang, G. L., & de Fiebre, C. M. (1997). 3-[2,4-Dimethoxybenzylidene]anabaseine (DMXB) selectively activates rat alpha7 receptors and improves memory-related behaviors in a mecamylamine-sensitive manner. *Brain Research*, *768*, 49–56.
Micoulaud-Franchi, J. A., Hetrick, W. P., Aramaki, M., Bolbecker, A., Boyer, L., Ystad, S., et al. (2014). Do schizophrenia patients with low P50-suppression report more perceptual anomalies with the sensory gating inventory? *Schizophrenia Research*, *157*, 157–162.

Mobascher, A., Brinkmeyer, J., Warbrick, T., Wels, C., Wagner, M., Grunder, G., et al. (2010). The P300 event-related potential and smoking—A population-based case–control study. *International Journal of Psychophysiology*, *77*, 166–175.

Mobascher, A., Warbrick, T., Brinkmeyer, J., Musso, F., Stoecker, T., Jon Shah, N., et al. (2012). Nicotine effects on anterior cingulate cortex in schizophrenia and healthy smokers as revealed by EEG-informed fMRI. *Psychiatry Research*, *204*, 168–177.

Mocking, R. J., Patrick Pflanz, C., Pringle, A., Parsons, E., McTavish, S. F., Cowen, P. J., et al. (2013). Effects of short-term varenicline administration on emotional and cognitive processing in healthy, non-smoking adults: A randomized, double-blind, study. *Neuropsychopharmacology*, *38*, 476–484.

Mohler, E. G., Franklin, S. R., Rueter, L. E., Fox, G. B., Decker, M. W., & Browman, K. E. (2010). ABT-594 improves performance in the 5-choice serial reaction time task under conditions of increased difficulty, sub-chronic dosing, and in poorly-performing subjects. *Pharmacology, Biochemistry, and Behavior*, *95*, 146–157.

Muhle, H., Steinich, I., von Spiczak, S., Franke, A., Weber, Y., Lerche, H., et al. (2010). A duplication in 1q21.3 in a family with early onset and childhood absence epilepsy. *Epilepsia*, *51*, 2453–2456.

Murphy, J. R., Rawdon, C., Kelleher, I., Twomey, D., Markey, P. S., Cannon, M., et al. (2013). Reduced duration mismatch negativity in adolescents with psychotic symptoms: Further evidence for mismatch negativity as a possible biomarker for vulnerability to psychosis. *BMC Psychiatry*, *13*, 45.

Myers, C. S., Robles, O., Kakoyannis, A. N., Sherr, J. D., Avila, M. T., Blaxton, T. A., et al. (2004). Nicotine improves delayed recognition in schizophrenic patients. *Psychopharmacology*, *174*, 334–340.

Naatanen, R., Gaillard, A. W., & Mantysalo, S. (1978). Early selective-attention effect on evoked potential reinterpreted. *Acta Psychologica*, *42*, 313–329.

Naatanen, R., Jacobsen, T., & Winkler, I. (2005). Memory-based or afferent processes in mismatch negativity (MMN): A review of the evidence. *Psychophysiology*, *42*, 25–32.

Nakauchi, S., Brennan, R. J., Boulter, J., & Sumikawa, K. (2007). Nicotine gates long-term potentiation in the hippocampal CA1 region via the activation of alpha2* nicotinic ACh receptors. *The European Journal of Neuroscience*, *25*, 2666–2681.

Neuhaus, A., Bajbouj, M., Kienast, T., Kalus, P., von Haebler, D., Winterer, G., et al. (2006). Persistent dysfunctional frontal lobe activation in former smokers. *Psychopharmacology*, *186*, 191–200.

Nieman, D. H., Koelman, J. H., Linszen, D. H., Bour, L. J., Dingemans, P. M., & Ongerboer de Visser, B. W. (2002). Clinical and neuropsychological correlates of the P300 in schizophrenia. *Schizophrenia Research*, *55*, 105–113.

O'Donnell, B. F., Faux, S. F., McCarley, R. W., Kimble, M. O., Salisbury, D. F., Nestor, P. G., et al. (1995). Increased rate of P300 latency prolongation with age in schizophrenia. Electrophysiological evidence for a neurodegenerative process. *Archives of General Psychiatry*, *52*, 544–549.

Olincy, A., Harris, J. G., Johnson, L. L., Pender, V., Kongs, S., Allensworth, D., et al. (2006). Proof-of-concept trial of an alpha7 nicotinic agonist in schizophrenia. *Archives of General Psychiatry*, *63*, 630–638.

Olney, J. W., & Farber, N. B. (1995). Glutamate receptor dysfunction and schizophrenia. *Archives of General Psychiatry*, *52*, 998–1007.

Oranje, B., Gispen-de Wied, C. C., Verbaten, M. N., & Kahn, R. S. (2002). Modulating sensory gating in healthy volunteers: The effects of ketamine and haloperidol. *Biological Psychiatry*, *52*, 887–895.

Orejarena, M. J., Herrera-Solis, A., Pons, S., Maskos, U., Maldonado, R., & Robledo, P. (2012). Selective re-expression of beta2 nicotinic acetylcholine receptor subunits in the ventral tegmental area of the mouse restores intravenous nicotine self-administration. *Neuropharmacology, 63*, 235–241.

Orr-Urtreger, A., Goldner, F. M., Saeki, M., Lorenzo, I., Goldberg, L., De Biasi, M., et al. (1997). Mice deficient in the alpha7 neuronal nicotinic acetylcholine receptor lack alpha-bungarotoxin binding sites and hippocampal fast nicotinic currents. *The Journal of Neuroscience, 17*, 9165–9171.

Parasuraman, R., Greenwood, P. M., Kumar, R., & Fossella, J. (2005). Beyond heritability: Neurotransmitter genes differentially modulate visuospatial attention and working memory. *Psychological Science, 16*, 200–207.

Patkar, A. A., Gopalakrishnan, R., Lundy, A., Leone, F. T., Certa, K. M., & Weinstein, S. P. (2002). Relationship between tobacco smoking and positive and negative symptoms in schizophrenia. *The Journal of Nervous and Mental Disease, 190*, 604–610.

Patterson, J. V., Hetrick, W. P., Boutros, N. N., Jin, Y., Sandman, C., Stern, H., et al. (2008). P50 sensory gating ratios in schizophrenics and controls: A review and data analysis. *Psychiatry Research, 158*, 226–247.

Paylor, R., Nguyen, M., Crawley, J. N., Patrick, J., Beaudet, A., & Orr-Urtreger, A. (1998). Alpha7 nicotinic receptor subunits are not necessary for hippocampal-dependent learning or sensorimotor gating: A behavioral characterization of Acra7-deficient mice. *Learning & Memory, 5*, 302–316.

Perkins, K. A., Lerman, C., Mercincavage, M., Fonte, C. A., & Briski, J. L. (2009). Nicotinic acetylcholine receptor beta2 subunit (CHRNB2) gene and short-term ability to quit smoking in response to nicotine patch. *Cancer Epidemiology, Biomarkers & Prevention, 18*, 2608–2612.

Perlman, G., Markin, A., & Iacono, W. G. (2013). P300 amplitude reduction is associated with early-onset and late-onset pathological substance use in a prospectively studied cohort of 14-year-old adolescents. *Psychophysiology*. http://dx.doi.org/10.1111/psyp.12081. [Epub ahead of print].

Petrovsky, N., Quednow, B. B., Ettinger, U., Schmechtig, A., Mossner, R., Collier, D. A., et al. (2010). Sensorimotor gating is associated with CHRNA3 polymorphisms in schizophrenia and healthy volunteers. *Neuropsychopharmacology, 35*, 1429–1439.

Pfefferbaum, A., Ford, J. M., White, P. M., & Roth, W. T. (1989). P3 in schizophrenia is affected by stimulus modality, response requirements, medication status, and negative symptoms. *Archives of General Psychiatry, 46*, 1035–1044.

Phillips, J. M., Ehrlichman, R. S., & Siegel, S. J. (2007). Mecamylamine blocks nicotine-induced enhancement of the P20 auditory event-related potential and evoked gamma. *Neuroscience, 144*, 1314–1323.

Picciotto, M. R., Zoli, M., Lena, C., Bessis, A., Lallemand, Y., Le Novere, N., et al. (1995). Abnormal avoidance learning in mice lacking functional high-affinity nicotine receptor in the brain. *Nature, 374*, 65–67.

Picciotto, M. R., Zoli, M., Rimondini, R., Lena, C., Marubio, L. M., Pich, E. M., et al. (1998). Acetylcholine receptors containing the beta2 subunit are involved in the reinforcing properties of nicotine. *Nature, 391*, 173–177.

Pichat, P., Bergis, O. E., Terranova, J. P., Urani, A., Duarte, C., Santucci, V., et al. (2007). SSR180711, a novel selective alpha7 nicotinic receptor partial agonist: (II) efficacy in experimental models predictive of activity against cognitive symptoms of schizophrenia. *Neuropsychopharmacology, 32*, 17–34.

Pineda, J. A., Herrera, C., Kang, C., & Sandler, A. (1998). Effects of cigarette smoking and 12-h abstention on working memory during a serial-probe recognition task. *Psychopharmacology, 139*, 311–321.

Polich, J. (2007). Updating P300: An integrative theory of P3a and P3b. *Clinical Neurophysiology, 118*, 2128–2148.

Polich, J., & Ochoa, C. J. (2004). Alcoholism risk, tobacco smoking, and P300 event-related potential. *Clinical Neurophysiology, 115*, 1374–1383.
Pons, S., Fattore, L., Cossu, G., Tolu, S., Porcu, E., McIntosh, J. M., et al. (2008). Crucial role of alpha4 and alpha6 nicotinic acetylcholine receptor subunits from ventral tegmental area in systemic nicotine self-administration. *The Journal of Neuroscience, 28*, 12318–12327.
Portugal, G. S., Kenney, J. W., & Gould, T. J. (2008). Beta2 subunit containing acetylcholine receptors mediate nicotine withdrawal deficits in the acquisition of contextual fear conditioning. *Neurobiology of Learning and Memory, 89*, 106–113.
Preskorn, S. H., Gawryl, M., Dgetluck, N., Palfreyman, M., Bauer, L. O., & Hilt, D. C. (2014). Normalizing effects of EVP-6124, an alpha-7 nicotinic partial agonist, on event-related potentials and cognition: A proof of concept, randomized trial in patients with schizophrenia. *Journal of Psychiatric Practice, 20*, 12–24.
Prickaerts, J., van Goethem, N. P., Chesworth, R., Shapiro, G., Boess, F. G., Methfessel, C., et al. (2012). EVP-6124, a novel and selective alpha7 nicotinic acetylcholine receptor partial agonist, improves memory performance by potentiating the acetylcholine response of alpha7 nicotinic acetylcholine receptors. *Neuropharmacology, 62*, 1099–1110.
Quisenaerts, C., Morrens, M., Hulstijn, W., de Bruijn, E., Timmers, M., Streffer, J., et al. (2014). The nicotinergic receptor as a target for cognitive enhancement in schizophrenia: Barking up the wrong tree? *Psychopharmacology, 231*, 543–550.
Radek, R. J., Robb, H. M., Stevens, K. E., Gopalakrishnan, M., & Bitner, R. S. (2012). Effects of the novel alpha7 nicotinic acetylcholine receptor agonist ABT-107 on sensory gating in DBA/2 mice: Pharmacodynamic characterization. *The Journal of Pharmacology and Experimental Therapeutics, 343*, 736–745.
Ramirez-Latorre, J., Yu, C. R., Qu, X., Perin, F., Karlin, A., & Role, L. (1996). Functional contributions of alpha5 subunit to neuronal acetylcholine receptor channels. *Nature, 380*, 347–351.
Revheim, N., Corcoran, C. M., Dias, E., Hellmann, E., Martinez, A., Butler, P. D., et al. (2014). Reading deficits in schizophrenia and individuals at high clinical risk: Relationship to sensory function, course of illness, and psychosocial outcome. *The American Journal of Psychiatry, 171*, 949–959.
Rezvani, A. H., Cauley, M. C., Johnson, E. C., Gatto, G. J., & Levin, E. D. (2012). Effects of AZD3480, a neuronal nicotinic acetylcholine receptor agonist, and donepezil on dizocilpine-induced attentional impairment in rats. *Psychopharmacology, 223*, 251–258.
Rezvani, A. H., Kholdebarin, E., Brucato, F. H., Callahan, P. M., Lowe, D. A., & Levin, E. D. (2009). Effect of R3487/MEM3454, a novel nicotinic alpha7 receptor partial agonist and 5-HT3 antagonist on sustained attention in rats. *Progress in Neuro-Psychopharmacology & Biological Psychiatry, 33*, 269–275.
Rigbi, A., Kanyas, K., Yakir, A., Greenbaum, L., Pollak, Y., Ben-Asher, E., et al. (2008). Why do young women smoke? V. Role of direct and interactive effects of nicotinic cholinergic receptor gene variation on neurocognitive function. *Genes, Brain, and Behavior, 7*, 164–172.
Ringel, T. M., Heidrich, A., Jacob, C. P., Pfuhlmann, B., Stoeber, G., & Fallgatter, A. J. (2004). Sensory gating deficit in a subtype of chronic schizophrenic patients. *Psychiatry Research, 125*, 237–245.
Rissling, A. J., Miyakoshi, M., Sugar, C. A., Braff, D. L., Makeig, S., & Light, G. A. (2014). Cortical substrates and functional correlates of auditory deviance processing deficits in schizophrenia. *NeuroImage: Clinical, 6*, 424–437.
Rissling, A. J., Park, S. H., Young, J. W., Rissling, M. B., Sugar, C. A., Sprock, J., et al. (2013). Demand and modality of directed attention modulate "pre-attentive" sensory processes in schizophrenia patients and nonpsychiatric controls. *Schizophrenia Research, 146*, 326–335.

Roh, S., Hoeppner, S. S., Schoenfeld, D., Fullerton, C. A., Stoeckel, L. E., & Evins, A. E. (2014). Acute effects of mecamylamine and varenicline on cognitive performance in non-smokers with and without schizophrenia. *Psychopharmacology, 231*, 765–775.

Roncarati, R., Scali, C., Comery, T. A., Grauer, S. M., Aschmi, S., Bothmann, H., et al. (2009). Procognitive and neuroprotective activity of a novel alpha7 nicotinic acetylcholine receptor agonist for treatment of neurodegenerative and cognitive disorders. *The Journal of Pharmacology and Experimental Therapeutics, 329*, 459–468.

Ross, S. A., Wong, J. Y., Clifford, J. J., Kinsella, A., Massalas, J. S., Horne, M. K., et al. (2000). Phenotypic characterization of an alpha 4 neuronal nicotinic acetylcholine receptor subunit knock-out mouse. *The Journal of Neuroscience, 20*, 6431–6441.

Roth, W. T., & Cannon, E. H. (1972). Some features of the auditory evoked response in schizophrenics. *Archives of General Psychiatry, 27*, 466–471.

Rudnick, N. D., Koehler, C., Picciotto, M. R., & Siegel, S. J. (2009). Role of beta2-containing nicotinic acetylcholine receptors in auditory event-related potentials. *Psychopharmacology, 202*, 745–751.

Rudnick, N. D., Strasser, A. A., Phillips, J. M., Jepson, C., Patterson, F., Frey, J. M., et al. (2010). Mouse model predicts effects of smoking and varenicline on event-related potentials in humans. *Nicotine & Tobacco Research, 12*, 589–597.

Sacco, K. A., Termine, A., Seyal, A., Dudas, M. M., Vessicchio, J. C., Krishnan-Sarin, S., et al. (2005). Effects of cigarette smoking on spatial working memory and attentional deficits in schizophrenia: Involvement of nicotinic receptor mechanisms. *Archives of General Psychiatry, 62*, 649–659.

Saccone, S. F., Hinrichs, A. L., Saccone, N. L., Chase, G. A., Konvicka, K., Madden, P. A., et al. (2007). Cholinergic nicotinic receptor genes implicated in a nicotine dependence association study targeting 348 candidate genes with 3713 SNPs. *Human Molecular Genetics, 16*, 36–49.

Saccone, N. L., Saccone, S. F., Hinrichs, A. L., Stitzel, J. A., Duan, W., Pergadia, M. L., et al. (2009). Multiple distinct risk loci for nicotine dependence identified by dense coverage of the complete family of nicotinic receptor subunit (CHRN) genes. *American Journal of Medical Genetics. Part B, Neuropsychiatric Genetics, 150B*, 453–466.

Saccone, N. L., Schwantes-An, T. H., Wang, J. C., Grucza, R. A., Breslau, N., Hatsukami, D., et al. (2010). Multiple cholinergic nicotinic receptor genes affect nicotine dependence risk in African and European Americans. *Genes, Brain, and Behavior, 9*, 741–750.

Saccone, N. L., Wang, J. C., Breslau, N., Johnson, E. O., Hatsukami, D., Saccone, S. F., et al. (2009). The CHRNA5-CHRNA3-CHRNB4 nicotinic receptor subunit gene cluster affects risk for nicotine dependence in African-Americans and in European-Americans. *Cancer Research, 69*, 6848–6856.

Salas, R., Cook, K. D., Bassetto, L., & De Biasi, M. (2004). The alpha3 and beta4 nicotinic acetylcholine receptor subunits are necessary for nicotine-induced seizures and hypolocomotion in mice. *Neuropharmacology, 47*, 401–407.

Salas, R., Orr-Urtreger, A., Broide, R. S., Beaudet, A., Paylor, R., & De Biasi, M. (2003). The nicotinic acetylcholine receptor subunit alpha 5 mediates short-term effects of nicotine in vivo. *Molecular Pharmacology, 63*, 1059–1066.

Salas, R., Pieri, F., & De Biasi, M. (2004). Decreased signs of nicotine withdrawal in mice null for the beta4 nicotinic acetylcholine receptor subunit. *The Journal of Neuroscience, 24*, 10035–10039.

Salas, R., Pieri, F., Fung, B., Dani, J. A., & De Biasi, M. (2003). Altered anxiety-related responses in mutant mice lacking the beta4 subunit of the nicotinic receptor. *The Journal of Neuroscience, 23*, 6255–6263.

Salas, R., Sturm, R., Boulter, J., & De Biasi, M. (2009). Nicotinic receptors in the habenulo-interpeduncular system are necessary for nicotine withdrawal in mice. *The Journal of Neuroscience, 29*, 3014–3018.

Salminen, O., Murphy, K. L., McIntosh, J. M., Drago, J., Marks, M. J., Collins, A. C., et al. (2004). Subunit composition and pharmacology of two classes of striatal presynaptic nicotinic acetylcholine receptors mediating dopamine release in mice. *Molecular Pharmacology, 65*, 1526–1535.

Sanchez-Morla, E. M., Santos, J. L., Aparicio, A., Garcia-Jimenez, M. A., Soria, C., & Arango, C. (2013). Neuropsychological correlates of P50 sensory gating in patients with schizophrenia. *Schizophrenia Research, 143*, 102–106.

Santos, J. L., Sanchez-Morla, E. M., Aparicio, A., Garcia-Jimenez, M. A., Villanueva, C., Martinez-Vizcaino, V., et al. (2010). P50 gating in deficit and nondeficit schizophrenia. *Schizophrenia Research, 119*, 183–190.

Saykin, A. J., Gur, R. C., Gur, R. E., Mozley, P. D., Mozley, L. H., Resnick, S. M., et al. (1991). Neuropsychological function in schizophrenia. Selective impairment in memory and learning. *Archives of General Psychiatry, 48*, 618–624.

Schlaepfer, I. R., Hoft, N. R., Collins, A. C., Corley, R. P., Hewitt, J. K., Hopfer, C. J., et al. (2008). The CHRNA5/A3/B4 gene cluster variability as an important determinant of early alcohol and tobacco initiation in young adults. *Biological Psychiatry, 63*, 1039–1046.

Semenova, S., Contet, C., Roberts, A. J., & Markou, A. (2012). Mice lacking the beta4 subunit of the nicotinic acetylcholine receptor show memory deficits, altered anxiety- and depression-like behavior, and diminished nicotine-induced analgesia. *Nicotine & Tobacco Research, 14*, 1346–1355.

Serreau, P., Chabout, J., Suarez, S. V., Naude, J., & Granon, S. (2011). Beta2-containing neuronal nicotinic receptors as major actors in the flexible choice between conflicting motivations. *Behavioural Brain Research, 225*, 151–159.

Shajahan, P. M., O'Carroll, R. E., Glabus, M. F., Ebmeier, K. P., & Blackwood, D. H. (1997). Correlation of auditory 'oddball' P300 with verbal memory deficits in schizophrenia. *Psychological Medicine, 27*, 579–586.

Shelley, A. M., Ward, P. B., Catts, S. V., Michie, P. T., Andrews, S., & McConaghy, N. (1991). Mismatch negativity: An index of a preattentive processing deficit in schizophrenia. *Biological Psychiatry, 30*, 1059–1062.

Sherr, J. D., Myers, C., Avila, M. T., Elliott, A., Blaxton, T. A., & Thaker, G. K. (2002). The effects of nicotine on specific eye tracking measures in schizophrenia. *Biological Psychiatry, 52*, 721–728.

Shiina, A., Shirayama, Y., Niitsu, T., Hashimoto, T., Yoshida, T., Hasegawa, T., et al. (2010). A randomised, double-blind, placebo-controlled trial of tropisetron in patients with schizophrenia. *Annals of General Psychiatry, 9*, 27.

Shim, J. C., Jung, D. U., Jung, S. S., Seo, Y. S., Cho, D. M., Lee, J. H., et al. (2012). Adjunctive varenicline treatment with antipsychotic medications for cognitive impairments in people with schizophrenia: A randomized double-blind placebo-controlled trial. *Neuropsychopharmacology, 37*, 660–668.

Shutara, Y., Koga, Y., Fujita, K., Takeuchi, H., Mochida, M., & Takemasa, K. (1996). An event-related potential study on the impairment of automatic processing of auditory input in schizophrenia. *Brain Topography, 8*, 285–289.

Siegel, C., Waldo, M., Mizner, G., Adler, L. E., & Freedman, R. (1984). Deficits in sensory gating in schizophrenic patients and their relatives. Evidence obtained with auditory evoked responses. *Archives of General Psychiatry, 41*, 607–612.

Silverman, M. A., Neale, M. C., Sullivan, P. F., Harris-Kerr, C., Wormley, B., Sadek, H., et al. (2000). Haplotypes of four novel single nucleotide polymorphisms in the nicotinic acetylcholine receptor beta2-subunit (CHRNB2) gene show no association with

smoking initiation or nicotine dependence. *American Journal of Medical Genetics*, *96*, 646–653.
Simosky, J. K., Stevens, K. E., Kem, W. R., & Freedman, R. (2001). Intragastric DMXB-A, an alpha7 nicotinic agonist, improves deficient sensory inhibition in DBA/2 mice. *Biological Psychiatry*, *50*, 493–500.
Smith, A. K., Edgar, J. C., Huang, M., Lu, B. Y., Thoma, R. J., Hanlon, F. M., et al. (2010). Cognitive abilities and 50- and 100-msec paired-click processes in schizophrenia. *The American Journal of Psychiatry*, *167*, 1264–1275.
Smith, D. M., Grant, B., Fisher, D. J., Borracci, G., Labelle, A., & Knott, V. J. (2013). Auditory verbal hallucinations in schizophrenia correlate with P50 gating. *Clinical Neurophysiology*, *124*, 1329–1335.
Smith, R. C., Lindenmayer, J. P., Davis, J. M., Cornwell, J., Noth, K., Gupta, S., et al. (2009). Cognitive and antismoking effects of varenicline in patients with schizophrenia or schizoaffective disorder. *Schizophrenia Research*, *110*, 149–155.
Smith, R. C., Singh, A., Infante, M., Khandat, A., & Kloos, A. (2002). Effects of cigarette smoking and nicotine nasal spray on psychiatric symptoms and cognition in schizophrenia. *Neuropsychopharmacology*, *27*, 479–497.
Smith, R. C., Warner-Cohen, J., Matute, M., Butler, E., Kelly, E., Vaidhyanathaswamy, S., et al. (2006). Effects of nicotine nasal spray on cognitive function in schizophrenia. *Neuropsychopharmacology*, *31*, 637–643.
Smucny, J., Olincy, A., Eichman, L. C., Lyons, E., & Tregellas, J. R. (2013). Early sensory processing deficits predict sensitivity to distraction in schizophrenia. *Schizophrenia Research*, *147*, 196–200.
Souza, V. B., Muir, W. J., Walker, M. T., Glabus, M. F., Roxborough, H. M., Sharp, C. W., et al. (1995). Auditory P300 event-related potentials and neuropsychological performance in schizophrenia and bipolar affective disorder. *Biological Psychiatry*, *37*, 300–310.
Spitz, M. R., Amos, C. I., Dong, Q., Lin, J., & Wu, X. (2008). The CHRNA5-A3 region on chromosome 15q24-25.1 is a risk factor both for nicotine dependence and for lung cancer. *Journal of the National Cancer Institute*, *100*, 1552–1556.
Stefansson, H., Rujescu, D., Cichon, S., Pietilainen, O. P., Ingason, A., Steinberg, S., et al. (2008). Large recurrent microdeletions associated with schizophrenia. *Nature*, *455*, 232–236.
Stephens, S. H., Franks, A., Berger, R., Palionyte, M., Fingerlin, T. E., Wagner, B., et al. (2012). Multiple genes in the 15q13-q14 chromosomal region are associated with schizophrenia. *Psychiatric Genetics*, *22*, 1–14.
Stevens, K. E., Freedman, R., Collins, A. C., Hall, M., Leonard, S., Marks, M. J., et al. (1996). Genetic correlation of inhibitory gating of hippocampal auditory evoked response and alpha-bungarotoxin-binding nicotinic cholinergic receptors in inbred mouse strains. *Neuropsychopharmacology*, *15*, 152–162.
Stevens, K. E., Kem, W. R., Mahnir, V. M., & Freedman, R. (1998). Selective alpha7-nicotinic agonists normalize inhibition of auditory response in DBA mice. *Psychopharmacology*, *136*, 320–327.
Stevens, K. E., & Wear, K. D. (1997). Normalizing effects of nicotine and a novel nicotinic agonist on hippocampal auditory gating in two animal models. *Pharmacology, Biochemistry, and Behavior*, *57*, 869–874.
Stoker, A. K., Marks, M. J., & Markou, A. (2014). Null mutation of the beta2 nicotinic acetylcholine receptor subunit attenuates nicotine withdrawal-induced anhedonia in mice. *European Journal of Pharmacology*, *753*, 146–150.
Stoker, A. K., Olivier, B., & Markou, A. (2012). Role of alpha7- and beta4-containing nicotinic acetylcholine receptors in the affective and somatic aspects of nicotine withdrawal: Studies in knockout mice. *Behavior Genetics*, *42*, 423–436.
Strik, W. K., Dierks, T., & Maurer, K. (1993). Amplitudes of auditory P300 in remitted and residual schizophrenics: Correlations with clinical features. *Neuropsychobiology*, *27*, 54–60.

Sudweeks, S. N., & Yakel, J. L. (2000). Functional and molecular characterization of neuronal nicotinic ACh receptors in rat CA1 hippocampal neurons. *The Journal of Physiology*, *527*(Pt 3), 515–528.

Sullivan, P. F., Neale, B. M., van den Oord, E., Miles, M. F., Neale, M. C., Bulik, C. M., et al. (2004). Candidate genes for nicotine dependence via linkage, epistasis, and bioinformatics. *American Journal of Medical Genetics. Part B, Neuropsychiatric Genetics*, *126B*, 23–36.

Sutton, S., Braren, M., Zubin, J., & John, E. R. (1965). Evoked-potential correlates of stimulus uncertainty. *Science*, *150*, 1187–1188.

Tabares-Seisdedos, R., & Rubenstein, J. L. (2009). Chromosome 8p as a potential hub for developmental neuropsychiatric disorders: Implications for schizophrenia, autism and cancer. *Molecular Psychiatry*, *14*, 563–589.

Taiminen, T. J., Salokangas, R. K., Saarijarvi, S., Niemi, H., Lehto, H., Ahola, V., et al. (1998). Smoking and cognitive deficits in schizophrenia: A pilot study. *Addictive Behaviors*, *23*, 263–266.

Tang, Y. L., George, T. P., Mao, P. X., Cai, Z. J., & Chen, Q. (2007). Cigarette smoking in Chinese male inpatients with schizophrenia: A cross-sectional analysis. *Journal of Psychiatric Research*, *41*, 43–48.

Tapper, A. R., McKinney, S. L., Nashmi, R., Schwarz, J., Deshpande, P., Labarca, C., et al. (2004). Nicotine activation of alpha4* receptors: Sufficient for reward, tolerance, and sensitization. *Science*, *306*, 1029–1032.

Thoma, R. J., Hanlon, F. M., Moses, S. N., Ricker, D., Huang, M., Edgar, C., et al. (2005). M50 sensory gating predicts negative symptoms in schizophrenia. *Schizophrenia Research*, *73*, 311–318.

Thorgeirsson, T. E., Geller, F., Sulem, P., Rafnar, T., Wiste, A., Magnusson, K. P., et al. (2008). A variant associated with nicotine dependence, lung cancer and peripheral arterial disease. *Nature*, *452*, 638–642.

Tidey, J. W., Colby, S. M., & Xavier, E. M. (2014). Effects of smoking abstinence on cigarette craving, nicotine withdrawal, and nicotine reinforcement in smokers with and without schizophrenia. *Nicotine & Tobacco Research*, *16*, 326–334.

Tietje, K. R., Anderson, D. J., Bitner, R. S., Blomme, E. A., Brackemeyer, P. J., Briggs, C. A., et al. (2008). Preclinical characterization of A-582941: A novel alpha7 neuronal nicotinic receptor agonist with broad spectrum cognition-enhancing properties. *CNS Neuroscience & Therapeutics*, *14*, 65–82.

Timmermann, D. B., Sandager-Nielsen, K., Dyhring, T., Smith, M., Jacobsen, A. M., Nielsen, E. O., et al. (2012). Augmentation of cognitive function by NS9283, a stoichiometry-dependent positive allosteric modulator of alpha2- and alpha4-containing nicotinic acetylcholine receptors. *British Journal of Pharmacology*, *167*, 164–182.

Toyomaki, A., Kusumi, I., Matsuyama, T., Kako, Y., Ito, K., & Koyama, T. (2008). Tone duration mismatch negativity deficits predict impairment of executive function in schizophrenia. *Progress in Neuro-Psychopharmacology & Biological Psychiatry*, *32*, 95–99.

Tregellas, J. R., Davalos, D. B., Rojas, D. C., Waldo, M. C., Gibson, L., Wylie, K., et al. (2007). Increased hemodynamic response in the hippocampus, thalamus and prefrontal cortex during abnormal sensory gating in schizophrenia. *Schizophrenia Research*, *92*, 262–272.

Turetsky, B. I., Bilker, W. B., Siegel, S. J., Kohler, C. G., & Gur, R. E. (2009). Profile of auditory information-processing deficits in schizophrenia. *Psychiatry Research*, *165*, 27–37.

Turetsky, B. I., Calkins, M. E., Light, G. A., Olincy, A., Radant, A. D., & Swerdlow, N. R. (2007). Neurophysiological endophenotypes of schizophrenia: The viability of selected candidate measures. *Schizophrenia Bulletin*, *33*, 69–94.

Turetsky, B. I., Dent, G., Jaeger, J., & Zukin, S. R. (2012). P50 amplitude reduction: A nicotinic receptor-mediated deficit in first-degree relatives of schizophrenia patients. *Psychopharmacology*, *221*, 39–52.

Turetsky, B. I., Dress, E. M., Braff, D. L., Calkins, M. E., Green, M. F., Greenwood, T. A., et al. (2014). The utility of P300 as a schizophrenia endophenotype and predictive biomarker: Clinical and socio-demographic modulators in COGS-2. *Schizophrenia Research, 163*(1-3), 53–62.

Turner, J. R., Castellano, L. M., & Blendy, J. A. (2010). Nicotinic partial agonists varenicline and sazetidine-a have differential effects on affective behavior. *The Journal of Pharmacology and Experimental Therapeutics, 334*, 665–672.

Ucok, A., Polat, A., Bozkurt, O., & Meteris, H. (2004). Cigarette smoking among patients with schizophrenia and bipolar disorders. *Psychiatry and Clinical Neurosciences, 58*, 434–437.

Umbricht, D., Keefe, R. S., Murray, S., Lowe, D. A., Porter, R., Garibaldi, G., et al. (2014). A randomized, placebo-controlled study investigating the nicotinic alpha7 agonist, RG3487, for cognitive deficits in schizophrenia. *Neuropsychopharmacology, 39*, 1568–1577.

Umbricht, D., & Krljes, S. (2005). Mismatch negativity in schizophrenia: A meta-analysis. *Schizophrenia Research, 76*, 1–23.

Umbricht, D., Schmid, L., Koller, R., Vollenweider, F. X., Hell, D., & Javitt, D. C. (2000). Ketamine-induced deficits in auditory and visual context-dependent processing in healthy volunteers: Implications for models of cognitive deficits in schizophrenia. *Archives of General Psychiatry, 57*, 1139–1147.

van Berckel, B. N., Oranje, B., van Ree, J. M., Verbaten, M. N., & Kahn, R. S. (1998). The effects of low dose ketamine on sensory gating, neuroendocrine secretion and behavior in healthy human subjects. *Psychopharmacology, 137*, 271–281.

Velligan, D., Brenner, R., Sicuro, F., Walling, D., Riesenberg, R., Sfera, A., et al. (2012). Assessment of the effects of AZD3480 on cognitive function in patients with schizophrenia. *Schizophrenia Research, 134*, 59–64.

Vianin, P., Posada, A., Hugues, E., Franck, N., Bovet, P., Parnas, J., et al. (2002). Reduced P300 amplitude in a visual recognition task in patients with schizophrenia. *NeuroImage, 17*, 911–921.

Vicens, P., Ribes, D., Torrente, M., & Domingo, J. L. (2011). Behavioral effects of PNU-282987, an alpha7 nicotinic receptor agonist, in mice. *Behavioural Brain Research, 216*, 341–348.

Voineskos, S., De Luca, V., Mensah, A., Vincent, J. B., Potapova, N., & Kennedy, J. L. (2007). Association of alpha4beta2 nicotinic receptor and heavy smoking in schizophrenia. *Journal of Psychiatry & Neuroscience, 32*, 412–416.

Wada, E., Wada, K., Boulter, J., Deneris, E., Heinemann, S., Patrick, J., et al. (1989). Distribution of alpha 2, alpha 3, alpha 4, and beta 2 neuronal nicotinic receptor subunit mRNAs in the central nervous system: A hybridization histochemical study in the rat. *The Journal of Comparative Neurology, 284*, 314–335.

Wallace, T. L., Callahan, P. M., Tehim, A., Bertrand, D., Tombaugh, G., Wang, S., et al. (2011). RG3487, a novel nicotinic alpha7 receptor partial agonist, improves cognition and sensorimotor gating in rodents. *The Journal of Pharmacology and Experimental Therapeutics, 336*, 242–253.

Walsh, T., McClellan, J. M., McCarthy, S. E., Addington, A. M., Pierce, S. B., Cooper, G. M., et al. (2008). Rare structural variants disrupt multiple genes in neurodevelopmental pathways in schizophrenia. *Science, 320*, 539–543.

Walters, C. L., Brown, S., Changeux, J. P., Martin, B., & Damaj, M. I. (2006). The beta2 but not alpha7 subunit of the nicotinic acetylcholine receptor is required for nicotine-conditioned place preference in mice. *Psychopharmacology, 184*, 339–344.

Wang, S., D van der Vaart, A., Xu, Q., Seneviratne, C., Pomerleau, O. F., Pomerleau, C. S., et al. (2014). Significant associations of CHRNA2 and CHRNA6 with nicotine dependence in European American and African American populations. *Human Genetics, 133*, 575–586.

Wang, F., Gerzanich, V., Wells, G. B., Anand, R., Peng, X., Keyser, K., et al. (1996). Assembly of human neuronal nicotinic receptor alpha5 subunits with alpha3, beta2, and beta4 subunits. *The Journal of Biological Chemistry*, *271*, 17656–17665.

Wang, J. C., Grucza, R., Cruchaga, C., Hinrichs, A. L., Bertelsen, S., Budde, J. P., et al. (2009). Genetic variation in the CHRNA5 gene affects mRNA levels and is associated with risk for alcohol dependence. *Molecular Psychiatry*, *14*, 501–510.

Wehner, J. M., Keller, J. J., Keller, A. B., Picciotto, M. R., Paylor, R., Booker, T. K., et al. (2004). Role of neuronal nicotinic receptors in the effects of nicotine and ethanol on contextual fear conditioning. *Neuroscience*, *129*, 11–24.

Weiss, R. B., Baker, T. B., Cannon, D. S., von Niederhausern, A., Dunn, D. M., Matsunami, N., et al. (2008). A candidate gene approach identifies the CHRNA5-A3-B4 region as a risk factor for age-dependent nicotine addiction. *PLoS Genetics*, *4*, e1000125.

Werkheiser, J. L., Sydserff, S., Hubbs, S. J., Ding, M., Eisman, M. S., Perry, D., et al. (2011). Ultra-low exposure to alpha-7 nicotinic acetylcholine receptor partial agonists elicits an improvement in cognition that corresponds with an increase in alpha-7 receptor expression in rodents: Implications for low dose clinical efficacy. *Neuroscience*, *186*, 76–87.

Wessel, J., McDonald, S. M., Hinds, D. A., Stokowski, R. P., Javitz, H. S., Kennemer, M., et al. (2010). Resequencing of nicotinic acetylcholine receptor genes and association of common and rare variants with the Fagerstrom test for nicotine dependence. *Neuropsychopharmacology*, *35*, 2392–2402.

Wiklund, A., Granon, S., Faure, P., Sundman, E., Changeux, J. P., & Eriksson, L. I. (2009). Object memory in young and aged mice after sevoflurane anaesthesia. *Neuroreport*, *20*, 1419–1423.

Wildeboer-Andrud, K. M., & Stevens, K. E. (2011). The smoking cessation drug varenicline improves deficient P20-N40 inhibition in DBA/2 mice. *Pharmacology, Biochemistry, and Behavior*, *100*, 17–24.

Williams, J. M., Anthenelli, R. M., Morris, C. D., Treadow, J., Thompson, J. R., Yunis, C., et al. (2012). A randomized, double-blind, placebo-controlled study evaluating the safety and efficacy of varenicline for smoking cessation in patients with schizophrenia or schizoaffective disorder. *The Journal of Clinical Psychiatry*, *73*, 654–660.

Williams, T. J., Nuechterlein, K. H., Subotnik, K. L., & Yee, C. M. (2011). Distinct neural generators of sensory gating in schizophrenia. *Psychophysiology*, *48*, 470–478.

Wing, V. C., Wass, C. E., Bacher, I., Rabin, R. A., & George, T. P. (2013). Varenicline modulates spatial working memory deficits in smokers with schizophrenia. *Schizophrenia Research*, *149*, 190–191.

Winterer, G., Gallinat, J., Brinkmeyer, J., Musso, F., Kornhuber, J., Thuerauf, N., et al. (2013). Allosteric alpha-7 nicotinic receptor modulation and P50 sensory gating in schizophrenia: A proof-of-mechanism study. *Neuropharmacology*, *64*, 197–204.

Winterer, G., Mittelstrass, K., Giegling, I., Lamina, C., Fehr, C., Brenner, H., et al. (2010). Risk gene variants for nicotine dependence in the CHRNA5-CHRNA3-CHRNB4 cluster are associated with cognitive performance. *American Journal of Medical Genetics. Part B, Neuropsychiatric Genetics*, *153B*, 1448–1458.

Winterer, G., Musso, F., Konrad, A., Vucurevic, G., Stoeter, P., Sander, T., et al. (2007). Association of attentional network function with exon 5 variations of the CHRNA4 gene. *Human Molecular Genetics*, *16*, 2165–2174.

Wu, E. Q., Birnbaum, H. G., Shi, L., Ball, D. E., Kessler, R. C., Moulis, M., et al. (2005). The economic burden of schizophrenia in the United States in 2002. *The Journal of Clinical Psychiatry*, *66*, 1122–1129.

Wynn, J. K., Sugar, C., Horan, W. P., Kern, R., & Green, M. F. (2010). Mismatch negativity, social cognition, and functioning in schizophrenia patients. *Biological Psychiatry*, *67*, 940–947.

Xu, W., Gelber, S., Orr-Urtreger, A., Armstrong, D., Lewis, R. A., Ou, C. N., et al. (1999). Megacystis, mydriasis, and ion channel defect in mice lacking the alpha3 neuronal nicotinic acetylcholine receptor. *Proceedings of the National Academy of Sciences of the United States of America, 96*, 5746–5751.

Yang, J., Wang, S., Yang, Z., Hodgkinson, C. A., Iarikova, P., Ma, J. Z., et al. (2014). The contribution of rare and common variants in 30 genes to risk nicotine dependence. *Molecular Psychiatry*. http://dx.doi.org/10.1038/mp.2014.156. [Epub ahead of print].

Yee, C. M., Nuechterlein, K. H., Morris, S. E., & White, P. M. (1998). P50 suppression in recent-onset schizophrenia: Clinical correlates and risperidone effects. *Journal of Abnormal Psychology, 107*, 691–698.

Yildirim, E., Connor, D. A., & Gould, T. J. (2014). ABT-089, but not ABT-107, ameliorates nicotine withdrawal-induced cognitive deficits in C57BL6/J mice. *Behavioural Pharmacology*. http://dx.doi.org/10.1097/FBP.0000000000000111.

Yoon, H. H., Iacono, W. G., Malone, S. M., & McGue, M. (2006). Using the brain P300 response to identify novel phenotypes reflecting genetic vulnerability for adolescent substance misuse. *Addictive Behaviors, 31*, 1067–1087.

Young, J. W., Crawford, N., Kelly, J. S., Kerr, L. E., Marston, H. M., Spratt, C., et al. (2007). Impaired attention is central to the cognitive deficits observed in alpha 7 deficient mice. *European Neuropsychopharmacology, 17*, 145–155.

Young, J. W., Finlayson, K., Spratt, C., Marston, H. M., Crawford, N., Kelly, J. S., et al. (2004). Nicotine improves sustained attention in mice: Evidence for involvement of the alpha7 nicotinic acetylcholine receptor. *Neuropsychopharmacology, 29*, 891–900.

Young, J. W., Meves, J. M., & Geyer, M. A. (2013). Nicotinic agonist-induced improvement of vigilance in mice in the 5-choice continuous performance test. *Behavioural Brain Research, 240*, 119–133.

Young, J. W., Meves, J. M., Tarantino, I. S., Caldwell, S., & Geyer, M. A. (2011). Delayed procedural learning in alpha7-nicotinic acetylcholine receptor knockout mice. *Genes, Brain, and Behavior, 10*, 720–733.

Young, J. W., Powell, S. B., Risbrough, V., Marston, H. M., & Geyer, M. A. (2009). Using the MATRICS to guide development of a preclinical cognitive test battery for research in schizophrenia. *Pharmacology & Therapeutics, 122*, 150–202.

Zhang, X. Y., Chen da, C., Tan, Y. L., Xiu, M. H., Cui, J., Hui, L., et al. (2014). Sociodemographic and clinical characteristics of heavy and non-heavy smokers among schizophrenia inpatients in a Chinese Han population. *Psychopharmacology, 231*, 305–314.

Zhang, X. Y., Chen, D. C., Xiu, M. H., Haile, C. N., He, S. C., Luo, X., et al. (2013). Cigarette smoking, psychopathology and cognitive function in first-episode drug-naive patients with schizophrenia: A case–control study. *Psychological Medicine, 43*, 1651–1660.

Zhang, H., Kranzler, H. R., Poling, J., & Gelernter, J. (2010). Variation in the nicotinic acetylcholine receptor gene cluster CHRNA5-CHRNA3-CHRNB4 and its interaction with recent tobacco use influence cognitive flexibility. *Neuropsychopharmacology, 35*, 2211–2224.

Zhang, X. Y., Tan, Y. L., Zhou, D. F., Haile, C. N., Wu, G. Y., Cao, L. Y., et al. (2007). Nicotine dependence, symptoms and oxidative stress in male patients with schizophrenia. *Neuropsychopharmacology, 32*, 2020–2024.

Ziedonis, D. M., Kosten, T. R., Glazer, W. M., & Frances, R. J. (1994). Nicotine dependence and schizophrenia. *Hospital & Community Psychiatry, 45*, 204–206.

Zoli, M., Lena, C., Picciotto, M. R., & Changeux, J. P. (1998). Identification of four classes of brain nicotinic receptors using beta2 mutant mice. *The Journal of Neuroscience, 18*, 4461–4472.

Zoli, M., Picciotto, M. R., Ferrari, R., Cocchi, D., & Changeux, J. P. (1999). Increased neurodegeneration during ageing in mice lacking high-affinity nicotine receptors. *The EMBO Journal, 18*, 1235–1244.

CHAPTER THREE

Neuronal α7 Nicotinic Receptors as a Target for the Treatment of Schizophrenia

Tanya L. Wallace*, Daniel Bertrand†,[1]
*SRI International, Menlo Park, California, USA
†HiQScreen Sàrl, Geneva, Switzerland
[1]Corresponding author: e-mail address: daniel.bertrand@hiqscreen.com

Contents

Abstract

Schizophrenia is a lifelong disease, the burden of which is often underestimated. Characterized by positive (e.g., hallucinations) and negative (e.g., avolition, amotivation) symptoms, schizophrenia is also accompanied with profound impairments in cognitive function that progress throughout the development of the disease. Although treatment with antipsychotic medications can effectively dampen some of the positive symptoms, these medications largely fail to reverse cognitive deficits or to mitigate negative symptoms. With a worldwide prevalence of approximately 1%, schizophrenia remains a large unmet medical need that stands to benefit greatly from (1) continued research to better understand the biological underpinnings of the disease and (2) the targeted development of novel therapeutics to improve the lives of those affected individuals.

International Review of Neurobiology, Volume 124
ISSN 0074-7742
http://dx.doi.org/10.1016/bs.irn.2015.08.003

Improvements in our understanding of the neuronal networks associated with schizophrenia as well as progress in identifying genetic risk factors and environmental conditions that may predispose individuals to developing the disease are advancing new strategies to study and treat it. Herein, we review the evidence that supports the role of α7 nicotinic acetylcholine receptors in the central nervous system and why these receptors constitute a promising target to treat some of the prominent symptoms of schizophrenia.

1. INTRODUCTION

Schizophrenia, then, is not like a flood, where one can envision watching all one's possessions being washed away. Nor like a cancer, where one can imagine a slowly growing tumor, relentlessly spreading from organ to organ and squeezing life from your body. No, schizophrenia is madness. Those who are afflicted act bizarrely, say strange things, withdraw from us, and may even try to hurt us. They are no longer the same person—they are mad! We don't understand why they say what they say and do what they do. We don't understand the disease process. Rather than a steadily growing tumor, which we can understand, it is as if the person has lost control of his/her brain.

***Excerpt from E. Fuller Torrey's, M.D., book* Surviving Schizophrenia: A Manual for Families, Patients and Providers**

With a manifestation in young adulthood, schizophrenia is often difficult to recognize from its first symptoms, but a pattern progressively emerges constituting the hallmark of a profoundly debilitating and lifelong illness. Illustrated by the description above, schizophrenia is characterized by thought disturbances, altered senses, abnormal social behavior, and an altered sense of self. Additional confusion may be apparent in the patient's mind by hearing voices, or other hallucinations, that are often interpreted as the presence of another "me" interfering with their own thoughts.

Given its early onset, the disease was initially referred as "démence précoce" or "adolescent insanity" by the German psychiatrist Emil Kraeplin (1856–1926). It was later the Swiss psychiatrist Paul Eugen Bleuler (1857–1939) who stated that it is not a single disease in the strict sense of the term but rather appears as a group of diseases that he termed "schizophrenia." He based his conclusions on the observation of the basic (obligatory, or negative) and accessory (supplementary, or positive) symptoms of the disorder. Generally occurring in late adolescence, the first manifestations of schizophrenia can be very different which render its diagnosis difficult (Cornblatt, Lenzenweger, Dworkin, & Erlenmeyer-Kimling, 1992; Davidson et al., 1999; Mednick, Parnas, & Schulsinger, 1987; Nuechterlein, 1983).

Schizophrenia is a severe and debilitating disorder that affects approximately 1% of the worldwide population. It is characterized by a constellation of symptoms that can include (i) positive symptoms, such as delusions and hallucinations (auditory and/or visual); (ii) negative symptoms, characterized by impairment in normal emotional responses and social behaviors, motivational deficits, affective flattening, and alogia among others; and (iii) cognitive impairments in executive functions and memory that are thought to contribute to social and/or occupational dysfunction.

Throughout history, patients suffering from schizophrenia were essentially excluded from normal life in society and, at the beginning of the last century, were secluded in psychiatric hospitals away from the general population. The introduction of the antipsychotic chlorpromazine to mitigate the positive symptoms of the disease in many patients marked a turning point in the treatment of the disease. While it can be hypothesized that the availability of pharmacological compounds that improve the patient's conditions shall rapidly refine our knowledge of the pathology, our understanding of the biological basis of schizophrenia is still only partial and can be attributed to at least three factors: (1) the broad spectrum of action of a compound such as chlorpromazine, making it difficult to conclude the primary mechanism(s) of action, (2) the complexity of the brain circuitry and networks involved in the disease, and (3) the etiology of the disease itself.

Individuals who have a relative diagnosed with schizophrenia have a 10% chance of developing the disease as compared to the 1% prevalence within the general population (see chapter "The role of nicotine in schizophrenia" by Robert E. Featherstone and Steven J. Siegel). Interactions between genetic risk factors and environmental factors, such as a viral infection during critical periods of development, as well as psychosocial factors (e.g., abuse, malnutrition) during childhood also have been associated with development of the disease (Tan, Callicott, & Weinberger, 2009). Thus, neurodevelopmental abnormalities are hypothesized to play a role in the pathophysiology of schizophrenia of predisposed individuals. In particular, abnormal development of prefrontal cortical neuronal circuitry and neurochemical transmitter systems that include acetylcholine, glutamate, and dopamine may then contribute to cognitive deficits that become evident at later stages of development.

Despite our lack of understanding of the cause(s) of schizophrenia, studies have shown differences in brain structure (e.g., less gray matter, larger ventricles) in some patients, as well as functional differences including malformed circuits (e.g., dorsolateral prefrontal cortex), and an imbalance in neurotransmitter systems (e.g., dopamine and glutamate) that are being studied for their involvement in producing the many symptoms of the disease.

As a major excitatory transmitter system in the brain, the cholinergic system has also been implicated in numerous neuropsychiatric, neurodevelopmental, and neurological disorders (Dani & Bertrand, 2007). More specifically for this chapter, the nicotinic receptors have been implicated in the pathophysiology of schizophrenia because of the high prevalence of smoking tobacco in patients with the disease, which is estimated between 65% and 85% and is two to five times higher than the general population (de Leon & Diaz, 2005). It has been hypothesized that schizophrenic patients smoke in attempt to self-medicate due, in part, to a dysregulation of the cholinergic system.

Two predominant pentameric nicotinic receptors have been identified in the brain, namely, the heteromeric α4β2 and the homomeric α7 forms. In this work, we will discuss the role of the α7 neuronal nicotinic acetylcholine receptors (nAChRs) in brain function and why these receptors are thought to represent an important target for the treatment of schizophrenia. A significant decrease in receptor binding using [^{125}I]-α-bungarotoxin (α-Btx), a selective ligand for the α7 receptor, was observed in the hippocampi (Freedman, Hall, Adler, & Leonard, 1995) and thalamic nuclei (Court et al., 1999) of schizophrenic patients in postmortem studies. In addition to binding studies, a reduction in protein expression of the α7 receptor in the frontal cortex (Guan, Zhang, Blennow, & Nordberg, 1999) and decreased mRNA levels of the *CHRNA7* gene, which encodes the α7 receptor, have been reported in individuals with schizophrenia (Mexal et al., 2010). Taken together, these data indicate altered α7 receptor levels are present in patients with schizophrenia and may contribute to some of the symptomatology of the disease. In particular, because of the recognized role of the cholinergic system in cognition, the investigation of the α7 receptor as a potential therapeutic target to improve cognitive function in patients afflicted with schizophrenia has received a great deal of attention.

2. THE α7 nAChRs

The α7 nAChRs belong to a category of ligand-gated ion channels thought to have appeared millions of years ago. The nAChRs are membrane-bound proteins comprised of both the ligand-binding site and the ion channel, which when open cause a change in the electrical properties of the pre- and/or postsynaptic cell, and facilitate neuronal transmission (Le Novère & Changeux, 1995). In their simplest and probably most widely expressed form, the α7 receptors result from the assembly of five identical (5) subunits around an axis of pseudo-symmetry that forms the ionic pore in its center. Each subunit spans four times the membrane with the second transmembrane domain (TM) lining the ionic pore in a barrel-like manner. The ligand-binding site resides at the interface between two adjacent subunits as schematized in Fig. 1. Although a crystal structure is still awaited, sequence homologies with other known pore-forming proteins, such as ELIC, GLIC, or the $5HT_3$ receptors (Bocquet et al., 2009; Hassaine et al., 2014; Hilf & Dutzler, 2008), are sufficient to confirm the quaternary organization of these

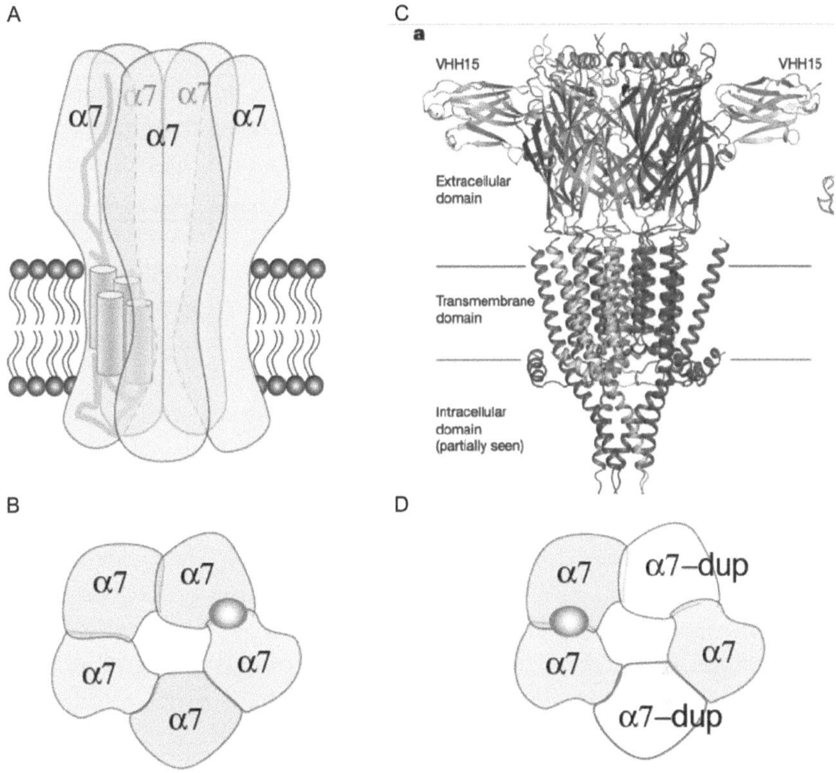

Figure 1 Structure of the receptor and possible assembly. (A) Schematic representation of the α7 receptor inserted in the membrane, with the five subunits arranged around the axis of symmetry. (B) Top view of the receptor with its organization laying in the center of the ionic pore. One binding site is represented for convenience. (C) Crystal structure of the $5HT_3$ receptor, a closely related receptor to the homomeric α7 nAChRs illustrating the protein arrangement. (D) Schematic representation of an α7: α7–dup (duplicate) combination in an heteromeric receptor complex to illustrate the difference in binding sites caused by the incorporation of the α7–dup subunit. *Panel (C): Reprinted by permission from Macmillan Publishers Ltd.:* Nature *(Hassaine et al., 2014), Copyright (2014).*

receptors. These data illustrate the topological features of the proteins with a large extracellular domain (also referred to as the ligand-binding domain [LBD]) and the four TMs with both the N- and C-terminal end of the protein facing the extracellular space. Given the high degree of homology and conservation of the key amino acids in the formation of these domains, it is even possible to exchange the LBD or TMs domain of one receptor with another to produce functional chimeras (Eiselé et al., 1993; Tillman, Seyoum, Mowrey, Xu, & Tang, 2014).

At rest, in the absence of bound ligand, the α7 receptor is closed and therefore impermeable to ions. Binding of the endogenous agonist acetylcholine (ACh) occurs in the extracellular domain, at the interface between two adjacent subunits, and stabilizes the receptor in the active open state. The transition from closed-to-open state is thought to result from a rotation and twist of the TM domains causing the opening in a blooming flower-like manner (Althoff, Hibbs, Banerjee, & Gouaux, 2014; Calimet et al., 2013).

Important functional features that characterize the α7 receptors are (1) a rapid desensitization, (2) a high calcium permeability, (3) a strong voltage dependence, and (4) a pharmacological profile distinct from other nAChRs (Alkondon & Albuquerque, 1993; Couturier et al., 1990; Séguéla, Wadiche, Dineley-Miller, Dani, & Patrick, 1993). Studies of the α7 receptor conducted using exogenous recombinant systems or from *in vitro* brain slice recordings have yielded very similar results and suggests that the observed properties are intrinsic to the receptors and not a peculiarity associated to the experimental conditions (Alkondon & Albuquerque, 1993; Alkondon, Pereira, & Albuquerque, 1998; Bloem, Poorthuis, & Mansvelder, 2014; Gopalakrishnan et al., 1995; Poorthuis, Bloem, Verhoog, & Mansvelder, 2013; Puchacz, Buisson, Bertrand, & Lukas, 1994).

Desensitization of the α7 receptor is the fastest of all known nAChR subtypes and indicates a phasic and transient physiological response following activation. An additional property that must be considered is that the activation and desensitization profiles of the α7 receptors display no significant overlap, which is observed by the absence of a window current, indicating that upon sustained exposure the α7 receptors will be desensitized (Dani & Bertrand, 2007). The rapid desensitization properties of the α7 receptor make it a challenging drug target in that activating the receptor with agonist compounds is likely to cause desensitization at high doses. Therefore, selecting the proper dose range for receptor activation is paramount to establish and maintain efficacy.

When open, the α7 receptors are permeable to cations, the influx of which causes a depolarization of the postsynaptic membrane in which they are located. These receptors display a high permeability to calcium independently of the species in which it was measured (Bertrand, Galzi, Devillers-Thiéry, Bertrand, & Changeux, 1993; Fayuk & Yakel, 2005; Fucile, 2004; Séguéla et al., 1993). Moreover, as reported in different studies, the calcium influx associated with the opening of the α7 receptors is sufficient to trigger the release of other neurotransmitters (see, for example, Albuquerque, Pereira, Braga, & Alkondon, 1998; Barik & Wonnacott, 2006; Dickinson, Kew, & Wonnacott, 2008; Gray, Rajan, Radcliffe, Yakehiro, & Dani, 1996; Ji & Dani, 2000) or to activate second messenger cascades (El Kouhen, Hu, Anderson, Li, & Gopalakrishnan, 2009; Gubbins, Gopalakrishnan, & Li, 2010). Site-directed mutagenesis conducted in the pore and extracellular domains has identified key residues that are responsible for this high calcium permeability (Bertrand et al., 1993; Colón-Sáez & Yakel, 2013). The conservation of the amino acids bordering the ionic pore across species ranging from avian to humans underscores the physiological relevance of the ionic selectivity of the α7 receptors, and the functional significance associated with activation of these receptors to cause both a depolarization of the membrane and an influx of calcium.

Upon activation, the amplitude of the response of the α7 receptors at different transmembrane voltages reveals a strong inward rectification. Whereas a large inward current is observed when the cell is maintained at a negative potential, the response becomes negligible for membrane potentials above −40 mV (Couturier et al., 1990; Gopalakrishnan et al., 1995; Séguéla et al., 1993). The reduction in current amplitude is larger than predicted by Ohms law, and the nonlinearity is attributed to the binding of molecules at the intracellular mouth of the α7 channels (Forster & Bertrand, 1995). As a physiological consequence, signaling mediated by these nAChRs depends upon the resting potential of the cell in which they are expressed.

Characterized by their unique sensitivity to the snake toxin α-Btx or methyllycaconitine (MLA), the α7 receptors display pharmacological properties that are significantly distinct from other nAChR subtypes. To date, several compounds displaying selective agonistic activity for the α7 receptors with EC_{50}s (half-maximal effective concentrations) in the micromolar range were discovered (reviewed in Hurst, Rollema, & Bertrand, 2012; Wallace et al., 2010).

3. THE HETEROMERIC α7-CONTAINING RECEPTORS

As functional receptors can readily be reconstituted by expression of the *CHRNA7* gene alone, it was concluded that the DNA sequence encodes for a protein that can yield homomeric receptors. Additional complexities must, however, be considered with the following alternatives. The identification of CHRFAM7A (which encodes for α7–dup) is a partial duplicate DNA sequence of *CHRNA7* that is localized on the same arm of chromosome 15 but encodes only for exons 6–10 of α7 raised the question of a possible assembly of the corresponding protein in a receptor complex (Gault et al., 1998). Heterologous expression of *CHRNA7* with *CHRFAM7A* revealed that these two proteins can assemble in a single receptor complex, but the presence of the truncated form impairs the functionality of the receptors and might act as a dominant negative (Araud et al., 2011; de Lucas-Cerrillo et al., 2011). Determination of the effects of the Type-II allosteric modulator revealed a larger potentiation in α7–dup-containing receptors, suggesting that these receptors might become functional under specific conditions (Araud et al., 2011).

Expression of *CHRNA7* with *CHRNB2* which encodes for the β2 nAChR subunit revealed that this protein can assemble with α7 to form functional heteromeric receptors (Khiroug et al., 2002; Murray et al., 2011; Zwart, Strotton, Ching, Astles, & Sher, 2014). Insertion of a different subunit in the receptor complex is expected to cause changes in critical regions such as the LBD or the ionic pore. This hypothesis was confirmed by the observation of a preferential inhibition of β2-containing receptors by the competitive inhibitor dihydro-beta-erythroïdine (DHβE) (Murray et al., 2011). Additional differences in the pharmacology of the α7 receptors were observed in the potency but not efficacy of partial agonists such as cytisine, EVP-6124, and RG3487 (Wallace, Porter, Neveu, & Bertrand, 2011; Zwart et al., 2014).

Results obtained from coexpression of CHRNA7 with CHRNB4 suggest that the β4 subunit can also assemble in functional heteromeric complexes (Criado et al., 2012). The α7β4 heteromers display distinct physiological and pharmacological properties resembling those of α7β2 with changes in the sensitivity to DHβE, maximal amplitude of the current evoked by partial agonists, and a slower desensitization time course.

Although coexpression of α7 with the β2 subunit appears to be limited to specific brain areas, such as the basal forebrain and the hippocampus, or the chromaffin cells for α7 with the β4 subunit, these data illustrate the presence of heteromeric complexes and add further complexity to the physiology and pharmacology of α7 receptors.

4. LOCALIZATION OF THE α7 RECEPTORS

The first information concerning the distribution of the α7 receptors in the brain was obtained from autoradiography of histological brain sections labeled with [^{125}I]α-Btx (Clarke, Schwartz, Paul, Pert, & Pert, 1985; Court, Martin-Ruiz, Graham, & Perry, 2000; Han et al., 2003; Tribollet, Bertrand, Marguerat, & Raggenbass, 2004). Data obtained from rat, monkey, and human brains revealed a strong labeling by α-Btx in the cortex, the hippocampus, and substantia nigra. As initial labeling experiments were conducted prior to the discovery of the α7 receptor itself, it was concluded that α-Btx and nicotine binding were similar in the substantia nigra pars compacta but differed in other areas such as in the hippocampus CA1 region indicating a difference in receptors specific for α-Btx and nicotine (Court & Clementi, 1995). A high density of α-Btx labeling was also observed in the cerebellum (Court et al., 2000). Autoradiograms obtained from monkey brains indicated that expression of the α7 receptor displays a higher density in layer I of the telencephalon, prefrontal, or auditory cortex, but is lower in layers V–VI of the telencephalon (Han et al., 2003). This illustrates that α7 receptors are highly expressed in a regionally specific manner within the brain. A good agreement was observed between the α-Btx labeling and the mRNA levels (Han et al., 2000). Subsequently, much of the early localization studies using toxins (e.g., α-Btx) have been confirmed through the use of α7-selective chemical entities such as A-585539, CHIBA-1001, AZ11637326, A-582941, A-844606, NS14492, or NS14490 (Anderson et al., 2008; Ettrup et al., 2011; Hashimoto et al., 2008; Maier et al., 2011; Rötering et al., 2014; Toyohara et al., 2010).

Marked subcellular organization of α7 receptors was demonstrated by electron microscopy conducted in the hippocampus and ventral tegmental areas (VTA) in the rat (Fabian-Fine et al., 2001; Jones & Wonnacott, 2004). A higher level of expression was observed in the hippocampus than VTA with labeling present at nearly every synapse of the CA1 region, at both pre and postsynaptic sites (Fabian-Fine et al., 2001). Furthermore, α7 receptors were equivalently expressed at γ-amino-butyric-acid (GABA) GABAergic and glutamatergic synapses as shown by double immunolabeling (Fabian-Fine et al., 2001). Postsynaptic receptors were distributed on dendritic spines and in the perisynaptic annulus. In contrast, expression of the α7 receptors in the VTA was both somatodendritic and postsynaptic. Most presynaptic receptors were on glutamatergic axon terminals and principally at extrasynaptic and perisynaptic loci (Jones & Wonnacott, 2004). The absence of α7 receptors on cholinergic synapses in the VTA suggests, however, a paracrine mode of activation in this brain area (Jones & Wonnacott, 2004).

Overall, α7 receptor localization within the brain is similar across mammalian species, although some important differences have been reported in brain regions associated with cognition and sensory processing that are worth noting. In particular, [^{125}I]-α-Btx binding has been identified in the lateral and medial geniculate areas of the thalamic nuclei in the rat, whereas in monkey and human postmortem brain samples, α7 receptors have been identified in the reticular nuclei of the thalamus (Cimino, Marini, Fornasari, Cattabeni, & Clementi, 1992; Breese et al., 1997; Spurden et al., 1997). The reticular nucleus of the thalamus receives excitatory input from the cortex and from within other nuclei in the thalamus and sends inhibitory projections to dorsal thalamic nuclei (Ferrarelli & Tononi, 2011), playing an important role in mediating attention and sensory gating, two processes that are disrupted in schizophrenia. The localization of the α7 receptors in the reticular nuclei of the thalamus suggests that they could contribute to modulating these processes and their dysregulation could be involved in the disruption of these processes in schizophrenia.

The pattern of α7 expression described above indicates that these receptors must contribute to specific and differential functions in given brain areas. Getting a better understanding of their contribution therefore requires precise functional studies at the cellular and network levels. Physiological experiments conducted in brain slices and in animal models are beginning to shine a new light on both the distribution and functionality of the α7 receptors.

The development of new molecules that can be radiolabeled and are sufficiently safe offers the possibility to examine the distribution of α7 receptors using positron emission tomography (PET) imaging in humans. To date, five molecules ([^{11}C]-CHIBA-1001, [^{11}C]-NS-14492, [^{11}C]-A-833834, [^{18}F]-AZ11637326, [^{18}F]-ASEM) have been developed for PET imaging (Ettrup et al., 2011; Hashimoto et al., 2008; Horti et al., 2013; Maier et al., 2011; Wong et al., 2014; Yin et al., 2013). Successful labeling of α7 receptors was already obtained in humans with [^{18}F]-ASEM (Wong et al., 2014). The first images obtained from human showed a distribution that is comparable to postmortem data with a high expression of α7 in the cortex and to a lesser extent in the white matter and corpus callosum. Moreover, as [^{18}F]-ASEM rapidly dissociates from the receptors, it is appropriate for interrogating neuropsychiatric disorders (Wong et al., 2014). In a study in which humans were administered either tropisetron (α7 agonist/$5HT_3$ antagonist) or ondansetron ($5HT_3$ antagonist) along with the CHIBA-1001 ligand, it was only in the tropesitron condition in which radiolabeling was diminished, presumably due to less available α7 receptors to bind the CHIBA-1001 molecule (Ishikawa & Hashimoto, 2011). Although the brain distribution of the [^{11}C]-CHIBA-1001 compound has not yet been evaluated in humans, data obtained from the brains of nonhuman primates show comparable distributions as those predicted from rodent brain studies. In addition, in an experiment attempting to reproduce "schizophrenia-like" conditions in nonhuman primates through the repeated administration of the *N*-methyl-D-aspartate (NMDA) receptor antagonist phencyclidine (0.3 mg/kg, twice a day; 13 days), a statistically significant reduction of α7 receptor labeling in the frontal cortex of these animals was observed (Hashimoto et al., 2008). Results already obtained with PET are showing promising strategies to investigate α7 receptors in different psychiatric conditions and are expected to provide additional answers about expression of the nAChRs in schizophrenia.

5. ALLOSTERIC MODULATION

The transition from a closed to an open or desensitized conformation of the α7 receptors is governed, as for any protein, by the energy barriers existing between the different states. Increasing or lowering these barriers will reduce or enhance the probability of transition between the considered states. As a consequence, the binding of a molecule at a site distinct from the ligand-binding site might affect the transition energy barriers and act as an allosteric modulator.

The proof of principle of such allosteric modulation was first provided by the observation that anthelmintic ivermectin potentiated the ACh-evoked current at α7 receptors (Krause et al., 1998). Evidence for allosteric modulation includes the lack of inward current, but increased sensitivity and maximal response following application of ivermectin. Following these initial findings, several allosteric modulators active at the human α7 receptor were identified and subsequently classified as Type-I and -II modulators (Bertrand & Gopalakrishnan, 2007; Grønlien et al., 2007). These two types are distinguished by their effect on desensitization kinetics. Type-I modulators, such as 5-hydroxyindole (5-HI) or NS-1738, potentiate the agonist-evoked currents and shift the dose–response to agonist to the left (i.e., increase the sensitivity of the agonist) without significant changes of the response decay time (Timmermann et al., 2007, Zwart et al., 2002). Type-II modulators such as PNU-120596 potentiate the agonist response in addition to slowing down the desensitization of the receptor (Hurst et al., 2005). Moreover, some Type-II modulators such as PNU-120596 are able to restore the response of receptors that are desensitized by sustained exposure to an agonist (Hurst et al., 2005).

Increasing the energy barrier between the resting and active state is expected to cause a shift to the right of the agonist sensitivity and to reduce the maximal amplitude of the response. The identification in mammals of endogenously expressed molecules that share a high degree of homology with the toxin α-Btx suggested that these proteins might interact with the nAChRs (Chimienti et al., 2003; Dessaud, Salaün, Gayet, Chabbert, & deLapeyrière, 2006; Miwa et al., 1999). The membrane glycosylphosphatidylinositol-anchored molecules such as Lynx1 were shown to inhibit the α7 receptors, likely through a negative allosteric effect (Ibañez-Tallon et al., 2004; Miwa et al., 1999, 2006; Tsetlin, 2014). On the contrary, the secreted mammalian Ly-6/uPAR-related protein 1 (SLURP-1) protein was shown to potentiate the α7 receptors (Chimienti et al., 2003). Finally, increased α7 activity was reported for another closely related protein from the protein/Ly-6/urokinase-type plasminogen activator receptor (*PATE*) gene family (Levitin et al., 2008). Altogether, this suggests that endogenous α7 ligands might serve to fine-tune these membrane-bound receptors.

6. THE α7 RECEPTORS IN BRAIN CIRCUITS

Although the psychotropic effects of nicotine were acknowledged a long time ago, its effects on α7 receptor central nervous system biology remained a puzzle for scientists. The lack of effects that nicotine exposure exhibited on α7 receptor-mediated brain activity either in extracellular or in intracellular recordings suggested that this alkaloid was not interacting directly with cholinergic neurotransmission. This was in sharp contrast to the electrophysiological findings observed with the high-affinity sites for nicotine (e.g., α4β2) or with the histological localization results obtained from α-Btx binding (Arimatsu, Seto, & Amano, 1978; Clarke et al., 1985).

Cloning and expression of the nAChR genes, especially the α7 subunit, shone a new light on the physiology of these receptors. In addition, studies of recombinant receptors allowed the characterization of the α7 receptors, and the identification of α-Btx and MLA as useful pharmacological tools to dissect their activity provided new hints on how to investigate these receptors (Drasdo, Caulfield, Bertrand, Bertrand, & Wonnacott, 1992).

While previous attempts to characterize nicotinic transmission were accomplished by the application of nicotine in the bath of the cells *in vitro* or by iontophoresis directly onto the cells, it was acknowledged that these methods were too slow and that the putative signals could be blurred by desensitization. Subsequently, the use of fast application inward-evoked currents, following either ACh or nicotine test pulses in rat hippocampal slices, demonstrated properties comparable to those observed in recombinant experiments (Alkondon & Albuquerque, 1993). In parallel, evidence for the activity of α7 receptors was provided by the slow application of nicotine but at a low concentration, which caused an increase in the frequency of miniature excitatory postsynaptic currents in the hippocampus that could be blocked by MLA or α-Btx (Gray et al., 1996). Collectively, these data renewed interest in investigating the effects of nicotine and other molecules at the α7 receptors.

Results obtained from *in vitro* recordings confirmed that α7 receptors are widely expressed in the central nervous system. The largest fraction of α7 receptors modulates neuronal activity, whereas a smaller remaining fraction mediates synaptic transmission (Albuquerque, Pereira, Alkondon, & Rogers, 2009; Dani & Bertrand, 2007). Recordings from distinct cortical layers in the prefrontal cortex revealed that in layers I–III, α7 receptors are expressed in interneurons, whereas in layers V, they are expressed both in interneurons and in the soma of pyramidal cells (Bloem et al., 2014; see Fig. 2). While these studies illustrate where α7 receptors are expressed in the cortex and begin to reveal their contribution in cell signaling, our understanding is still too limited to fully apprehend the complexity of cholinergic neurotransmission in the prefrontal cortex.

To get one step further, it is necessary to examine the role of α7 nAChRs in different cell types and, moreover, as a function of their subcellular localization. When inserted in the postsynaptic complex (schematized in Fig. 3A), α7 nAChRs can mediate excitatory neurotransmission such as that observed in the hippocampus (Hefft, Hulo, Bertrand, & Muller, 1999). However, the majority of α7 nAChRs are expressed in different cell compartments including presynaptic or perisynaptic areas. Receptors expressed

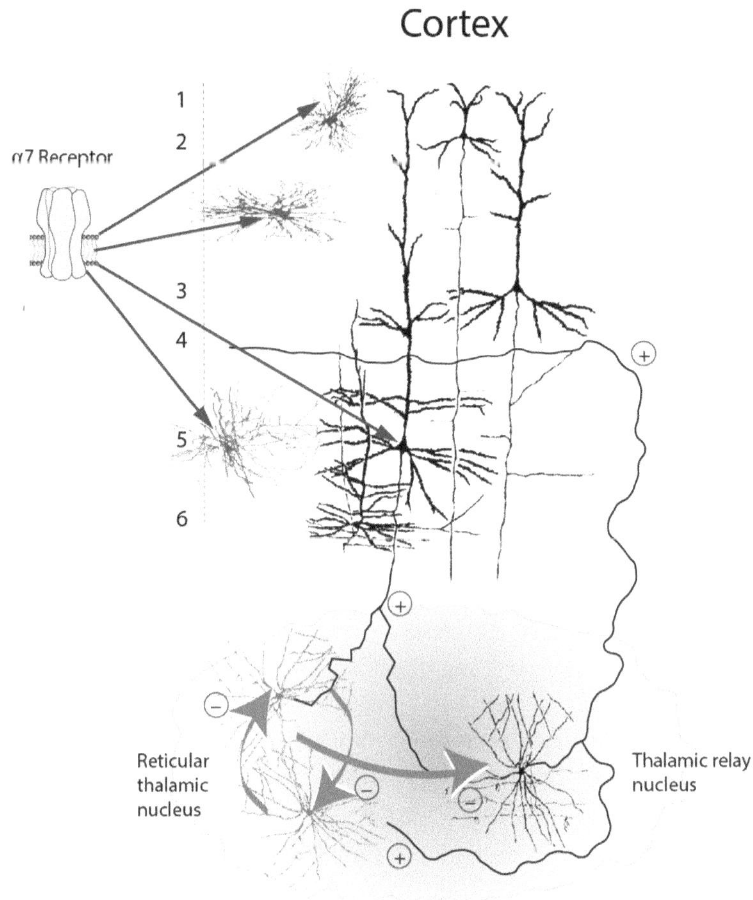

Figure 2 Localization of α7 receptors in the prefrontal cortex. Distribution of the α7 nicotinic receptor in the cortical layers of the prefrontal cortex with a schematic representation of the thalamo-cortical feedback.

in the presynaptic bouton (as schematized in Fig. 3B) can modulate the release of the neurotransmitter contained in the synaptic vesicle. The high permeability of α7 nAChRs causes an influx of calcium, which increases the release of neurotransmitter and, in the case of an inhibitory synapse, such as, illustrated in this figure, will potentiate the inhibition. On the contrary, activation of receptors expressed perisynaptically, such as schematized in Fig. 3C, can block the propagation of the action potential and therefore inhibit neurotransmission. Activation of α7 receptors inserted on neuronal spines, such as illustrated in Fig. 3D, can modulate the activity of the main

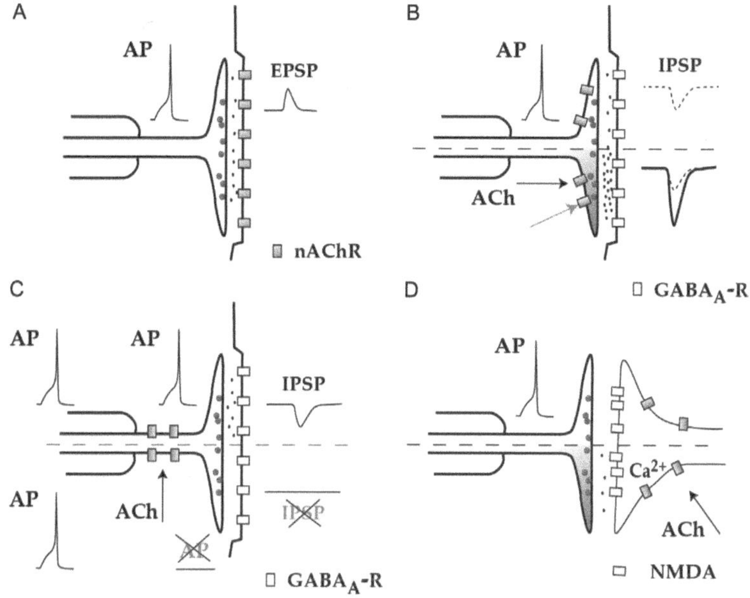

Figure 3 Expression of α7 receptors in different subcellular compartments. Schematic representation of the α7 localization in the different synaptic endings with their respective effects. (A) α7 receptors expressed postsynaptically can mediate excitatory neurotransmission. (B) α7 receptors expressed presynaptically can cause an increase of the intracellular calcium concentration (represented in the lower part of the scheme) and cause an increase in the release of neurotransmitter. (C) α7 receptors expressed perisynaptically can modulate the electrical activity and propagation of the action potential (lower part indicating the effects of α7 stimulation). (D) α7 receptors expressed in the dendritic spine can modulate the activity of NMDA receptors. The calcium entry caused by α7 stimulation enhances the NMDA response. Abbreviations: ACh, acetylcholine; AP, action potential; ESPC, excitatory postsynaptic potentials; IPSC, inhibitory postsynaptic potentials; GABA, gamma-amino-butyric acid receptor; NMDA, *N*-methyl-D-aspartate receptor).

receptors that mediate neurotransmission as it was proposed in the prefrontal cortex (Arnsten, Wang, & Paspalas, 2012).

Expression of α7 receptors on the main dendrite of the pyramidal cell can affect the process known as back propagation of the action potential from the soma that, in turn, impacts the efficacy of signal transmission from the apical tuft. Activation of receptors expressed in the vicinity of the axon hillock can modify the firing threshold or might prevent the triggering of the action potential.

As interneurons are known to express the α7 nAChRs, this provides a further level of complexity of modulation that these receptors can provide. Namely, activation of α7 receptors on an inhibitory interneuron might exacerbate or prevent this cell effect by mechanisms that are described above.

Altogether, given their broad degree of expression and peculiar properties, such as the high calcium permeability, the α7 nAChRs provide a broad possibility of modulation by the cholinergic system and may explain why these receptors are highly conserved throughout evolution.

7. α7 RECEPTORS AND SCHIZOPHRENIA

The high prevalence of smoking in schizophrenic patients led to the investigation of the nicotinic receptors in the disease. Later, the ability to study the α7 receptors using selective venomous toxins (e.g., α-Btx) demonstrated that α7 receptors were localized in brain regions with recognized involvement in cognition such as the cortex and hippocampus. Impairments in cognitive processes such as executive function mediated by the prefrontal cortex, including working memory and cognitive flexibility, as well as attention, precede psychotic symptoms and are a predictor of functional outcome (Caspi, 2003; Green, Kern, Braff, & Mintz, 2000).

8. α7 AND NMDA RECEPTOR INTERACTIONS IN SCHIZOPHRENIA

A growing body of evidence suggests an interaction of the glutamatergic and cholinergic systems, at the receptor level (α7 nicotinic receptors for the cholinergic system and NMDA receptors (NMDAR) for the glutamatergic system), may underlie some of the cognitive impairments of schizophrenia. In particular, working memory, as well as other executive functions, can be profoundly impaired in schizophrenia. As the described basis of abstract thought, an inability to reflect salient information in the absence of sensory stimulation could contribute to the ineffective cognitive function (Arnsten et al., 2012). Moreover, working memory underlies other cognitive processes, such as long-term memory and language, which also can become impaired with age and disease (Arnsten et al., 2012; Bizon, Foster, Alexander, & Glisky, 2012).

It is the persistent firing of a network of interconnected glutamatergic pyramidal neurons (also known as "delay cells") within layer III of the dorsolateral (dl) region of the PFC that subserves visuospatial working memory (Goldman-Rakic, 1995, 1996; Wang et al., 2011). Activation of delay cells enables continuous updating of information to maintain and flexibly manipulate information that is no longer present in the environment. NMDAR are integral to maintaining persistent firing in dlPFC neurons and keeping abstract information "online" during the delay periods between stimulus presentation and action.

Hypofunction of the NMDAR system has been implicated in the etiology of schizophrenia (Moghaddam & Krystal, 2012). NMDAR are critical for memory function, and inhibition of these receptors either pharmacologically or genetically (Park et al., 2014; Zorumski & Izumi, 2012) reproduces psychotomimetic properties and cognitive impairments of schizophrenia in animals and in humans (Javitt, 2007). Interestingly, the administration of nicotine and, perhaps more intriguing, selective α7 receptor agonists reverses NMDA receptor-induced working memory deficits as assessed in rodents (Hashimoto, Fujita, Ishima, Hagiwara, & Iyo, 2006; Levin, Bettegowda, Weaver, & Christopher, 1998; Lewis, Curley, Glausier, & Volk, 2012; Pichat et al., 2007; Thomsen, Christensen, Hansen, Redrobe, & Mikkelsen, 2009). In addition, neurochemical changes following NMDA receptor antagonism are consistent with some of the most reproducible findings in schizophrenic brains including deficits in GABAergic neurotransmission due to a reduction of glutamic acid decarboxylase 67 (GAD67) and parvalbumin content (Lewis et al., 2012). Improvement in working memory deficits observed after α7 agonist administration occurs prior to the restoration of neurochemical alterations (e.g., GAD67) following NMDA receptor antagonism.

Acetylcholine also plays an important role in working memory function and depletion of ACh from the prefrontal cortex impairs spatial working memory (Croxson, Kyriazis, & Baxter, 2011). Iontophoresis of α7 receptor agonists onto dorsolateral prefrontal cortical neurons increases delay cell firing, whereas iontophoresis of MLA inhibited neuron firing (Yang et al., 2013). Moreover, iontophoresis of α7 agonists such as PHA543613 was able to restore firing following NMDAR antagonism (Yang et al., 2013). Localization of α7 receptors and NMDAR adjacent to one another suggests that activation of α7 receptors may depolarize the spine, allowing the removal of the Mg^{2+} block within the NMDA receptor. Activation of the NMDAR then becomes possible, which is sufficient to trigger action potentials NMDAR.

Additional evidence of an interaction of α7 receptors and NMDAR comes from genetic studies. Microdeletion of the *CHRNA7* gene has been associated with cortical dysfunction in schizophrenia as well as other neurological and neuropsychiatric disorders (Stefansson et al., 2008). More specifically, gene deletion of the α7 receptor in mice results in reduced NMDAR electrophysiological currents as well as a reduction in NMDAR expression and glutamatergic synapse formation (Lin, Hsu, Baumann, Coulter, & Lynch, 2014). As a coagonist of the NMDAR, D-serine has been of interest in schizophrenia. Many studies have demonstrated perturbation or deletion of the D-serine synthetic enzyme (serine racemate) not only decreases D-serine production as would be expected but also reduces glutamatergic transmission in the forebrain and yields schizophrenia-like behaviors in mice (Puhl et al., 2014; Van Horn, Sild, & Ruthazer, 2013). These data suggest a tight regulation of NMDAR by the α7 receptor, which could be involved in the disruption in cortical-mediated functions underlying schizophrenia.

9. α7 RECEPTORS AS A DRUG TARGET

Results obtained from the ensemble of animal models, human postmortem analyses, and functional imaging studies suggest that schizophrenia results from an imbalance of the neuronal network activity with a determinant contribution of both the glutamatergic and cholinergic systems. Given their brain distribution, cellular localization, and physiological properties, the α7 nAChRs can be seen as an ideal target for the development of compounds with therapeutic activity aiming for the treatment of neurological diseases.

Different approaches can be considered in the search of active compounds that would restore the balance between the excitatory glutamatergic and cholinergic functions.

9.1 The Agonist Approach

The first set of attempts made to create compounds that displayed selective agonist properties for the α7 receptor led to the identification of 2,4-dimethoxybenzylidene anabaseine (DMXB, also referenced as GTS-21) (Hunter, de Fiebre, Papke, Kem, & Meyer, 1994). While GTS-21 acts as a partial agonist at the human α7 receptors, compounds showing better selectivity were subsequently identified following the observation of the preferential activity of molecules containing a quinuclidine moiety (Bodnar et al., 2005; Feuerbach, Nozulak, Lingenhoehl, McAllister, & Hoyer, 2007; Marrero et al., 2010; Mazurov et al., 2012; Prickaerts et al., 2011; Wallace et al., 2010). Alternate scaffolds showing selective agonistic activities at the α7 receptors were also identified (Biton et al., 2007; Ghiron et al., 2010; Haydar et al., 2009).

Experiments conducted with these molecules in animal models ranging from mice to monkeys revealed their procognitive properties, effects that were antagonized with MLA, supporting the hypothesis of direct activity at the α7 nAChRs. While some of these compounds showed a complex pharmacology with activation of the α7 nAChR and blockade of the closely related serotonergic $5HT_3$ receptors, the inhibition observed with MLA established that the procognitive effects were mediated by the α7 receptors. Interestingly, improvements in memory performance and cognitive tasks were observed almost independently of the fact that some molecules acted as partial agonists, while others were full agonists. A common trend, however, was that improvement of cognitive performance was observed over a rather narrow range of concentrations in the form of an inverted U-shaped dose–response curve (Fig. 4), i.e., increasing the concentration of an α7 agonist above a threshold value results in a loss of drug effect.

Despite promising and converging results obtained in animal models and early clinical trials, success in larger, controlled clinical trials has been limited. For example, while a proof of concept for DMXB-A (GTS-21) was

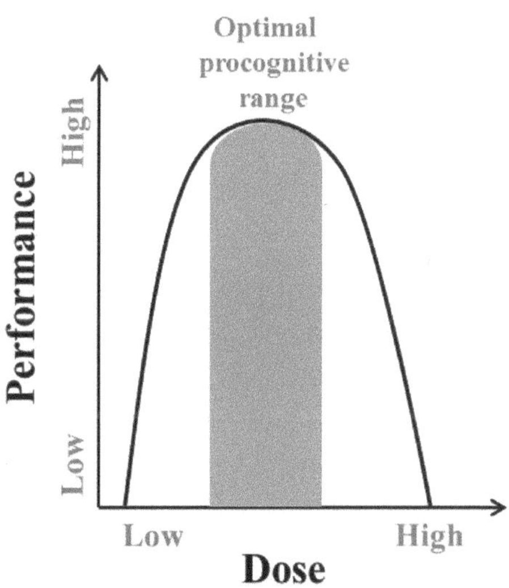

Figure 4 Representation of "inverted U" dose–response function. α7 receptor agonists exhibit procognitive properties in animals and humans within a narrow dose range, often yielding "inverted U" shaped dose–response relationships in which lower and higher doses are ineffective. It is presumed that at higher doses, α7 receptors become desensitized and the cognitive-enhancing properties of the compound often are lost.

obtained in schizophrenic patients, subsequent results in a phase II clinical trial were not conclusive (Freedman et al., 2008; Olincy et al., 2006). Molecules were abandoned due to side effects, such as cardiovascular events for PNU-292987, insufficient efficacy, or other undocumented reasons. To date, two α7 agonists are in clinical trials include encenicline (EVP-6124) and ABT-126.

Based on the physiological properties of the α7 receptors, it is predicted that sustained exposure to an agonist in conditions comparable to those observed during systemic administration of the molecule will rapidly desensitize the receptors. This inhibition of the α7 receptors most likely accounts for the inverted U observed *in vivo* and the loss of effect observed during exposure to the highest concentrations of the agonist.

When taking into account that schizophrenic patients tend to be heavy smokers, it should be remembered that the effects of the drug under investigation should combine with the effects caused by nicotine inhaled with the cigarette smoke. Consequently, a shift in the dose–effect curve would be predicted further limiting the efficacious range for the compound and perhaps could explain the limited success observed in clinical trials thus far. In addition, while preclinical experiments almost exclusively have been conducted with the investigational molecule in isolation, most patients undergoing clinical trials are following their experimental treatments with antipsychotic drugs (and perhaps other psychotropic agents). To better reproduce the patient's therapeutic conditions, it therefore would be necessary to coexpose the α7 receptors to the classical antipsychotic medication and the compound under evaluation during preclinical testing.

9.2 The Allosteric Modulation Approach

Potentiation of α7 receptor activity by an allosteric modulator represents an alternative and perhaps more physiological approach to increase the effects of these receptors. One important advantage of the allosteric modulator is increased temporal resolution in that these molecules are expected to have minimal or no intrinsic activity but should increase (or decrease) the receptor response only when the endogenous ligand is present. Additionally, because of the high incidence of smoking in schizophrenic patients, an allosteric modulator approach has the benefit of not competing with nicotine at the orthosteric site of the receptor, thereby maximizing the therapeutic potential of α7 receptor activation.

Experiments conducted in different animal models have shown that both Type-I and Type-II positive allosteric modulators exert procognitive activity (Eskildsen et al., 2014; Sahdeo et al., 2014; Thomsen, El-Sayed, & Mikkelsen, 2011; Thomsen, Hansen, Timmerman, & Mikkelsen, 2010). Moreover, additive effects of the positive allosteric modulator with an agonist have been demonstrated in rats and nonhuman primates. Enhanced cognitive performance was observed following the combined administration of the Type-II modulator PNU-120596 and the cholinesterase inhibitor donepezil at doses that were ineffective alone, but were efficacious when administered together. These effects were mediated through the α7 receptor as they could be blocked by MLA (Callahan, Hutchings, Kille, Chapman, & Terry, 2013).

Experiments conducted *in vitro* and *in vivo* suggest that kynurenic acid is a potent inhibitor at the α7 receptors and that excessive production of this molecule is likely to affect the cholinergic neurotransmission (Hilmas et al., 2001; reviewed in Albuquerque & Schwarcz, 2013). Following these observations, it was thought that an excess of kynurenic acid in the frontal lobe might inhibit the α7 receptors causing a cholinergic glutamatergic imbalance leading to schizophrenia. Attempts to restore the cholinergic function by exploiting the allosteric modulation of α7 receptors with galantamine were subsequently conducted in animals and humans. Offspring born from rats fed with kynurenic acid L-kynurenine showed cognitive deficits that were reversed by acute administration of galantamine (Alexander et al., 2013). Further testing of this hypothesis was completed in a clinical trial in which individuals with schizophrenia were treated with a combination of a dietary source of CDP-choline and galantamine (Deutsch et al., 2013). However, in this study, no significant effect was noted on either the negative symptoms (primary endpoint) or the cognitive impairments, which were assessed as well. Interestingly, this treatment combination did improve overall functioning in patients, suggesting further investigation of the α7 agonist/PAM approach may be warranted.

The identification of endogenous proteins that modulate the activity of α7 receptors, such as Lynx-1 or SLURP-1, is another factor that must be taken into account. Dysfunction of these proteins and their corresponding pathways also may impact the α7 receptor function.

9.3 α7 Receptor Priming

The procognitive effects of α7 agonists generally display a dose–response activity corresponding to an inverted U (Fig. 4) in which the decline of cognitive effects at higher doses is attributed to the desensitization of the α7 nAChRs. Moreover, comparison between *in vivo* and *in vitro* data reveals that cognitive effects observed *in vivo* occur at concentrations causing no detectable activation of the α7 receptors *in vitro*. Electrophysiological recordings conducted with sustained exposure to low concentration of the partial agonist EVP-6124 shed a new light on this contradictory results. Responses evoked by brief ACh test pulses were potentiated by sub-nanomolar concentration of EVP-6124, which corresponds to the active concentration range observed in animal models (Prickaerts et al., 2011). This mechanism is now referred as *priming* and is thought to occur by the occupancy of the ligand on the receptors that facilitate the activation by ACh. Interestingly, potentiation of other nAChRs by antagonists was previously reported for heteromeric receptors suggesting of a common mechanism of action (Cachelin & Rust, 1994; Smulders et al., 2005; Zwart, van Kleef, Gotti, Smulders, & Vijverberg, 2000).

Enhancement of the ACh-evoked currents caused by *priming* is an interesting alternative to the pure agonistic activity and can be considered closer to receptor modulation. The main advantage of this mechanism is that *priming* is observed at low concentrations of compound that enhances the endogenous cholinergic activity. The intrinsic inverted U response associated with the agonist activity causes a self-limitation in the efficacious concentration range that should protect patients from overdosing. Moreover, as potentiation of the endogenous cholinergic tone by the agonist is observed at concentrations that are not sufficient to stimulate the α7 receptors alone, this may protect from undesired side effects associated with receptor activation.

10. EFFECTS OF α7-SELECTIVE COMPOUNDS ON BEHAVIOR

10.1 Clinical Trials: Cognition

The therapeutic potential of α7 nAChR activation is currently under investigation. In a phase II, double-blinded, placebo-controlled clinical trial that included smokers and nonsmokers, EVP-6124 (encenicline), which exhibits α7 agonist and $5HT_3$ antagonist properties, improved cognitive performance in schizophrenic patients stable on antipsychotic medication, as assessed by the CogState test battery. This positive result has culminated in the initiation of a 26-week phase III study (*COGNITIV*) investigating the cognitive-enhancing potential of EVP-6124 in schizophrenic patients (clinicaltrials.gov trial identification number: NCT01714661). The primary outcome measure will be a change from baseline in the MATRICS Consensus Cognitive Battery (MCCB), which failed to show significance in the phase II study despite positive results in the CogState test battery.

In addition, in a smaller trial, the antiemetic drug, tropesitron, which exhibits mixed α7 agonist and $5HT_3$ receptor antagonist properties was shown to improve sustained attention and to significantly improve sensory gating in a randomized, double-blind, placebo-controlled study in schizophrenic patients (Shiina et al., 2010). However, not all studies with α7 agonists have shown improvements in cognitive performance in schizophrenic patients. Whereas initial reports of the α7 partial agonist DMXB-A (also known as GTS-21) showed improvement of the cognitive performance in healthy volunteers (Kitagawa et al., 2003) and in a cohort of nonsmoking schizophrenic patients using the Repeatable Battery for the Assessment of Neuropsychological Status (Freedman, 2013), a larger phase II study with GTS-21 using the MATRICS test battery as the primary endpoint did not replicate these initial findings in a schizophrenic population. Similarly, a lack of effect on cognitive measures has been noted with other compounds (e.g., RG3487) (Umbricht et al., 2014); therefore, the cognitive-enhancing potential of α7 agonists remains to be confirmed in a schizophrenic patient population.

10.2 Clinical Trials: Negative Symptoms

Whereas the procognitive properties in schizophrenic patients have yielded mixed results in early clinical development studies, EVP-6124, GTS-21, RG3487, and tropesitron all have been reported to improve negative symptoms in schizophrenic patients. This is an interesting finding in that negative symptoms of schizophrenia are independent of cognitive deficits, although like the cognitive impairments, both have few, if any, therapeutic options. Current evidence suggests that negative symptoms can be categorized into two subdomains that include (1) diminished expression (e.g., alogia, blunted affect, asociality) and (2) amotivation (e.g., avolition, anhedonia) (Foussias, Mann, Zakzanis, van Reekum, & Remington, 2009) and affect up to 58% of patients with chronic schizophrenia (Bobes, Arango, Garcia-Garcia, Rejas, & CLAMORS Study Collaborative Group, 2010; Galderisi, Bucci, Üçok, & Peuskens, 2012; Mäkinen et al., 2010). Persistent negative symptoms are associated with a poorer prognosis and an overall lower quality of life in patients. Therefore, effectively treating the negative symptoms of schizophrenia is expected to decrease the burden of disease for these patients and to improve functional outcome (Bowie & Harvey, 2008; Foussias et al., 2009; Gur et al., 2007).

The underlying neurobiological substrates for negative symptoms are not well understood, although dysfunction in cortico-striatal-thalamo-cortico loops may play an important role in the manifestation of some of the key symptoms (i.e., motivation, goal-directed behaviors, working memory) (Barch & Dowd, 2010; Dowd & Barch, 2010; Gold, Waltz, Prentice, Morris, & Heerey, 2008). The localization of the α7 receptor in brain regions such as cortex and thalamus suggests a role for them in mediating this function. The development of animal models to better study negative symptoms of schizophrenia is an important area of research that may help to elucidate the involvement of the α7 receptor in mediating these symptoms.

11. CONCLUSIONS

Schizophrenia is a multifaceted disease with limited treatment options. The classification of schizophrenic symptoms into different domains (i.e., positive and negative symptoms, cognitive impairments) provides a starting point to analyze the dysfunctional neuronal systems and to examine the biology from cellular processing, neurotransmission, and brain structures underlying the disease. In this review, we summarized our current knowledge of the α7 nicotinic receptor and its role as a potential therapeutic to treat some (i.e., cognitive deficits, negative symptoms), but not all (i.e., positive symptoms), aspects of schizophrenia. Whereas some of the key characteristics of the α7 receptor (e.g., localization, function) make it an ideal drug target candidate, other properties (e.g., rapid desensitization) make it challenging to study and to translate effects into clinical studies. Continued investigation of the biology of the α7 receptor as well as pending results from ongoing clinical studies with selective agonists certainly will aid in our understanding of this interesting receptor system and its potential as a therapeutic agent.

REFERENCES

Albuquerque, E. X., Pereira, E. F., Alkondon, M., & Rogers, S. W. (2009). Mammalian nicotinic acetylcholine receptors: From structure to function. *Physiological Reviews*, *89*(1), 73–120.

Albuquerque, E. X., Pereira, E. F., Braga, M. F., & Alkondon, M. (1998). Contribution of nicotinic receptors to the function of synapses in the central nervous system: The action of choline as a selective agonist of alpha 7 receptors. *Journal of Physiology, Paris*, *92*(3–4), 309–316.

Albuquerque, E. X., & Schwarcz, R. (2013). Kynurenic acid as an antagonist of α7 nicotinic acetylcholine receptors in the brain: Facts and challenges. *Biochemical Pharmacology*, *85*(8), 1027–1032.

Alexander, K. S., Pocivavsek, A., Wu, H. Q., Pershing, M. L., Schwarcz, R., & Bruno, J. P. (2013). Early developmental elevations of brain kynurenic acid impair cognitive flexibility in adults: Reversal with galantamine. *Neuroscience*, *238*, 19–28.

Alkondon, M., & Albuquerque, E. X. (1993). Diversity of nicotinic acetylcholine receptors in rat hippocampal neurons. I. Pharmacological and functional evidence for distinct structural subtypes. *The Journal of Pharmacology and Experimental Therapeutics*, *265*(3), 1455–1473.

Alkondon, M., Pereira, E. F., & Albuquerque, E. X. (1998). Alpha-bungarotoxin- and methyllycaconitine-sensitive nicotinic receptors mediate fast synaptic transmission in interneurons of rat hippocampal slices. *Brain Research*, *810*(1–2), 257–263.

Althoff, T., Hibbs, R. E., Banerjee, S., & Gouaux, E. (2014). X-ray structures of GluCl in apo states reveal a gating mechanism of cys-loop receptors. *Nature*, *512*(7514), 333–337.

Anderson, D. J., Bunnelle, W., Surber, B., Du, J., Surowy, C., Tribollet, E., et al. (2008). [3H]A-585539 [(1S,4S)-2,2-dimethyl-5-(6-phenylpyridazin-3-yl)-5-aza-2-azoniabicyclo[2.2.1]heptane], a novel high-affinity alpha7 neuronal nicotinic receptor agonist: Radioligand binding characterization to rat and human brain. *The Journal of Pharmacology and Experimental Therapeutics*, *324*(1), 179–187.

Araud, T., Graw, S., Berger, R., Lee, M., Neveu, E., Bertrand, D., et al. (2011). The chimeric gene CHRFAM7A, a partial duplication of the CHRNA7 gene, is a dominant negative regulator of α7*nAChR function. *Biochemical Pharmacology*, *82*(8), 904–914.

Arimatsu, Y., Seto, A., & Amano, T. (1978). Localization of alpha-bungarotoxin binding sites in mouse brain by light and electron microscopic autoradiography. *Brain Research*, *147*(1), 165–169.

Arnsten, A. F., Wang, M. J., & Paspalas, C. D. (2012). Neuromodulation of thought: Flexibilities and vulnerabilities in prefrontal cortical network synapses. *Neuron*, *76*(1), 223–239.

Barch, D. M., & Dowd, E. C. (2010). Goal representations and motivational drive in schizophrenia: The role of prefrontal-striatal interactions. *Schizophrenia Bulletin*, *36*(5), 919–934.

Barik, J., & Wonnacott, S. (2006). Indirect modulation by alpha7 nicotinic acetylcholine receptors of noradrenaline release in rat hippocampal slices: Interaction with glutamate and GABA systems and effect of nicotine withdrawal. *Molecular Pharmacology*, *69*(2), 618–628.

Bertrand, D., Galzi, J. L., Devillers-Thiéry, A., Bertrand, S., & Changeux, J. P. (1993). Mutations at two distinct sites within the channel domain M2 alter calcium permeability of neuronal alpha 7 nicotinic receptor. *Proceedings of the National Academy of Sciences of the United States of America*, *90*(15), 6971–6975.

Bertrand, D., & Gopalakrishnan, M. (2007). Allosteric modulation of nicotinic acetylcholine receptors. *Biochemical Pharmacology*, *74*(8), 1155–1163.

Biton, B., Bergis, O. E., Galli, F., Nedelec, A., Lochead, A. W., Jegham, S., et al. (2007). SSR180711, a novel selective alpha7 nicotinic receptor partial agonist: (1) Binding and functional profile. *Neuropsychopharmacology: Official Publication of the American College of Neuropsychopharmacology*, *32*(1), 1–16.

Bizon, J. L., Foster, T. C., Alexander, G. E., & Glisky, E. L. (2012). Characterizing cognitive aging of working memory and executive function in animal models. *Frontiers in Aging Neuroscience*, *4*, 19.

Bloem, B., Poorthuis, R. B., & Mansvelder, H. D. (2014). Cholinergic modulation of the medial prefrontal cortex: The role of nicotinic receptors in attention and regulation of neuronal activity. *Frontiers in Neural Circuits*, *8*, 17.

Bobes, J., Arango, C., Garcia-Garcia, M., Rejas, J., & CLAMORS Study Collaborative Group (2010). Prevalence of negative symptoms in outpatients with schizophrenia spectrum disorders treated with antipsychotics in routine clinical practice: Findings from the CLAMORS study. *The Journal of Clinical Psychiatry*, *71*(3), 280–286.

Bocquet, N., Nury, H., Baaden, M., Le Poupon, C., Changeux, J. P., Delarue, M., et al. (2009). X-ray structure of a pentameric ligand-gated ion channel in an apparently open conformation. *Nature*, *457*(7225), 111–114.

Bodnar, A. L., Cortes-Burgos, L. A., Cook, K. K., Dinh, D. M., Groppi, V. E., Hajos, M., et al. (2005). Discovery and structure-activity relationship of quinuclidine benzamides as agonists of alpha7 nicotinic acetylcholine receptors. *Journal of Medicinal Chemistry*, *48*(4), 905–908.

Bowie, C. R., & Harvey, P. D. (2008). Communication abnormalities predict functional outcomes in chronic schizophrenia: Differential associations with social and adaptive functions. *Schizophrenia Research*, *103*(1–3), 240–247.

Breese, C. R., Adams, C., Logel, J., Drebing, C., Rollins, Y., Barnhart, M., et al. (1997). Comparison of the regional expression of nicotinic acetylcholine receptor alpha7 MRNA and [125I]-alpha-bungarotoxin binding in human postmortem brain. *The Journal of Comparative Neurology*, *387*(3), 385–398.

Cachelin, A. B., & Rust, G. (1994). Unusual pharmacology of (+)-tubocurarine with rat neuronal nicotinic acetylcholine receptors containing beta 4 subunits. *Molecular Pharmacology*, *46*(6), 1168–1174.

Calimet, N., Simoes, M., Changeux, J. P., Karplus, M., Taly, A., & Cecchini, M. (2013). A gating mechanism of pentameric ligand-gated ion channels. *Proceedings of the National Academy of Sciences of the United States of America*, *110*(42), E3987–E3996.

Callahan, P. M., Hutchings, E. J., Kille, N. J., Chapman, J. M., & Terry, A. V. (2013). Positive allosteric modulator of alpha 7 nicotinic-acetylcholine receptors, PNU-120596 augments the effects of donepezil on learning and memory in aged rodents and non-human primates. *Neuropharmacology*, *67*, 201–212.

Caspi, A. (2003). Cognitive performance in schizophrenia patients assessed before and following the first psychotic episode. *Schizophrenia Research*, *65*(2–3). http://dx.doi.org/10.1016/s0920-9964(03)00056-2.

Chimienti, F., Hogg, R. C., Plantard, L., Lehmann, C., Brakch, N., Fischer, J., et al. (2003). Identification of SLURP-1 as an epidermal neuromodulator explains the clinical phenotype of Mal de Meleda. *Human Molecular Genetics*, *12*(22), 3017–3024.

Cimino, M., Marini, P., Fornasari, D., Cattabeni, F., & Clementi, F. (1992). Distribution of nicotinic receptors in cynomolgus monkey brain and ganglia: Localization of alpha 3 subunit MRNA, alpha-bungarotoxin and nicotine binding sites. *Neuroscience*, *51*(1), 77–86.

Clarke, P. B., Schwartz, R. D., Paul, S. M., Pert, C. B., & Pert, A. (1985). Nicotinic binding in rat brain: Autoradiographic comparison of [3H]acetylcholine, [3H]nicotine, and [125I]-alpha-bungarotoxin. *The Journal of Neuroscience: The Official Journal of the Society for Neuroscience*, *5*(5), 1307–1315.

Colón-Sáez, J. O., & Yakel, J. L. (2013). A mutation in the extracellular domain of the α7 nAChR reduces calcium permeability. *Pflügers Archiv/European Journal of Physiology*, *466*, 1571–1579.

Cornblatt, B. A., Lenzenweger, M. F., Dworkin, R. H., & Erlenmeyer-Kimling, L. (1992). Childhood attentional dysfunctions predict social deficits in unaffected adults at risk for schizophrenia. *The British Journal of Psychiatry*, *161*(Suppl. 18), 59–64.

Court, J., & Clementi, F. (1995). Distribution of nicotinic subtypes in human brain. *Alzheimer Disease and Associated Disorders*, *9*(Suppl. 2), 6–14.

Court, J. A., Martin-Ruiz, C., Graham, A., & Perry, E. (2000). Nicotinic receptors in human brain: Topography and pathology. *Journal of Chemical Neuroanatomy*, *20*(3–4), 281–298.

Court, J., Spurden, D., Lloyd, S., McKeith, I., Ballard, C., Cairns, N., et al. (1999). Neuronal nicotinic receptors in dementia with Lewy bodies and schizophrenia: Alpha-bungarotoxin and nicotine binding in the thalamus. *Journal of Neurochemistry*, *73*(4), 1590–1597.

Couturier, S., Bertrand, D., Matter, J. M., Hernandez, M. C., Bertrand, S., Millar, N., et al. (1990). A neuronal nicotinic acetylcholine receptor subunit (alpha 7) is developmentally regulated and forms a homo-oligomeric channel blocked by alpha-BTX. *Neuron*, *5*(6), 847–856.

Criado, M., Valor, L. M., Mulet, J., Gerber, S., Sala, S., & Sala, F. (2012). Expression and functional properties of α7 acetylcholine nicotinic receptors are modified in the presence of other receptor subunits. *Journal of Neurochemistry*, *123*(4), 504–514.

Croxson, P. L., Kyriazis, D. A., & Baxter, M. G. (2011). Cholinergic modulation of a specific memory function of prefrontal cortex. *Nature Neuroscience*, *14*(12), 1510–1512.

Dani, J. A., & Bertrand, D. (2007). Nicotinic acetylcholine receptors and nicotinic cholinergic mechanisms of the central nervous system. *Annual Review of Pharmacology and Toxicology*, *47*, 699–729.

Davidson, M., Reichenberg, A., Rabinowitz, J., Weiser, M., Kaplan, Z., & Mark, M. (1999). Behavioral and intellectual markers for schizophrenia in apparently healthy male adolescents. *The American Journal of Psychiatry*, *156*(9), 1328–1335.

de Leon, J., & Diaz, F. J. (2005). A meta-analysis of worldwide studies demonstrates an association between schizophrenia and tobacco smoking behaviors. *Schizophrenia Research*, *76*(2–3), 135–157.

de Lucas-Cerrillo, A. M., Maldifassi, M. C., Arnalich, F., Renart, J., Atienza, G., Serantes, R., et al. (2011). Function of partially duplicated human α77 nicotinic receptor subunit CHRFAM7A gene: Potential implications for the cholinergic anti-inflammatory response. *The Journal of Biological Chemistry*, *286*(1), 594–606.

Dessaud, E., Salaün, D., Gayet, O., Chabbert, M., & deLapeyrière, O. (2006). Identification of lynx2, a novel member of the ly-6/neurotoxin superfamily, expressed in neuronal subpopulations during mouse development. *Molecular and Cellular Neurosciences*, *31*(2), 232–242.

Deutsch, S. I., Schwartz, B. L., Schooler, N. R., Brown, C. H., Rosse, R. B., & Rosse, S. M. (2013). Targeting alpha-7 nicotinic neurotransmission in schizophrenia: A novel agonist strategy. *Schizophrenia Research*, *148*(1–3), 138–144.

Dickinson, J. A., Kew, J. N., & Wonnacott, S. (2008). Presynaptic alpha 7- and beta 2-containing nicotinic acetylcholine receptors modulate excitatory amino acid release from rat prefrontal cortex nerve terminals via distinct cellular mechanisms. *Molecular Pharmacology*, *74*(2), 348–359.

Dowd, E. C., & Barch, D. M. (2010). Anhedonia and emotional experience in schizophrenia: Neural and behavioral indicators. *Biological Psychiatry*, *67*(10), 902–911.

Drasdo, A., Caulfield, M., Bertrand, D., Bertrand, S., & Wonnacott, S. (1992). Methyllycaconitine: A novel nicotinic antagonist. *Molecular and Cellular Neurosciences*, *3*(3), 237–243.

Eiselé, J. L., Bertrand, S., Galzi, J. L., Devillers-Thiéry, A., Changeux, J. P., & Bertrand, D. (1993). Chimaeric nicotinic-serotonergic receptor combines distinct ligand binding and channel specificities. *Nature*, *366*(6454), 479–483.

El Kouhen, R., Hu, M., Anderson, D. J., Li, J., & Gopalakrishnan, M. (2009). Pharmacology of alpha7 nicotinic acetylcholine receptor mediated extracellular signal-regulated kinase signalling in PC12 cells. *British Journal of Pharmacology*, *156*(4), 638–648.

Eskildsen, J., Redrobe, J. P., Sams, A. G., Dekermendjian, K., Laursen, M., Boll, J. B., et al. (2014). Discovery and optimization of lu AF58801, a novel, selective and brain penetrant positive allosteric modulator of alpha-7 nicotinic acetylcholine receptors: Attenuation of subchronic phencyclidine (PCP)-induced cognitive deficits in rats following oral administration. *Bioorganic & Medicinal Chemistry Letters*, *24*(1), 288–293.

Ettrup, A., Mikkelsen, J. D., Lehel, S., Madsen, J., Nielsen, E. Ø., Palner, M., et al. (2011). 11C-NS14492 as a novel PET radioligand for imaging cerebral alpha7 nicotinic acetylcholine receptors: In vivo evaluation and drug occupancy measurements. *Journal of Nuclear Medicine: Official Publication, Society of Nuclear Medicine, 52*(9), 1449–1456.

Fabian-Fine, R., Skehel, P., Errington, M. L., Davies, H. A., Sher, E., Stewart, M. G., et al. (2001). Ultrastructural distribution of the alpha7 nicotinic acetylcholine receptor subunit in rat hippocampus. *The Journal of Neuroscience: The Official Journal of the Society for Neuroscience, 21*(20), 7993–8003.

Fayuk, D., & Yakel, J. L. (2005). Ca2+ permeability of nicotinic acetylcholine receptors in rat hippocampal CA1 interneurones. *The Journal of Physiology, 566*(Pt. 3), 759–768.

Ferrarelli, F., & Tononi, G. (2011). The thalamic reticular nucleus and schizophrenia. *Schizophrenia Bulletin. 37*(2). http://dx.doi.org/10.1093/schbul/sbq142.

Feuerbach, D., Nozulak, J., Lingenhoehl, K., McAllister, K., & Hoyer, D. (2007). JN403, in vitro characterization of a novel nicotinic acetylcholine receptor alpha7 selective agonist. *Neuroscience Letters, 416*(1), 61–65.

Forster, I., & Bertrand, D. (1995). Inward rectification of neuronal nicotinic acetylcholine receptors investigated by using the homomeric alpha 7 receptor. *Proceedings. Biological Sciences/The Royal Society, 260*(1358), 139–148.

Foussias, G., Mann, S., Zakzanis, K. K., van Reekum, R., & Remington, G. (2009). Motivational deficits as the central link to functioning in schizophrenia: A pilot study. *Schizophrenia Research, 115*(2–3), 333–337.

Freedman, R. (2013). α7-nicotinic acetylcholine receptor agonists for cognitive enhancement in schizophrenia. *Annual Review of Medicine, 65*, 245–261.

Freedman, R., Hall, M., Adler, L. E., & Leonard, S. (1995). Evidence in postmortem brain tissue for decreased numbers of hippocampal nicotinic receptors in schizophrenia. *Biological Psychiatry, 38*(1), 22–33.

Freedman, R., Olincy, A., Buchanan, R. W., Harris, J. G., Gold, J. M., Johnson, L., et al. (2008). Initial phase 2 trial of a nicotinic agonist in schizophrenia. *The American Journal of Psychiatry, 165*(8), 1040–1047.

Fucile, S. (2004). Ca2+ permeability of nicotinic acetylcholine receptors. *Cell Calcium, 35*(1), 1–8.

Galderisi, S., Bucci, P., Üçok, A., & Peuskens, J. (2012). No gender differences in social outcome in patients suffering from schizophrenia. *European Psychiatry: The Journal of the Association of European Psychiatrists, 27*(6), 406–408.

Gault, J., Robinson, M., Berger, R., Drebing, C., Logel, J., Hopkins, J., et al. (1998). Genomic organization and partial duplication of the human alpha7 neuronal nicotinic acetylcholine receptor gene (CHRNA7). *Genomics, 52*(2), 173–185.

Ghiron, C., Haydar, S. N., Aschmies, S., Bothmann, H., Castaldo, C., Cocconcelli, G., et al. (2010). Novel alpha-7 nicotinic acetylcholine receptor agonists containing a urea moiety: Identification and characterization of the potent, selective, and orally efficacious agonist 1-[6-(4-fluorophenyl)pyridin-3-yl]-3-(4-piperidin-1-ylbutyl) urea (SEN34625/WYE-103914). *Journal of Medicinal Chemistry, 53*(11), 4379–4389.

Gold, J. M., Waltz, J. A., Prentice, K. J., Morris, S. E., & Heerey, E. A. (2008). Reward processing in schizophrenia: A deficit in the representation of value. *Schizophrenia Bulletin, 34*(5), 835–847.

Goldman-Rakic, P. S. (1995). Cellular basis of working memory. *Neuron, 14*(3), 477–485.

Goldman-Rakic, P. S. (1996). Memory: Recording experience in cells and circuits: Diversity in memory research. *Proceedings of the National Academy of Sciences of the United States of America, 93*(24), 13435–13437.

Gopalakrishnan, M., Buisson, B., Touma, E., Giordano, T., Campbell, J. E., Hu, I. C., et al. (1995). Stable expression and pharmacological properties of the human alpha 7 nicotinic acetylcholine receptor. *European Journal of Pharmacology, 290*(3), 237–246.

Gray, R., Rajan, A. S., Radcliffe, K. A., Yakehiro, M., & Dani, J. A. (1996). Hippocampal synaptic transmission enhanced by low concentrations of nicotine. *Nature, 383*(6602), 713–716.

Green, M. F., Kern, R. S., Braff, D. L., & Mintz, J. (2000). Neurocognitive deficits and functional outcome in schizophrenia: Are we measuring the "right stuff"? *Schizophrenia Bulletin*, *26*(1), 119–136.

Grønlien, J. H., Håkerud, M., Ween, H., Thorin-Hagene, K., Briggs, C. A., Gopalakrishnan, M., et al. (2007). Distinct profiles of alpha7 nAChR positive allosteric modulation revealed by structurally diverse chemotypes. *Molecular Pharmacology*, *72*(3), 715–724.

Guan, Z. Z., Zhang, X., Blennow, K., & Nordberg, A. (1999). Decreased protein level of nicotinic receptor alpha7 subunit in the frontal cortex from schizophrenic brain. *Neuroreport*, *10*(8), 1779–1782.

Gubbins, E. J., Gopalakrishnan, M., & Li, J. (2010). Alpha7 nAChR-mediated activation of MAP kinase pathways in PC12 cells. *Brain Research*, *1328*, 1–11.

Gur, R. E., Calkins, M. E., Gur, R. C., Horan, W. P., Nuechterlein, K. H., Seidman, L. J., et al. (2007). The consortium on the genetics of schizophrenia: Neurocognitive endophenotypes. *Schizophrenia Bulletin*, *33*(1), 49–68.

Han, Z. Y., Le Novère, N., Zoli, M., Hill, J. A., Champtiaux, N., & Changeux, J. P. (2000). Localization of nAChR subunit mRNAs in the brain of Macaca mulatta. *The European Journal of Neuroscience*, *12*(10), 3664–3674.

Han, Z. Y., Zoli, M., Cardona, A., Bourgeois, J. P., Changeux, J. P., & Le Novère, N. (2003). Localization of [3H]nicotine, [3H]cytisine, [3H]epibatidine, and [125I]alpha-bungarotoxin binding sites in the brain of Macaca mulatta. *The Journal of Comparative Neurology*, *461*(1), 49–60.

Hashimoto, K., Fujita, Y., Ishima, T., Hagiwara, H., & Iyo, M. (2006). Phencyclidine-induced cognitive deficits in mice are improved by subsequent subchronic administration of tropisetron: Role of alpha7 nicotinic receptors. *European Journal of Pharmacology*, *553*(1–3), 191–195.

Hashimoto, K., Nishiyama, S., Ohba, H., Matsuo, M., Kobashi, T., Takahagi, M., et al. (2008). [11C]CHIBA-1001 as a novel PET ligand for alpha7 nicotinic receptors in the brain: A PET study in conscious monkeys. *PLoS One*, *3*(9), e3231.

Hassaine, G., Deluz, C., Grasso, L., Wyss, R., Tol, M. B., Hovius, R., et al. (2014). X-ray structure of the mouse serotonin 5-HT3 receptor. *Nature*, *512*, 276–281.

Haydar, S. N., Ghiron, C., Bettinetti, L., Bothmann, H., Comery, T. A., Dunlop, J., et al. (2009). SAR and biological evaluation of SEN12333/WAY-317538: Novel alpha 7 nicotinic acetylcholine receptor agonist. *Bioorganic & Medicinal Chemistry*, *17*(14), 5247–5258.

Hefft, S., Hulo, S., Bertrand, D., & Muller, D. (1999). Synaptic transmission at nicotinic acetylcholine receptors in rat hippocampal organotypic cultures and slices. *The Journal of Physiology*, *515*(Pt. 3), 769–776.

Hilf, R. J., & Dutzler, R. (2008). X-ray structure of a prokaryotic pentameric ligand-gated ion channel. *Nature*, *452*(7185), 375–379.

Hilmas, C., Pereira, E. F., Alkondon, M., Rassoulpour, A., Schwarcz, R., & Albuquerque, E. X. (2001). The brain metabolite kynurenic acid inhibits alpha7 nicotinic receptor activity and increases non-alpha7 nicotinic receptor expression: Physiopathological implications. *The Journal of Neuroscience: The Official Journal of the Society for Neuroscience*, *21*(19), 7463–7473.

Horti, A. G., Ravert, H. T., Gao, Y., Holt, D. P., Bunnelle, W. H., Schrimpf, M. R., et al. (2013). Synthesis and evaluation of new radioligands [(11)C]A-833834 and [(11)C]A-752274 for positron-emission tomography of α7-nicotinic acetylcholine receptors. *Nuclear Medicine and Biology*, *40*(3), 395–402.

Hunter, B. E., de Fiebre, C. M., Papke, R. L., Kem, W. R., & Meyer, E. M. (1994). A novel nicotinic agonist facilitates induction of long-term potentiation in the rat hippocampus. *Neuroscience Letters*, *168*(1–2), 130–134.

Hurst, R. S., Hajós, M., Raggenbass, M., Wall, T. M., Higdon, N. R., Lawson, J. A., et al. (2005). A novel positive allosteric modulator of the alpha7 neuronal nicotinic

acetylcholine receptor: In vitro and in vivo characterization. *The Journal of Neuroscience: The Official Journal of the Society for Neuroscience*, *25*(17), 4396–4405.
Hurst, R., Rollema, H., & Bertrand, D. (2012). Nicotinic acetylcholine receptors: From basic science to therapeutics. *Pharmacology & Therapeutics*, *137*, 22–54.
Ibañez-Tallon, I., Wen, H., Miwa, J. M., Xing, J., Tekinay, A. B., Ono, F., et al. (2004). Tethering naturally occurring peptide toxins for cell autonomous modulation of ion channels and receptors in vivo. *Neuron*, *43*(3), 305–311.
Ishikawa, M., & Hashimoto, K. (2011). α7 nicotinic acetylcholine receptor as a potential therapeutic target for schizophrenia. *Current Pharmaceutical Design*, *17*(2), 121–129.
Javitt, D. C. (2007). Glutamate and schizophrenia: Phencyclidine, n-methyl-D-aspartate receptors, and dopamine-glutamate interactions. *International Review of Neurobiology*, *78*, 69–108.
Ji, D., & Dani, J. A. (2000). Inhibition and disinhibition of pyramidal neurons by activation of nicotinic receptors on hippocampal interneurons. *Journal of Neurophysiology*, *83*(5), 2682–2690.
Jones, I. W., & Wonnacott, S. (2004). Precise localization of alpha7 nicotinic acetylcholine receptors on glutamatergic axon terminals in the rat ventral tegmental area. *The Journal of Neuroscience: The Official Journal of the Society for Neuroscience*, *24*(50), 11244–11252.
Khiroug, S. S., Harkness, P. C., Lamb, P. W., Sudweeks, S. N., Khiroug, L., Millar, N. S., et al. (2002). Rat nicotinic ach receptor α7 and β2 subunits co-assemble to form functional heteromeric nicotinic receptor channels. *The Journal of Physiology*, *540*(Pt. 2), 425.
Kitagawa, H., Takenouchi, T., Azuma, R., Wesnes, K. A., Kramer, W. G., Clody, D. E., et al. (2003). Safety, pharmacokinetics, and effects on cognitive function of multiple doses of GTS-21 in healthy, male volunteers. *Neuropsychopharmacology: Official Publication of the American College of Neuropsychopharmacology*, *28*(3), 542–551.
Krause, R. M., Buisson, B., Bertrand, S., Corringer, P. J., Galzi, J. L., Changeux, J. P., et al. (1998). Ivermectin: A positive allosteric effector of the alpha7 neuronal nicotinic acetylcholine receptor. *Molecular Pharmacology*, *53*(2), 283–294.
Le Novère, N., & Changeux, J. P. (1995). Molecular evolution of the nicotinic acetylcholine receptor: An example of multigene family in excitable cells. *Journal of Molecular Evolution*, *40*(2), 155–172.
Levin, E. D., Bettegowda, C., Weaver, T., & Christopher, N. C. (1998). Nicotine-dizocilpine interactions and working and reference memory performance of rats in the radial-arm maze. *Pharmacology, Biochemistry, and Behavior*, *61*(3), 335–340.
Levitin, F., Weiss, M., Hahn, Y., Stern, O., Papke, R. L., Matusik, R., et al. (2008). PATE gene clusters code for multiple, secreted TFP/ly-6/upar proteins that are expressed in reproductive and neuron-rich tissues and possess neuromodulatory activity. *The Journal of Biological Chemistry*, *283*(24), 16928–16939.
Lewis, D. A., Curley, A. A., Glausier, J. R., & Volk, D. W. (2012). Cortical parvalbumin interneurons and cognitive dysfunction in schizophrenia. *Trends in Neurosciences*, *35*(1), 57–67.
Lin, H., Hsu, F. C., Baumann, B. H., Coulter, D. A., & Lynch, D. R. (2014). Cortical synaptic NMDA receptor deficits in α7 nicotinic acetylcholine receptor gene deletion models: Implications for neuropsychiatric diseases. *Neurobiology of Disease*, *63*, 129–140.
Maier, D. L., Hill, G., Ding, M., Tuke, D., Einstein, E., Gurley, D., et al. (2011). Pre-clinical validation of a novel alpha-7 nicotinic receptor radiotracer, [(3)H]AZ11637326: Target localization, biodistribution and ligand occupancy in the rat brain. *Neuropharmacology*, *61*(1–2), 161–171.
Mäkinen, J., Miettunen, J., Jääskeläinen, E., Veijola, J., Isohanni, M., & Koponen, H. (2010). Negative symptoms and their predictors in schizophrenia within the Northern Finland 1966 Birth Cohort. *Psychiatry Research*, *178*(1), 121–125.

Marrero, M. B., Lucas, R., Salet, C., Hauser, T. A., Mazurov, A., Lippiello, P. M., et al. (2010). An alpha7 nicotinic acetylcholine receptor-selective agonist reduces weight gain and metabolic changes in a mouse model of diabetes. *The Journal of Pharmacology and Experimental Therapeutics*, *332*(1), 173–180.

Mazurov, A. A., Kombo, D. C., Hauser, T. A., Miao, L., Dull, G., Genus, J. F., et al. (2012). Discovery of (2S,3R)-N-[2-(pyridin-3-ylmethyl)-1-azabicyclo[2.2.2]oct-3-yl]benzo[b]furan-2-carboxamide (TC-5619), a selective α7 nicotinic acetylcholine receptor agonist, for the treatment of cognitive disorders. *Journal of Medicinal Chemistry*, *55*(22), 9793–9809.

Mednick, S. A., Parnas, J., & Schulsinger, F. (1987). The Copenhagen high-risk project, 1962-86. *Schizophrenia Bulletin*, *13*(3), 485–495.

Mexal, S., Berger, R., Logel, J., Ross, R. G., Freedman, R., & Leonard, S. (2010). Differential regulation of alpha7 nicotinic receptor gene (CHRNA7) expression in schizophrenic smokers. *Journal of Molecular Neuroscience: MN*, *40*(1–2), 185–195.

Miwa, J. M., Ibanez-Tallon, I., Crabtree, G. W., Sánchez, R., Sali, A., Role, L. W., et al. (1999). Lynx1, an endogenous toxin-like modulator of nicotinic acetylcholine receptors in the mammalian CNS. *Neuron*, *23*(1), 105–114.

Miwa, J. M., Stevens, T. R., King, S. L., Caldarone, B. J., Ibanez-Tallon, I., Xiao, C., et al. (2006). The prototoxin lynx1 acts on nicotinic acetylcholine receptors to balance neuronal activity and survival in vivo. *Neuron*, *51*(5), 587–600.

Moghaddam, B., & Krystal, J. H. (2012). Capturing the angel in "angel dust": Twenty years of translational neuroscience studies of NMDA receptor antagonists in animals and humans. *Schizophrenia Bulletin*, *38*(5), 942–949.

Murray, T. A., Bertrand, D., Papke, R. L., George, A. A., Pantoja, R., Srinivasan, R., et al. (2011). α7β2 nAChRs assemble and function, and are activated primarily via their α7-α7 interfaces. *Molecular Pharmacology*, *81*(2), 175–188.

Nuechterlein, K. H. (1983). Signal detection in vigilance tasks and behavioral attributes among offspring of schizophrenic mothers and among hyperactive children. *Journal of Abnormal Psychology*, *92*(1), 4–28.

Olincy, A., Harris, J. G., Johnson, L. L., Pender, V., Kongs, S., Allensworth, D., et al. (2006). Proof-of-concept trial of an alpha7 nicotinic agonist in schizophrenia. *Archives of General Psychiatry*, *63*(6), 630–638.

Park, P., Volianskis, A., Sanderson, T. M., Bortolotto, Z. A., Jane, D. E., Zhuo, M., et al. (2014). NMDA receptor-dependent long-term potentiation comprises a family of temporally overlapping forms of synaptic plasticity that are induced by different patterns of stimulation. *Philosophical Transactions of the Royal Society of London. Series B, Biological Sciences*, *369*(1633), 20130131.

Pichat, P., Bergis, O. E., Terranova, J. P., Urani, A., Duarte, C., Santucci, V., et al. (2007). SSR180711, a novel selective alpha7 nicotinic receptor partial agonist: (II) efficacy in experimental models predictive of activity against cognitive symptoms of schizophrenia. *Neuropsychopharmacology: Official Publication of the American College of Neuropsychopharmacology*, *32*(1), 17–34.

Poorthuis, R. B., Bloem, B., Verhoog, M. B., & Mansvelder, H. D. (2013). Layer-specific interference with cholinergic signaling in the prefrontal cortex by smoking concentrations of nicotine. *The Journal of Neuroscience: The Official Journal of the Society for Neuroscience*, *33*(11), 4843–4853.

Prickaerts, J., van Goethem, N. P., Chesworth, R., Shapiro, G., Boess, F. G., Methfessel, C., et al. (2011). EVP-6124, a novel and selective α7 nicotinic acetylcholine receptor partial agonist, improves memory performance by potentiating the acetylcholine response of α7 nicotinic acetylcholine receptors. *Neuropharmacology*, *62*(2), 1099–1110.

Puchacz, E., Buisson, B., Bertrand, D., & Lukas, R. J. (1994). Functional expression of nicotinic acetylcholine receptors containing rat α7 subunits in human SH-SY5Y neuroblastoma cells. *FEBS Letters*, *354*(2), 155–159.

Puhl, M. D., Mintzopoulos, D., Jensen, J. E., Gillis, T. E., Konopaske, G. T., Kaufman, M. J., et al. (2014). In vivo magnetic resonance studies reveal neuroanatomical and neurochemical abnormalities in the serine racemase knockout mouse model of schizophrenia. *Neurobiology of Disease*, *73C*, 269–274.

Rötering, S., Deuther-Conrad, W., Cumming, P., Donat, C. K., Scheunemann, M., Fischer, S., et al. (2014). Imaging of α7 nicotinic acetylcholine receptors in brain and cerebral vasculature of juvenile pigs with [(18)F]NS14490. *EJNMMI Research*, *4*, 43.

Sahdeo, S., Wallace, T., Hirakawa, R., Knoflach, F., Bertrand, D., Maag, H., et al. (2014). Characterization of RO5126946, a novel α7 nicotinic acetylcholine receptor-positive allosteric modulator. *The Journal of Pharmacology and Experimental Therapeutics*, *350*(2), 455–468.

Séguéla, P., Wadiche, J., Dineley-Miller, K., Dani, J. A., & Patrick, J. W. (1993). Molecular cloning, functional properties, and distribution of rat brain alpha 7: A nicotinic cation channel highly permeable to calcium. *The Journal of Neuroscience: The Official Journal of the Society for Neuroscience*, *13*(2), 596–604.

Shiina, A., Shirayama, Y., Niitsu, T., Hashimoto, T., Yoshida, T., Hasegawa, T., et al. (2010). A randomised, double-blind, placebo-controlled trial of tropisetron in patients with schizophrenia. *Annals of General Psychiatry*, *9*, 27.

Spurden, D. P., Court, J. A., Lloyd, S., Oakley, A., Perry, R., Pearson, C., et al. (1997). Nicotinic receptor distribution in the human thalamus: Autoradiographical localization of [3H]nicotine and [125I] alpha-bungarotoxin binding. *Journal of Chemical Neuroanatomy*, *13*(2), 105–113.

Smulders, C. J., Zwart, R., Bermudez, I., van Kleef, R. G., Groot-Kormelink, P. J., & Vijverberg, H. P. (2005). Cholinergic drugs potentiate human nicotinic alpha4beta2 acetylcholine receptors by a competitive mechanism. *European Journal of Pharmacology*, *509*(2–3), 97–108.

Stefansson, H., Rujescu, D., Cichon, S., Ingason, A., Steinberg, S., Fossdal, R., et al. (2008). Large recurrent microdeletions associated with schizophrenia. *Nature*, *455*(7210), 232–236.

Tan, H. Y., Callicott, J. H., & Weinberger, D. R. (2009). Prefrontal cognitive systems in schizophrenia: Towards human genetic brain mechanisms. *Cognitive Neuropsychiatry*, *14*(4–5), 277–298.

Thomsen, M. S., Christensen, D. Z., Hansen, H. H., Redrobe, J. P., & Mikkelsen, J. D. (2009). Alpha(7) nicotinic acetylcholine receptor activation prevents behavioral and molecular changes induced by repeated phencyclidine treatment. *Neuropharmacology*, *56*(6–7), 1001–1009.

Thomsen, M. S., El-Sayed, M., & Mikkelsen, J. D. (2011). Differential immediate and sustained memory enhancing effects of alpha7 nicotinic receptor agonists and allosteric modulators in rats. *PLoS One*, *6*(11), e27014.

Thomsen, M. S., Hansen, H. H., Timmerman, D. B., & Mikkelsen, J. D. (2010). Cognitive improvement by activation of alpha7 nicotinic acetylcholine receptors: From animal models to human pathophysiology. *Current Pharmaceutical Design*, *16*(3), 323–343.

Tillman, T. S., Seyoum, E., Mowrey, D. D., Xu, Y., & Tang, P. (2014). ELIC-α7 nAChR chimeras reveal a prominent role of the extracellular-transmembrane domain interface in allosteric modulation. *The Journal of Biological Chemistry*, *289*, 13851–13857.

Timmermann, D. B., Grønlien, J. H., Kohlhaas, K. L., Nielsen, E. Ø., Dam, E., Jørgensen, T. D., et al. (2007). An allosteric modulator of the alpha7 nicotinic acetylcholine receptor possessing cognition-enhancing properties in vivo. *The Journal of Pharmacology and Experimental Therapeutics*, *323*(1), 294–307.

Toyohara, J., Ishiwata, K., Sakata, M., Wu, J., Nishiyama, S., Tsukada, H., et al. (2010). In vivo evaluation of alpha7 nicotinic acetylcholine receptor agonists [11C]A-582941 and [11C]A-844606 in mice and conscious monkeys. *PLoS One*, *5*(2), e8961.

Tribollet, E., Bertrand, D., Marguerat, A., & Raggenbass, M. (2004). Comparative distribution of nicotinic receptor subtypes during development, adulthood and aging: An autoradiographic study in the rat brain. *Neuroscience*, *124*(2), 405–420.

Tsetlin, V. I. (2014). Three-finger snake neurotoxins and Ly6 proteins targeting nicotinic acetylcholine receptors: Pharmacological tools and endogenous modulators. *Trends in Pharmacological Sciences*, *36*, 109–123.

Umbricht, D., Keefe, R. S., Murray, S., Lowe, D. A., Porter, R., Garibaldi, G., et al. (2014). A randomized, placebo-controlled study investigating the nicotinic α7 agonist, RG3487, for cognitive deficits in schizophrenia. *Neuropsychopharmacology: Official Publication of the American College of Neuropsychopharmacology*, *39*, 1568–1577.

Van Horn, M. R., Sild, M., & Ruthazer, E. S. (2013). D-Serine as a gliotransmitter and its roles in brain development and disease. *Frontiers in Cellular Neuroscience*, 7, 39.

Wallace, T. L., Callahan, P. M., Tehim, A., Bertrand, D., Tombaugh, G., Wang, S., et al. (2010). RG3487, a novel nicotinic alpha7 receptor partial agonist, improves cognition and sensorimotor gating in rodents. *The Journal of Pharmacology and Experimental Therapeutics*, *336*(1), 242–253.

Wallace, C., Porter, P., Neveu, E., & Bertrand, D. (2011). Effects of RG3487 at the α7β2 nicotinic acetylcholine receptor expressed in Xenopus oocytes. *Biochemical Pharmacology*, *82*, 1026.

Wang, M., Gamo, N. J., Yang, Y., Jin, L. E., Wang, X. J., Laubach, M., et al. (2011). Neuronal basis of age-related working memory decline. *Nature*, *476*(7359), 210–213.

Wong, D. F., Kuwabara, H., Pomper, M., Holt, D. P., Brasic, J. R., George, N., et al. (2014). Human brain imaging of α7 nAChR with [(18)F]ASEM: A new PET radiotracer for neuropsychiatry and determination of drug occupancy. *Molecular Imaging and Biology*, *16*(5), 730–738.

Yang, Y., Paspalas, C. D., Jin, L. E., Picciotto, M. R., Arnsten, A. F., & Wang, M. (2013). Nicotinic α7 receptors enhance NMDA cognitive circuits in dorsolateral prefrontal cortex. *Proceedings of the National Academy of Sciences of the United States of America*, *110*(29), 12078–12083.

Yin, L., Zhao, Q., Li, L., Zhang, S. L., Chen, X. Q., Ma, C., et al. (2013). An experimental study on (131)I-CHIBA-1001: A radioligand for α7 nicotinic acetylcholine receptors. *PLoS One*, *8*(7), e70188.

Zorumski, C. F., & Izumi, Y. (2012). NMDA receptors and metaplasticity: Mechanisms and possible roles in neuropsychiatric disorders. *Neuroscience and Biobehavioral Reviews*, *36*(3), 989–1000.

Zwart, R., De Filippi, G., Broad, L. M., McPhie, G. I., Pearson, K. H., Baldwinson, T., et al. (2002). 5-Hydroxyindole potentiates human alpha 7 nicotinic receptor-mediated responses and enhances acetylcholine-induced glutamate release in cerebellar slices. *Neuropharmacology*, *43*(3), 374–384.

Zwart, R., Strotton, M., Ching, J., Astles, P. C., & Sher, E. (2014). Unique pharmacology of heteromeric α7β2 nicotinic acetylcholine receptors expressed in Xenopus laevis oocytes. *European Journal of Pharmacology*, *726*, 77–86.

Zwart, R., van Kleef, R. G., Gotti, C., Smulders, C. J., & Vijverberg, H. P. (2000). Competitive potentiation of acetylcholine effects on neuronal nicotinic receptors by acetylcholinesterase-inhibiting drugs. *Journal of Neurochemistry*, *75*(6), 2492–2500.

CHAPTER FOUR

Role of the Neuregulin Signaling Pathway in Nicotine Dependence and Co-morbid Disorders

Miranda L. Fisher*, Anu Loukola†, Jaakko Kaprio†,‡,§, Jill R. Turner*,1

*Department of Drug Discovery and Biomedical Sciences, South Carolina College of Pharmacy, University of South Carolina, Columbia, South Carolina, USA

†Department of Public Health, Faculty of Medicine, University of Helsinki, Helsinki, Finland

‡National Institute for Health and Welfare, Helsinki, Finland

§Institute for Molecular Medicine Finland FIMM, University of Helsinki, Helsinki, Finland

[1]Corresponding author: e-mail address: jiturner@sccp.sc.edu

Contents

Abstract

Smoking is currently the leading cause of preventable death in the United States and is responsible for over four million deaths annually worldwide. Therefore, there is a vast clinical unmet need with regards to therapeutics targeting smoking cessation. This is even more apparent when examining smokers co-morbid with psychiatric illness, as rates of smoking in this population are ~4× higher than in the general population.

International Review of Neurobiology, Volume 124
ISSN 0074-7742
http://dx.doi.org/10.1016/bs.irn.2015.07.006

Examining common genetic and molecular signaling pathways impinging upon both smoking behavior and psychiatric illness will lead to a better understanding of co-morbid disorders and potential development of novel therapeutics. Studies have implicated the Neuregulin Signaling Pathway in the pathophysiology of a number of psychiatric illnesses. Additionally, recent studies have also shown an association between the Neuregulin Signaling Pathway and smoking behaviors. This review outlines basic mechanisms of the Neuregulin Signaling Pathway and how it may be exploited for precision medicine approaches in treating nicotine dependence and mental illness.

1. INTRODUCTION

Tobacco smoking is still the leading cause of preventable death in the United States even years after the discovery of multiple smoking cessation therapies. The main addictive component in cigarette smoke is nicotine (United States. Department of the Army. Office of the Surgeon General, 1988), which drives the reinforcement behind smoking behavior. With global smoking-related mortality reaching nearly six million deaths annually ("WHO urges more countries to require large, graphic health warnings on tobacco packaging: the WHO report on the global tobacco epidemic, 2011 examines antitobacco mass-media campaigns," WHO, 2011), there is a high demand for targeted therapeutics that successfully aid smokers to quit. Several smoking cessation pharmacotherapies are available, including nicotine replacement therapy, prescription medication such as bupropion (originally designed as an antidepressant) and the nicotinic acetylcholine receptor (nAChR) partial agonist varenicline (Cummings & Mahoney, 2006; Jorenby et al., 2006); however, the success rate of such therapies after 1 year is at best only 20–25% (Gonzales et al., 2006). In comparison, approximately 3% of individuals trying to quit without any pharmacotherapies are still abstinent after 6 months (Hughes et al., 1992). The majority of smokers would like to quit and are aware of the risks of smoking, but are unable to do so. The positive reinforcing effect of nicotine is an important determinant of cessation failure; however, it is not the only factor that should be taken into account. The significant aversive withdrawal symptoms that occur during abstinence are also considered a major determinant of high relapse rates (Le Foll & Goldberg, 2009).

Withdrawal symptoms are relatively well-characterized and include both cognitive and affective symptoms. These symptoms primarily include depressed mood states, anxiety, irritability, concentration difficulties, and craving (Hughes, 2007). It is suggested that withdrawal symptom severity is a more valid indicator of smoking cessation outcome than nicotine intake or dependence (West, Hajek, & Belcher, 1989). Of these aversive withdrawal symptoms, a common affective symptom is depression. Interestingly, depressed mood is also associated with nicotine dependence, but it is not known whether depression predisposes an individual to begin smoking or whether depression develops during the course of nicotine dependence.

Broadly, nicotine dependence is highly co-morbid with several psychiatric illnesses and other substance use disorders, which further complicates smoking cessation. However, the relationship between nicotine use and mental disorders is still elusive and debatable (Moylan, Jacka, Pasco, & Berk, 2012). It has been suggested that nicotine is used in an effort to self-medicate symptoms occurring in psychiatric illnesses such as schizophrenia (Royal College of Physicians of London & Royal College of Psychiatrists, 2013), i.e., smoking would primarily be a consequence of the psychiatric disease. The second possible explanation for the co-morbidity is that smoking is itself a cause of psychiatric illness; the evidence for this is variable and depends on conditions being examined. For example, growing evidence supports the causal role of smoking in the etiology of depression (Breslau, Peterson, Schultz, Chilcoat, & Andreski, 1998; Kendler et al., 1993; Pasco et al., 2008). However, evidence must come from well-conducted prospective epidemiological studies, within-family studies, or Mendelian randomization studies using genetic markers to test causality, as randomized clinical trials cannot be used to test this hypothesis. The third potential reason for the co-morbidity may be that there are underlying genetic factors in common to specific mental disorders and smoking-related phenotypes, including nicotine dependence and withdrawal. A prime candidate for this third explanation is the co-morbidity observed between nicotine dependence and schizophrenia.

Within the co-morbid population, overall prevalence of smoking in schizophrenia patients is higher than in patients with other psychiatric conditions (Dickerson et al., 2013). Strikingly, high smoking prevalences, 60–90%, have been reported in schizophrenia patients (Dickerson et al., 2013; Matthews, Wilson, & Mitchell, 2011; Zabala et al., 2009), compared to the approximately 18% prevalence rate in the general U.S. population (Jamal et al., 2014). In addition to being more frequently current smokers, schizophrenia patients typically smoke more, are more likely nicotine dependent, and are less likely to succeed in quitting (Addington &

el-Guebaly, 1998; George et al., 2002; Ziedonis & George, 1997). However, the association of smoking with schizophrenia is not universal. For example, among Chinese women with schizophrenia, the prevalence of smoking was only slightly higher than in the general Chinese population (Hou et al., 2011; Xu et al., 2014). However, this finding could reasonably be due to a greater percentage of smokers in the Chinese population (28.1%) (Li, Hsia, & Yang, 2011) as compared to the ~20% of Americans. Nonetheless, the strength and consistency of the association over the Western world suggests that there may also be an underlying biological basis for it. Furthermore, given the differences in genetic architecture between major human ancestry groups, the findings in Chinese patients do not exclude a genetic contribution in European ancestry populations.

The neurodevelopmental theory of schizophrenia suggests that genetic and/or environmental factors negatively affect brain development during critical neural development milestones (Rapoport, Addington, Frangou, & Psych, 2005). These in turn are responsible for the biochemical alterations observed in people diagnosed with the disease (Marenco, Weinberger, & Schreurs, 2003). Breaking down the symptom profile of schizophrenia into several disease-relevant endophenotypes has enabled investigation of the role of specific risk genes that impact behavioral and biological components of this disease phenotype (Braff, Freedman, Schork, & Gottesman, 2007; Waddington et al., 2007; Walters & Owen, 2007). For example, linkage and association studies have resulted in several candidate genes such as *DTNBP1, DISC1, NRG1,* and *NRG3.* One of the most promising susceptibility genes for schizophrenia is *NRG3* due to the observation that structural and polymorphic variations of this gene are associated with a wide spectrum of neurodevelopmental disorders with phenotypes encompassing developmental delay, impairment of cognition, and autism (Balciuniene et al., 2007). This genetic variation is due to recurrent microdeletions of chromosome 10q22-q23 that involve the *NRG3* gene and also shows linkage to schizophrenia in Ashkenazi Jewish and Han Chinese populations (Fallin et al., 2003; Faraone et al., 2006). A noncoding genetic variation in *NRG3* has also been observed as a putative risk factor for schizophrenia (Chen et al., 2009; Morar et al., 2011; Sonuga-Barke et al., 2008; Wang et al., 2008). Additionally, genetic association studies show multiple genes and epistatic locus interactions (Benzel et al., 2007) within the NRG–ErbB signaling pathway that increases the risk for schizophrenia. These multiple genes encode for NRG3, NRG1, ERBB4, and AKT1, suggesting this signaling cascade may represent a pathogenic network occurring in schizophrenia.

While it is difficult to evaluate the possible therapeutic effects of nicotine in mental disorders, it may be more approachable to view these co-morbidities through the lens of genetics. For example, genes encoding for the Neuregulin Signaling Pathway have been consistently implicated in the etiology of schizophrenia (Li, Collier, & He, 2006; Munafo, Attwood, & Flint, 2008; Munafo, Thiselton, Clark, & Flint, 2006) and these same genes have recently also been implicated in smoking behavior (Loukola et al., 2014; Turner et al., 2014). Therefore, examining this pathway for possible alterations both in psychiatric illness and in nicotine dependence and cessation outcomes may aid in identifying a common link for these co-morbid disorders.

2. KNOWN MECHANISMS OF NEUREGULIN–ErbB SIGNALING

2.1 Overview

Recently, neuregulins (NRGs) have been studied as molecular links between several co-morbid disorders such as nicotine dependence, schizophrenia, attention deficit hyperactivity disorder (ADHD), and depression. This family of epidermal growth factor (EGF)-like proteins is widely expressed within the central nervous system (CNS) and has been implicated in a variety of processes, including neural development and brain activity homeostasis (for review, see Mei & Nave, 2014). While this review focuses on their effects in the CNS, NRGs signal through receptor tyrosine kinases of the ErbB family to achieve cell-to-cell interactions throughout the body, including breast and heart tissue (Yarden & Sliwkowski, 2001), where they have broad impact on cellular function and signaling. The *NRG* gene family encodes for NRG1-6, and each gene gives rise to multiple splice variants. NRG1 was the first ligand to be discovered in the brain for its function in biological processes such as activation of ErbB receptors, stimulation of Schwann cell growth, and induction of acetylcholine receptor expression (Falls, 2003; Mei & Xiong, 2008). NRG1 was also found to be a key regulator in neurotransmitter function, myelination, and synaptic plasticity related to drugs of abuse and schizophrenia (Law, 2014). However, since then, other members of the NRG family have been identified for various functions in the CNS (Carraway et al., 1997; Chang, Riese, Gilbert, Stern, & McMahan, 1997; Harari et al., 1999; Howard, Panchal, McCarthy, & Ashworth, 2005; Kanemoto et al., 2001; Kinugasa et al., 2004; Uchida et al., 1999; Watanabe et al., 1995; Zhang et al., 1997).

2.2 Neuregulin Binding and ErbB Dimerization

Neuregulins are produced as transmembrane bound precursors (Massague & Pandiella, 1993; Fig. 1(1)). The intracellular domain of NRG1 is released after proteolytic cleavage and is translocated to the nucleus of the presynaptic neuron, where it influences processes such as apoptosis (Bao, Wolpowitz, Role, & Talmage, 2003); this cascade of events is called *back signaling*. NRGs also interact with and activate ErbB receptors (ErbB1–4), resulting in

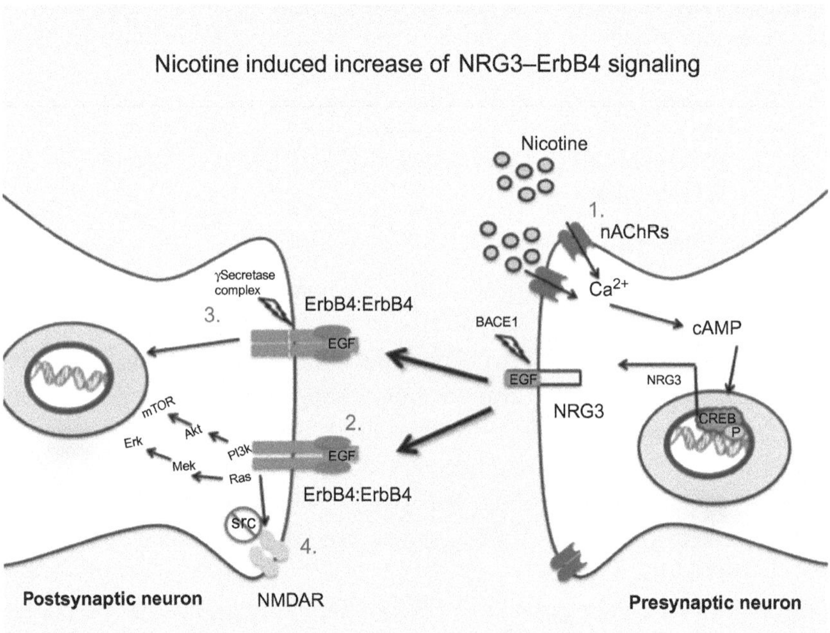

Figure 1 (1) Nicotine binds to the nicotinic acetylcholine receptor, causing a conformational change that opens the receptor's ion channel and allowing entry of Ca^{2+} and Na^{+}. The influx of these cations further activates voltage-dependent calcium channels, allowing more Ca^{2+} to enter, increasing the production of second messenger cyclic AMP (cAMP). These increases in cAMP lead to the activation of the transcription factor CREB, inducing increased expression of *NRG3*. (2) The *NRG3* EGF-like domain is then proteolytically cleaved by BACE1 and binds to the ErbB4 receptor. Upon binding of NRG3, conformational changes increase the affinity for another ErbB molecule, thus leading to homo- or heterodimerization. This dimerization results in activation of ErbB receptor tyrosine kinases and other intracellular signaling pathways referred to as *canonical forward signaling*. (3) An alternative pathway results from the cleavage of the intracellular domain of ErbB4 by a gamma-secretase complex and subsequent translocation to the nucleus to regulate gene transcription, also known as *noncanonical forward signaling*. (4) NRG–ErbB4 signaling can also directly suppress Src-mediated enhancement of synaptic NMDAR function.

activation of intracellular signaling pathways (such as ERK-, PI3K-, and Akt-mediated signaling) within the postsynaptic cell; this cascade of events is called *canonical forward signaling*, which has been shown to modulate neuronal migration and differentiation (Falls, 2003), as well as to play a role in the stimulation or inhibition of processes such as apoptosis, adhesion, proliferation, differentiation, and migration (Yarden & Sliwkowski, 2001; Fig. 1 (2)). The extracellular EGF domain of NRG binds to the ErbB receptor and initiates conformational changes in the receptor molecule, thereby increasing the affinity for another ErbB molecule and leading to homo- or heterodimerization (i.e., ErbB1–ErbB1 or Erb1–ErbB4) (Olayioye, Neve, Lane, & Hynes, 2000). This recruitment of specific ErbB molecules seems to be driven in part by the activating NRG. For example, NRG3 binds exclusively to ErbB4 receptors, but this can either be ErbB4 homodimers or be ErbB4:ErbB2 heterodimers (Zhang et al., 1997). Unlike recruitment of the dimer, however, the recruited phosphorylated ErbB partner determines the functional nature of signaling, irrespective of the ErbB ligand. The receptor dimerization activates the tyrosine kinase domain and allows it to phosphorylate tyrosine residues in the cytoplasmic region of the ErbB partner. The phosphorylated tyrosine residues then recruit various adaptors/effectors that induce specific intracellular signaling cascades, which appear to be subtype dependent. For example, ErbB4 mainly links to the Ras–MAPK and PI3k–Akt pathways (Muraoka-Cook, Feng, Strunk, & Earp, 2008; Ortega et al., 2012), and this signaling is considered to be important in many neural developmental processes, including circuitry generation, neurotransmission, and synaptic plasticity (Mei & Nave, 2014). A third mechanism of action for ErbB is the *noncanonical forward signaling*, where upon binding of NRG to the ErbB receptor, the C-terminal intracellular domain of ErbB is released by proteolytic cleavage and translocated to the nucleus where it can regulate gene transcription (Lee et al., 2002; Ni, Murphy, Golde, & Carpenter, 2001; Fig. 1(3)).

2.3 Effects of Alternative Splicing of the ErbB4 Receptor

In addition to dimerization of ErbB receptors, alternative splicing of the various ErbB receptors increases the system complexity by selectively shunting activation of intracellular signaling cascades. For example, in the human genome, alternative splicing of the *ERBB4* gene at exon 15/16 and exon 26 produces multiple ERBB4 variants (JM-a/b/c/d and CYT-1/2) (Veikkolainen et al., 2011; Zeng et al., 2009). These splice variants can have distinct effects. For example, the CYT-1 variant can recruit the p85 regulatory adapter to preferentially activate PI3k signaling. Additionally, this same splice variant is susceptible to proteolytic cleavage by TNF-alpha converting enzyme (TACE) and gamma-secretase (Sundvall et al., 2010; Vidal, Naresh, Marrero, & Jones, 2005), producing an 80-kDa intracellular fragment (ERBB4-ICD), which interacts with the transcription factor STAT4 and migrates to the nucleus, acting as a molecular chaperone (Sundvall et al., 2010; Vidal et al., 2005).

2.4 Neuregulin–ErbB4 Effects on NMDA Receptors

ErbB4 also contains a PDZ-binding motif at the carboxyl terminal and is anchored to the postsynaptic density protein 95 (PSD95) in neurons (Huang et al., 2000). Even when ErbB4 is phosphorylated by another partner, or proteolytically cleaved to produce ErbB4-ICD, the signal is only minimally transported to the soma or translocated to the nucleus (Lee et al., 2002; Ni et al., 2001). Instead, the interaction with the scaffolding protein PSD95 allows ErbB4 receptors to closely interact with ionotropic glutamate receptors (NMDARs), thereby enhancing this signaling within the postsynaptic compartments (Garcia, Vasudevan, & Buonanno, 2000). A recent study from Pitcher and colleagues (2011) demonstrated a new mechanism by which NRG–ErbB4 activation results in NMDA hypofunction (Fig. 1(4)). This constrained activity allowed ErbB4 activation to trigger dephosphorylation of the NMDAR, resulting in reduced function of the NMDAR. Dysregulation of glutamatergic transmission has been implicated in schizophrenia, mainly because of psychotomimetic effects of NMDA receptor antagonists (Coyle & Tsai, 2004). Therefore, these findings represent a new pathway by which NMDAR and ErbB4 interaction could underlie schizophrenic pathophysiology. However, whether and how this mechanism is altered in nicotine dependence is currently unknown.

2.5 Acetylcholine Receptor Inducing Activity: Modulation of nAChR Expression

One potential way that NRG–ErbB signaling and nicotine dependence may overlap is through modulation of nAChR expression. Some members of the NRG family are shown to stimulate nAChR synthesis and clustering in cultured chick and rat myotubes (Falls, Rosen, Corfas, Lane, & Fischbach, 1993), and thus are called "acetylcholine receptor inducing activity" (ARIA) proteins. This observation has now been extended to the CNS, where studies have demonstrated that NRG1 activity results in an increase in synaptic expression of α7-containing nAChRs (Hancock, Canetta, Role, & Talmage, 2008; Yang, Kuo, Devay, Yu, & Role, 1998). Thus, a direct association between NRG and cholinergic signaling exists at the level of nAChR expression. These studies are particularly intriguing in light of reported deficits in α7-homopentameric nAChRs in schizophrenia patients (Leonard et al., 1996, 2002). However, these phenomena have only been recently evaluated in nicotine dependence and cessation phenotypes.

3. NRG3: RELEVANCE IN SMOKING BEHAVIOR AND CO-MORBID DISORDERS

3.1 Potential Role of NRG3 in Nicotine Withdrawal and Smoking Cessation Outcome

Recent research showcasing the translational utility of cross-species models identified potential mechanisms and functional outcomes associated with NRG3–ErbB4 signaling during nicotine withdrawal (Loukola et al., 2014; Turner et al., 2014). Turner and colleagues (2014) evaluated molecular adaptations to nicotine withdrawal in discrete brain regions implicated in both cognitive and affective withdrawal symptoms. These studies investigated chromatin alterations and transcriptional control of CREB target genes following chronic nicotine exposure and 24-h withdrawal using next-generation sequencing. This coupling of CREB chromatin immunoprecipitation and high-throughput sequencing (ChIP-seq) resulted in the identification of a novel molecular target for nicotine dependence, NRG3, with increased expression detected in response to chronic nicotine exposure and withdrawal. To directly evaluate whether NRG3–ErbB4 signaling could impact smoking cessation behaviors, the authors utilized both genetic and pharmacological tools to block NRG3–ErbB4 signaling during chronic nicotine treatment and withdrawal. They observed that a co-occurring induction of NRG3 during early withdrawal is associated with increased anxiety-like behavior in mice. However, if this increased NRG3 signaling is blunted, either in NRG3 hypomorphic mice ($NRG3^{ska}$) or in wild-type mice treated with an ErbB4 inhibitor (afatinib), the anxiety behaviors observed during withdrawal were also blunted, suggesting a relationship between changes in NRG3 signaling and behavior. While the precise mechanism by which NRG3 impacts these withdrawal behaviors is unknown, these studies encouraged further scrutiny of NRG3's role in smoking cessation outcomes. Therefore, in order to evaluate the clinical relevance of this finding in human smokers, Turner and colleagues (2014) examined genetic polymorphisms in *NRG3* and identified single nucleotide polymorphisms (SNPs) that significantly associated with reduced smoking cessation rates at both 6 weeks and 6 months.

3.2 NRG3 in Schizophrenia

Structural and polymorphic variation of *NRG3* is associated with not only schizophrenia but also a broad spectrum of neurodevelopmental disorders. Previous fine mapping of the 10q22-23 schizophrenia locus identified significant association between delusion severity and polymorphisms on intron 1 of *NRG3* (Morar et al., 2011). Following these findings, Kao and colleagues (2010) examined *NRG3* in 400 postmortem prefrontal cortical tissue samples from schizophrenia patients and controls, evaluating the association between disease state, genetic risk variants, and *NRG3* expression levels. Alternative splicing enables one gene to encode multiple proteins and is often regulated in a tissue-specific and developmental manner (Cotton et al., 2008; Kampa et al., 2004). Using RNA expression profiling, Kao and colleagues revealed that *NRG3* expression is developmentally regulated and increased in schizophrenia (Kao et al., 2010). Furthermore, *NRG3* undergoes complex splicing, leading to many distinct isoforms, all of which have an EGF-like bioactive domain, a transmembrane domain, and a complete cytoplasmic tail (Kao et al., 2010). Hatzimanolis and colleagues (2013) hypothesized that more than one damaging variant in the NRG signaling pathway genes may be needed to cause schizophrenia. They scrutinized all known genes within the NRG signaling pathway and detected an aggregation of predicted damaging variants in a subset of individuals showing unique phenotypic properties. Further, their data supports the notion that damaging variants in the NRG signaling pathway may underlie the heterogeneity of schizophrenia, which is observed in both as phenotypic variability and as genetic complexity.

3.3 NRG3: Possible Mechanisms Underlying Co-morbidity

While evaluation of SNPs common to both nicotine dependence and schizophrenia is unfortunately lacking to date, one potential way NRG3 may be contributing to smoking behavior as well as to co-morbid disorders, such as schizophrenia, is via its role in impulsivity. One shared distinct deficit among co-morbid disorders such as addiction, ADHD, and schizophrenia is impulse control. Impulsivity is suggested to be a prominent, heritable symptom among psychiatric disorders (American Psychiatric Association &

American Psychiatric Association. Task Force on DSM-IV, 1994) and can manifest in a variety of impulsive behaviors, which can be observed for example in computerized response tasks (Loos et al., 2014). A facet of impulsivity is impulsive action, which can be broadly defined as the inability to withhold from making a response. Genetic mapping of impulsive action in mice has revealed a locus on chromosome 14, which is homologous to the human 10q22-q23 schizophrenia-susceptibility locus encompassing *NRG3* (Loos et al., 2014). To confirm its influence on impulsive action, congenic mice carrying the impulsivity locus (Impu1) showed that increased impulsivity was associated with increased *Nrg3* expression in the medial prefrontal cortex (mPFC), a region known for its role in drug abuse-related behaviors. Loos and colleagues (2014) also showed that viral overexpression of *Nrg3* in the mPFC increased impulsivity, whereas loss-of-function mutant mice showed decreased impulsivity (Loos et al., 2014). Although the level of *NRG3* expression appears to influence levels of inhibitory control, the specific mechanism how NRG3 signaling impacts impulsivity and how this relates to nicotine dependence and schizophrenia is unknown.

4. ERBB4: RELEVANCE IN SMOKING BEHAVIOR AND CO-MORBID DISORDERS

4.1 Association Between ERBB4 and Nicotine Dependence

Recently, *ERBB4* was shown to be associated with nicotine dependence. Loukola and colleagues (2014) performed a genome-wide association study on 1114 adult twins ascertained for heavy smoking from a population-based Finnish Twin Cohort study. With 17 smoking-related phenotypes available, the authors were able to comprehensively portray the multiple dimensions of smoking behavior, such as smoking initiation, amount smoked, and nicotine dependence. By employing a convergent approach, the authors gathered multiple independent lines of evidence supporting the association between *ERBB4* and nicotine dependence defined by DSM-IV (American Psychiatric Association, 1994). The initial association detected in the Finnish twin sample was replicated in an independent Australian twin family sample of 4425 individuals. Further, *ERBB4* is located within a regular smoking linkage locus previously identified in the Finnish twin families (Loukola et al., 2008) and within a smoking quantity locus highlighted in a linkage meta-analysis (Han, Gelernter, Luo, & Yang, 2010). These results provided novel evidence for the involvement of ErbB4 in nicotine dependence (Loukola et al., 2014).

4.2 ErbB4–NMDA Receptor Interactions in Schizophrenia and Possible Relevance for Co-morbidity with Nicotine Dependence

While a valid animal model of schizophrenia has been difficult to construct due to the polygenetic nature of the disease, genetic mouse models resulting in increased activation of NRG–ErbB4 signaling have aided understanding of the disease. For example, Paterson and Law (2014) recently investigated the effects of *Nrg3* overexpression with regards to activation of the ErbB4–Akt signaling pathway. They found that excessive ErbB4 activation during development had life-long consequences on discrete behavioral phenotypes and posited that this enhanced signaling impacts early neonatal brain development and influences circuitry that is involved in behaviors related to anxiety and sociability (Paterson & Law, 2014). Further studies by Del Pino and colleagues (2013) examined schizophrenia-like phenotypes in ErbB4-floxed mutant mice. They found that deletion of *ErbB4* from two main types of fast-spiking neurons (chandelier and basket cells) caused disruption in the synchrony of cortical regions. This functional deficit was found to be associated with increased locomotor activity, abnormal emotional and social responses, and impaired cognitive function, thus leading to the conclusion that dysfunction of cortical fast-spiking interneurons might be central to the etiology of schizophrenia (Del Pino et al., 2013). However, these dual observations may be due to the close proximity of and interaction between NMDA and ErbB4 receptors (Garcia et al., 2000). As discussed earlier, ErbB4 activation can result in reduced NMDA receptor function. However, the effects of chronic ErbB4 inhibition on NMDA receptors are unknown, especially during development, but NMDA receptor hypofunction has been suggested to underlie some schizophrenic traits. In line with this, phencyclidine and ketamine, two anesthetics that induce schizophrenia-like symptoms, are in fact NMDAR channel blockers (Anis, Berry, Burton, & Lodge, 1983; Javitt & Zukin, 1991). Additionally, current animal models of NMDAR hypofunction via genetic down-regulation of NMDARs result in traits resembling schizophrenia (Belforte et al., 2010; Mohn, Gainetdinov, Caron, & Koller, 1999). Therefore, these findings represent a new pathway by which NMDAR and ErbB4 receptor interaction could underlie schizophrenic pathophysiology. However, whether and how this mechanism is altered in nicotine dependence is currently unknown, but may hold relevance both for understanding co-morbidity and for developing new treatments.

5. SUMMARY

With continual technological advancements, genetic studies have helped scientists identify common genetic variation within the human population that may underlie nicotine dependence and co-morbid disorders, such as schizophrenia. For example, SNPs on genes encoding the NRG–ErbB signaling pathway have been shown to influence nicotine dependence and withdrawal (Turner et al., 2014), as well as the pathophysiology of schizophrenia (Badner & Gershon, 2002; Gurling et al., 2001), providing researchers new insight into the potential benefits of examining the NRG–ErbB4 pathway for novel therapeutic targets not only for smoking cessation but also for treating symptoms seen in schizophrenia as well. Furthermore, due to such high demand for novel therapeutics targeted at treating co-morbid disorders such as tobacco smoking and schizophrenia, understanding common cellular processes that link these disorders is worth investigating and the NRG–ErbB pathway may represent a promising place to start.

REFERENCES

Addington, J., & el-Guebaly, N. (1998). Group treatment for substance abuse in schizophrenia. *Canadian Journal of Psychiatry*, *43*(8), 843–845.

American Psychiatric Association. (1994). *Diagnostic criteria from DSM-IV*. Washington, DC: The Association.

American Psychiatric Association & American Psychiatric Association. Task Force on DSM-IV. (1994). *Diagnostic and statistical manual of mental disorders: DSM-IV* (4th ed.). Washington, DC: American Psychiatric Association.

Anis, N. A., Berry, S. C., Burton, N. R., & Lodge, D. (1983). The dissociative anaesthetics, ketamine and phencyclidine, selectively reduce excitation of central mammalian neurones by N-methyl-aspartate. *British Journal of Pharmacology*, *79*(2), 565–575.

Badner, J. A., & Gershon, E. S. (2002). Meta-analysis of whole-genome linkage scans of bipolar disorder and schizophrenia. *Molecular Psychiatry*, 7(4), 405–411. http://dx.doi.org/10.1038/sj.mp.4001012.

Balciuniene, J., Feng, N., Iyadurai, K., Hirsch, B., Charnas, L., Bill, B. R., et al. (2007). Recurrent 10q22-q23 deletions: A genomic disorder on 10q associated with cognitive and behavioral abnormalities. *American Journal of Human Genetics*, *80*(5), 938–947. http://dx.doi.org/10.1086/513607.

Bao, J., Wolpowitz, D., Role, L. W., & Talmage, D. A. (2003). Back signaling by the Nrg-1 intracellular domain. *The Journal of Cell Biology*, *161*(6), 1133–1141. http://dx.doi.org/10.1083/jcb.200212085.

Belforte, J. E., Zsiros, V., Sklar, E. R., Jiang, Z., Yu, G., Li, Y., et al. (2010). Postnatal NMDA receptor ablation in corticolimbic interneurons confers schizophrenia-like phenotypes. *Nature Neuroscience*, *13*(1), 76–83. http://dx.doi.org/10.1038/nn.2447.

Benzel, I., Bansal, A., Browning, B. L., Galwey, N. W., Maycox, P. R., McGinnis, R., et al. (2007). Interactions among genes in the ErbB-Neuregulin signalling network are associated with increased susceptibility to schizophrenia. *Behavioral and Brain Functions*, *3*, 31. http://dx.doi.org/10.1186/1744-9081-3-31.

Braff, D. L., Freedman, R., Schork, N. J., & Gottesman, I. I. (2007). Deconstructing schizophrenia: An overview of the use of endophenotypes in order to understand a complex disorder. *Schizophrenia Bulletin*, *33*(1), 21–32. http://dx.doi.org/10.1093/schbul/sbl049.

Breslau, N., Peterson, E. L., Schultz, L. R., Chilcoat, H. D., & Andreski, P. (1998). Major depression and stages of smoking. A longitudinal investigation. *Archives of General Psychiatry*, *55*(2), 161–166.

Carraway, K. L., 3rd, Weber, J. L., Unger, M. J., Ledesma, J., Yu, N., Gassmann, M., et al. (1997). Neuregulin-2, a new ligand of ErbB3/ErbB4-receptor tyrosine kinases. *Nature*, *387*(6632), 512–516. http://dx.doi.org/10.1038/387512a0.

Chang, H., Riese, D. J., 2nd, Gilbert, W., Stern, D. F., & McMahan, U. J. (1997). Ligands for ErbB-family receptors encoded by a neuregulin-like gene. *Nature*, *387*(6632), 509–512. http://dx.doi.org/10.1038/387509a0.

Chen, P. L., Avramopoulos, D., Lasseter, V. K., McGrath, J. A., Fallin, M. D., Liang, K. Y., et al. (2009). Fine mapping on chromosome 10q22-q23 implicates Neuregulin 3 in schizophrenia. *American Journal of Human Genetics*, *84*(1), 21–34. http://dx.doi.org/10.1016/j.ajhg.2008.12.005.

Cotton, R. G., Auerbach, A. D., Axton, M., Barash, C. I., Berkovic, S. F., Brookes, A. J., et al. (2008). Genetics. The Human Variome Project. *Science*, *322*(5903), 861–862. http://dx.doi.org/10.1126/science.1167363.

Coyle, J. T., & Tsai, G. (2004). NMDA receptor function, neuroplasticity, and the pathophysiology of schizophrenia. *International Review of Neurobiology*, *59*, 491–515. http://dx.doi.org/10.1016/S0074-7742(04)59019-0.

Cummings, K. M., & Mahoney, M. (2006). Current and emerging treatment approaches for tobacco dependence. *Current Oncology Reports*, *8*(6), 475–483.

Del Pino, I., Garcia-Frigola, C., Dehorter, N., Brotons-Mas, J. R., Alvarez-Salvado, E., Martinez de Lagran, M., et al. (2013). Erbb4 deletion from fast-spiking interneurons causes schizophrenia-like phenotypes. *Neuron*, *79*(6), 1152–1168. http://dx.doi.org/10.1016/j.neuron.2013.07.010.

Dickerson, F., Stallings, C. R., Origoni, A. E., Vaughan, C., Khushalani, S., Schroeder, J., et al. (2013). Cigarette smoking among persons with schizophrenia or bipolar disorder in routine clinical settings, 1999–2011. *Psychiatric Services*, *64*(1), 44–50. http://dx.doi.org/10.1176/appi.ps.201200143.

Fallin, M. D., Lasseter, V. K., Wolyniec, P. S., McGrath, J. A., Nestadt, G., Valle, D., et al. (2003). Genomewide linkage scan for schizophrenia susceptibility loci among Ashkenazi Jewish families shows evidence of linkage on chromosome 10q22. *American Journal of Human Genetics*, *73*(3), 601–611. http://dx.doi.org/10.1086/378158.

Falls, D. L. (2003). Neuregulins: Functions, forms, and signaling strategies. *Experimental Cell Research*, *284*(1), 14–30.

Falls, D. L., Rosen, K. M., Corfas, G., Lane, W. S., & Fischbach, G. D. (1993). ARIA, a protein that stimulates acetylcholine receptor synthesis, is a member of the neu ligand family. *Cell*, *72*(5), 801–815.

Faraone, S. V., Hwu, H. G., Liu, C. M., Chen, W. J., Tsuang, M. M., Liu, S. K., et al. (2006). Genome scan of Han Chinese schizophrenia families from Taiwan: Confirmation of linkage to 10q22.3. *The American Journal of Psychiatry*, *163*(10), 1760–1766. http://dx.doi.org/10.1176/appi.ajp.163.10.1760.

Garcia, R. A., Vasudevan, K., & Buonanno, A. (2000). The neuregulin receptor ErbB-4 interacts with PDZ-containing proteins at neuronal synapses. *Proceedings of the National*

Academy of Sciences of the United States of America, 97(7), 3596–3601. http://dx.doi.org/10.1073/pnas.070042497.

George, T. P., Vessicchio, J. C., Termine, A., Bregartner, T. A., Feingold, A., Rounsaville, B. J., et al. (2002). A placebo controlled trial of bupropion for smoking cessation in schizophrenia. *Biological Psychiatry, 52*(1), 53–61.

Gonzales, D., Rennard, S. I., Nides, M., Oncken, C., Azoulay, S., Billing, C. B., et al. (2006). Varenicline, an alpha4beta2 nicotinic acetylcholine receptor partial agonist, vs sustained-release bupropion and placebo for smoking cessation: A randomized controlled trial. *JAMA, 296*(1), 47–55. http://dx.doi.org/10.1001/jama.296.1.47.

Gurling, H. M. D., Kalsi, G., Brynjolfson, J., Sigmundsson, T., Sherrington, R., Mankoo, B. S., et al. (2001). Genomewide genetic linkage analysis confirms the presence of susceptibility loci for schizophrenia, on chromosomes 1q32.2, 5q33.2, and 8p21-22 and provides support for linkage to schizophrenia, on chromosomes 11q23.3-24 and 20q12.1-11.23. *American Journal of Human Genetics, 68*(3), 661–673. http://dx.doi.org/10.1086/318788.

Han, S., Gelernter, J., Luo, X., & Yang, B. Z. (2010). Meta-analysis of 15 genome-wide linkage scans of smoking behavior. *Biological Psychiatry, 67*(1), 12–19. http://dx.doi.org/10.1016/j.biopsych.2009.08.028.

Hancock, M. L., Canetta, S. E., Role, L. W., & Talmage, D. A. (2008). Presynaptic type III neuregulin1-ErbB signaling targets alpha7 nicotinic acetylcholine receptors to axons. *The Journal of General Physiology, 131*(6), i4. http://dx.doi.org/10.1085/JGP1316OIA4.

Harari, D., Tzahar, E., Romano, J., Shelly, M., Pierce, J. H., Andrews, G. C., et al. (1999). Neuregulin-4: A novel growth factor that acts through the ErbB-4 receptor tyrosine kinase. *Oncogene, 18*(17), 2681–2689. http://dx.doi.org/10.1038/sj.onc.1202631.

Hatzimanolis, A., McGrath, J. A., Wang, R., Li, T., Wong, P. C., Nestadt, G., et al. (2013). Multiple variants aggregate in the neuregulin signaling pathway in a subset of schizophrenia patients. *Translational Psychiatry, 3*. http://dx.doi.org/10.1038/tp.2013.33. e264.

Hou, Y. Z., Xiang, Y. T., Yan, F., Ungvari, G. S., Dickerson, F., Chiu, H. F., et al. (2011). Cigarette smoking in community-dwelling patients with schizophrenia in China. *Journal of Psychiatric Research, 45*(12), 1551–1556. http://dx.doi.org/10.1016/j.jpsychires.2011.07.011.

Howard, B., Panchal, H., McCarthy, A., & Ashworth, A. (2005). Identification of the scaramanga gene implicates Neuregulin3 in mammary gland specification. *Genes & Development, 19*(17), 2078–2090. http://dx.doi.org/10.1101/gad.338505.

Huang, Y. Z., Won, S., Ali, D. W., Wang, Q., Tanowitz, M., Du, Q. S., et al. (2000). Regulation of neuregulin signaling by PSD-95 interacting with ErbB4 at CNS synapses. *Neuron, 26*(2), 443–455.

Hughes, J. R. (2007). Effects of abstinence from tobacco: Valid symptoms and time course. *Nicotine & Tobacco Research, 9*(3), 315–327. http://dx.doi.org/10.1080/14622200701188919.

Hughes, J. R., Gulliver, S. B., Fenwick, J. W., Valliere, W. A., Cruser, K., Pepper, S., et al. (1992). Smoking cessation among self-quitters. *Health Psychology, 11*(5), 331–334.

Jamal, A., Agaku, I. T., O'Connor, E., King, B. A., Kenemer, J. B., & Neff, L. (2014). Current cigarette smoking among adults—United States, 2005–2013. *Morbidity And Mortality Weekly Report, 63*(47), 1108–1112.

Javitt, D. C., & Zukin, S. R. (1991). Recent advances in the phencyclidine model of schizophrenia. *The American Journal of Psychiatry, 148*(10), 1301–1308.

Jorenby, D. E., Hays, J. T., Rigotti, N. A., Azoulay, S., Watsky, E. J., Williams, K. E., et al. (2006). Efficacy of varenicline, an alpha4beta2 nicotinic acetylcholine receptor partial agonist, vs placebo or sustained-release bupropion for smoking cessation: A randomized controlled trial. *JAMA, 296*(1), 56–63. http://dx.doi.org/10.1001/jama.296.1.56.

Kampa, D., Cheng, J., Kapranov, P., Yamanaka, M., Brubaker, S., Cawley, S., et al. (2004). Novel RNAs identified from an in-depth analysis of the transcriptome of human chromosomes 21 and 22. *Genome Research*, *14*(3), 331–342. http://dx.doi.org/10.1101/gr.2094104.

Kanemoto, N., Horie, M., Omori, K., Nishino, N., Kondo, M., Noguchi, K., et al. (2001). Expression of TMEFF1 mRNA in the mouse central nervous system: Precise examination and comparative studies of TMEFF1 and TMEFF2. *Brain Research. Molecular Brain Research*, *86*(1–2), 48–55.

Kao, W. T., Wang, Y., Kleinman, J. E., Lipska, B. K., Hyde, T. M., Weinberger, D. R., et al. (2010). Common genetic variation in Neuregulin 3 (NRG3) influences risk for schizophrenia and impacts NRG3 expression in human brain. *Proceedings of the National Academy of Sciences of the United States of America*, *107*(35), 15619–15624. http://dx.doi.org/10.1073/pnas.1005410107.

Kendler, K. S., Neale, M. C., MacLean, C. J., Heath, A. C., Eaves, L. J., & Kessler, R. C. (1993). Smoking and major depression. A causal analysis. *Archives of General Psychiatry*, *50*(1), 36–43.

Kinugasa, Y., Ishiguro, H., Tokita, Y., Oohira, A., Ohmoto, H., & Higashiyama, S. (2004). Neuroglycan C, a novel member of the neuregulin family. *Biochemical and Biophysical Research Communications*, *321*(4), 1045–1049. http://dx.doi.org/10.1016/j.bbrc.2004.07.066.

Law, A. J. (2014). Genetic mouse models of neuregulin 1: Gene dosage effects, isoform-specific functions, and relevance to schizophrenia. *Biological Psychiatry*, *76*(2), 89–90. http://dx.doi.org/10.1016/j.biopsych.2014.05.016.

Le Foll, B., & Goldberg, S. R. (2009). Effects of nicotine in experimental animals and humans: An update on addictive properties. *Handbook of Experimental Pharmacology*, *192*, 335–367. http://dx.doi.org/10.1007/978-3-540-69248-5_12.

Lee, H. J., Jung, K. M., Huang, Y. Z., Bennett, L. B., Lee, J. S., Mei, L., et al. (2002). Presenilin-dependent gamma-secretase-like intramembrane cleavage of ErbB4. *The Journal of Biological Chemistry*, *277*(8), 6318–6323. http://dx.doi.org/10.1074/jbc.M110371200.

Leonard, S., Adams, C., Breese, C. R., Adler, L. E., Bickford, P., Byerley, W., et al. (1996). Nicotinic receptor function in schizophrenia. *Schizophrenia Bulletin*, *22*(3), 431–445.

Leonard, S., Gault, J., Hopkins, J., Logel, J., Vianzon, R., Short, M., et al. (2002). Association of promoter variants in the alpha7 nicotinic acetylcholine receptor subunit gene with an inhibitory deficit found in schizophrenia. *Archives of General Psychiatry*, *59*(12), 1085–1096.

Li, D., Collier, D. A., & He, L. (2006). Meta-analysis shows strong positive association of the neuregulin 1 (NRG1) gene with schizophrenia. *Human Molecular Genetics*, *15*(12), 1995–2002. http://dx.doi.org/10.1093/hmg/ddl122.

Li, Q., Hsia, J., & Yang, G. (2011). Prevalence of smoking in China in 2010. *The New England Journal of Medicine*, *364*(25), 2469–2470. http://dx.doi.org/10.1056/NEJMc1102459.

Loos, M., Mueller, T., Gouwenberg, Y., Wijnands, R., van der Loo, R. J., Neuro Bsik Mouse Phenomics Consortium, et al. (2014). Neuregulin-3 in the mouse medial prefrontal cortex regulates impulsive action. *Biological Psychiatry*, *76*(8), 648–655. http://dx.doi.org/10.1016/j.biopsych.2014.02.011.

Loukola, A., Broms, U., Maunu, H., Widen, E., Heikkila, K., Siivola, M., et al. (2008). Linkage of nicotine dependence and smoking behavior on 10q, 7q and 11p in twins with homogeneous genetic background. *The Pharmacogenomics Journal*, *8*(3), 209–219. http://dx.doi.org/10.1038/sj.tpj.6500464.

Loukola, A., Wedenoja, J., Keskitalo-Vuokko, K., Broms, U., Korhonen, T., Ripatti, S., et al. (2014). Genome-wide association study on detailed profiles of smoking behavior and nicotine dependence in a twin sample. *Molecular Psychiatry, 19*(5), 615–624. http://dx.doi.org/10.1038/mp.2013.72.

Marenco, S., Weinberger, D. R., & Schreurs, B. G. (2003). Single-cue delay and trace classical conditioning in schizophrenia. *Biological Psychiatry, 53*(5), 390–402.

Massague, J., & Pandiella, A. (1993). Membrane-anchored growth factors. *Annual Review of Biochemistry, 62*, 515–541. http://dx.doi.org/10.1146/annurev.bi.62.070193.002503.

Matthews, A. M., Wilson, V. B., & Mitchell, S. H. (2011). The role of antipsychotics in smoking and smoking cessation. *CNS Drugs, 25*(4), 299–315. http://dx.doi.org/10.2165/11588170-000000000-00000.

Mei, L., & Nave, K. A. (2014). Neuregulin-ERBB signaling in the nervous system and neuropsychiatric diseases. *Neuron, 83*(1), 27–49. http://dx.doi.org/10.1016/j.neuron.2014.06.007.

Mei, L., & Xiong, W. C. (2008). Neuregulin 1 in neural development, synaptic plasticity and schizophrenia. *Nature Reviews. Neuroscience, 9*(6), 437–452. http://dx.doi.org/10.1038/nrn2392.

Mohn, A. R., Gainetdinov, R. R., Caron, M. G., & Koller, B. H. (1999). Mice with reduced NMDA receptor expression display behaviors related to schizophrenia. *Cell, 98*(4), 427–436.

Morar, B., Dragovic, M., Waters, F. A., Chandler, D., Kalaydjieva, L., & Jablensky, A. (2011). Neuregulin 3 (NRG3) as a susceptibility gene in a schizophrenia subtype with florid delusions and relatively spared cognition. *Molecular Psychiatry, 16*(8), 860–866. http://dx.doi.org/10.1038/mp.2010.70.

Moylan, S., Jacka, F. N., Pasco, J. A., & Berk, M. (2012). Cigarette smoking, nicotine dependence and anxiety disorders: A systematic review of population-based, epidemiological studies. *BMC Medicine, 10*, 123. http://dx.doi.org/10.1186/1741-7015-10-123.

Munafo, M. R., Attwood, A. S., & Flint, J. (2008). Neuregulin 1 genotype and schizophrenia. *Schizophrenia Bulletin, 34*(1), 9–12. http://dx.doi.org/10.1093/schbul/sbm129.

Munafo, M. R., Thiselton, D. L., Clark, T. G., & Flint, J. (2006). Association of the NRG1 gene and schizophrenia: A meta-analysis. *Molecular Psychiatry, 11*(6), 539–546. http://dx.doi.org/10.1038/sj.mp.4001817.

Muraoka-Cook, R. S., Feng, S. M., Strunk, K. E., & Earp, H. S., 3rd. (2008). ErbB4/HER4: Role in mammary gland development, differentiation and growth inhibition. *Journal of Mammary Gland Biology and Neoplasia, 13*(2), 235–246. http://dx.doi.org/10.1007/s10911-008-9080-x.

Ni, C. Y., Murphy, M. P., Golde, T. E., & Carpenter, G. (2001). Gamma-secretase cleavage and nuclear localization of ErbB-4 receptor tyrosine kinase. *Science, 294*(5549), 2179–2181. http://dx.doi.org/10.1126/science.1065412.

Olayioye, M. A., Neve, R. M., Lane, H. A., & Hynes, N. E. (2000). The ErbB signaling network: Receptor heterodimerization in development and cancer. *The EMBO Journal, 19*(13), 3159–3167. http://dx.doi.org/10.1093/emboj/19.13.3159.

Ortega, M. C., Bribian, A., Peregrin, S., Gil, M. T., Marin, O., & de Castro, F. (2012). Neuregulin-1/ErbB4 signaling controls the migration of oligodendrocyte precursor cells during development. *Experimental Neurology, 235*(2), 610–620. http://dx.doi.org/10.1016/j.expneurol.2012.03.015.

Pasco, J. A., Williams, L. J., Jacka, F. N., Ng, F., Henry, M. J., Nicholson, G. C., et al. (2008). Tobacco smoking as a risk factor for major depressive disorder: Population-based study. *The British Journal of Psychiatry, 193*(4), 322–326. http://dx.doi.org/10.1192/bjp.bp.107.046706.

Paterson, C., & Law, A. J. (2014). Transient overexposure of neuregulin 3 during early postnatal development impacts selective behaviors in adulthood. *PLoS One, 9*(8). http://dx.doi.org/10.1371/journal.pone.0104172, e104172.

Pitcher, G. M., Kalia, L. V., Ng, D., Goodfellow, N. M., Yee, K. T., Lambe, E. K., et al. (2011). Schizophrenia susceptibility pathway neuregulin 1-ErbB4 suppresses Src upregulation of NMDA receptors. *Nature Medicine*, *17*(4), 470–478. http://dx.doi.org/10.1038/nm.2315.

Rapoport, J. L., Addington, A. M., Frangou, S., & Psych, M. R. (2005). The neurodevelopmental model of schizophrenia: Update 2005. *Molecular Psychiatry*, *10*(5), 434–449. http://dx.doi.org/10.1038/sj.mp.4001642.

Royal College of Physicians of London and Royal College of Psychiatrists. (2013). *Smoking and mental health: A joint report*. London: Royal College of Physicians: Royal College of Psychiatrists.

Sonuga-Barke, E. J. S., Lasky-Su, J., Neale, B. M., Oades, R., Chen, W., Franke, B., et al. (2008). Does parental expressed emotion moderate genetic effects in ADHD? An exploration using a genome wide association scan. *American Journal of Medical Genetics. Part B, Neuropsychiatric Genetics: The Official Publication of the International Society of Psychiatric Genetics*, *147B*(8), 1359–1368. http://dx.doi.org/10.1002/Ajmg.B.30860.

Sundvall, M., Veikkolainen, V., Kurppa, K., Salah, Z., Tvorogov, D., van Zoelen, E. J., et al. (2010). Cell death or survival promoted by alternative isoforms of ErbB4. *Molecular Biology of the Cell*, *21*(23), 4275–4286. http://dx.doi.org/10.1091/mbc.E10-04-0332.

Turner, J. R., Ray, R., Lee, B., Everett, L., Xiang, J., Jepson, C., et al. (2014). Evidence from mouse and man for a role of neuregulin 3 in nicotine dependence. *Molecular Psychiatry*, *19*(7), 801–810. http://dx.doi.org/10.1038/mp.2013.104.

Uchida, T., Wada, K., Akamatsu, T., Yonezawa, M., Noguchi, H., Mizoguchi, A., et al. (1999). A novel epidermal growth factor-like molecule containing two follistatin modules stimulates tyrosine phosphorylation of erbB-4 in MKN28 gastric cancer cells. *Biochemical and Biophysical Research Communications*, *266*(2), 593–602. http://dx.doi.org/10.1006/bbrc.1999.1873.

U.S. Department of Health and Human Services (1988). The Health Consequences of Smoking. Nikotine addiction. A report of the Surgeon General. Washington, D.C.

Veikkolainen, V., Vaparanta, K., Halkilahti, K., Iljin, K., Sundvall, M., & Elenius, K. (2011). Function of ERBB4 is determined by alternative splicing. *Cell Cycle*, *10*(16), 2647–2657.

Vidal, G. A., Naresh, A., Marrero, L., & Jones, F. E. (2005). Presenilin-dependent gamma-secretase processing regulates multiple ERBB4/HER4 activities. *The Journal of Biological Chemistry*, *280*(20), 19777–19783. http://dx.doi.org/10.1074/jbc.M412457200.

Waddington, J. L., Corvin, A. P., Donohoe, G., O'Tuathaigh, C. M., Mitchell, K. J., & Gill, M. (2007). Functional genomics and schizophrenia: Endophenotypes and mutant models. *The Psychiatric Clinics of North America*, *30*(3), 365–399. http://dx.doi.org/10.1016/j.psc.2007.04.011.

Walters, J. T., & Owen, M. J. (2007). Endophenotypes in psychiatric genetics. *Molecular Psychiatry*, *12*(10), 886–890. http://dx.doi.org/10.1038/sj.mp.4002068.

Wang, Y. C., Chen, J. Y., Chen, M. L., Chen, C. H., Lai, I. C., Chen, T. T., et al. (2008). Neuregulin 3 genetic variations and susceptibility to schizophrenia in a Chinese population. *Biological Psychiatry*, *64*(12), 1093–1096. http://dx.doi.org/10.1016/j.biopsych.2008.07.012.

Watanabe, E., Maeda, N., Matsui, F., Kushima, Y., Noda, M., & Oohira, A. (1995). Neuroglycan C, a novel membrane-spanning chondroitin sulfate proteoglycan that is restricted to the brain. *The Journal of Biological Chemistry*, *270*(45), 26876–26882.

West, R. J., Hajek, P., & Belcher, M. (1989). Severity of withdrawal symptoms as a predictor of outcome of an attempt to quit smoking. *Psychological Medicine*, *19*(4), 981–985.

WHO. (2011). WHO urges more countries to require large, graphic health warnings on tobacco packaging: The WHO report on the global tobacco epidemic, 2011 examines anti-tobacco mass-media campaigns. *Central European Journal of Public Health*, *19*(3), 133, 151.

Xu, Y. M., Chen, H. H., Li, F., Deng, F., Liu, X. B., Yang, H. C., et al. (2014). Prevalence and correlates of cigarette smoking among Chinese schizophrenia inpatients receiving antipsychotic mono-therapy. *PLoS One, 9*(2). http://dx.doi.org/10.1371/journal.pone.0088478. e88478.

Yang, X., Kuo, Y., Devay, P., Yu, C., & Role, L. (1998). A cysteine-rich isoform of neuregulin controls the level of expression of neuronal nicotinic receptor channels during synaptogenesis. *Neuron, 20*(2), 255–270.

Yarden, Y., & Sliwkowski, M. X. (2001). Untangling the ErbB signalling network. *Nature Reviews. Molecular Cell Biology, 2*(2), 127–137. http://dx.doi.org/10.1038/35052073.

Zabala, A., Eguiluz, J. I., Segarra, R., Enjuto, S., Ezcurra, J., Gonzalez Pinto, A., et al. (2009). Cognitive performance and cigarette smoking in first-episode psychosis. *European Archives of Psychiatry and Clinical Neuroscience, 259*(2), 65–71. http://dx.doi.org/10.1007/s00406-008-0835-6.

Zeng, N., Liu, L., McCabe, M. G., Jones, D. T., Ichimura, K., & Collins, V. P. (2009). Real-time quantitative polymerase chain reaction (qPCR) analysis with fluorescence resonance energy transfer (FRET) probes reveals differential expression of the four ERBB4 juxtamembrane region variants between medulloblastoma and pilocytic astrocytoma. *Neuropathology and Applied Neurobiology, 35*(4), 353–366. http://dx.doi.org/10.1111/j.1365-2990.2008.01001.x.

Zhang, D. X., Sliwkowski, M. X., Mark, M., Frantz, G., Akita, R., Sun, Y., et al. (1997). Neuregulin-3 (NRG3): A novel neural tissue-enriched protein that binds and activates ErbB4. *Proceedings of the National Academy of Sciences of the United States of America, 94*(18), 9562–9567. http://dx.doi.org/10.1073/Pnas.94.18.9562.

Ziedonis, D. M., & George, T. P. (1997). Schizophrenia and nicotine use: Report of a pilot smoking cessation program and review of neurobiological and clinical issues. *Schizophrenia Bulletin, 23*(2), 247–254.

CHAPTER FIVE

Effective Cessation Strategies for Smokers with Schizophrenia

A. Eden Evins[*,†,1], **Corinne Cather**[*,†]
[*]Center for Addiction Medicine, Massachusetts General Hospital, Boston, Massachusetts, USA
[†]Department of Psychiatry, Harvard Medical School, Boston, Massachusetts, USA
[1]Corresponding author: e-mail address: a_eden_evins@hms.harvard.edu

Contents

Abstract

Despite availability of smoking cessation aids with proven efficacy and tolerability in those with schizophrenia, addiction to tobacco-derived nicotine remains highly prevalent among those with schizophrenia spectrum disorders. While smokers with serious mental illness have been excluded from most large nicotine-dependence treatment studies, and these treatments are woefully underutilized for smokers with psychiatric illness, a growing evidence base is available to guide cessation treatment for smokers with schizophrenia. The aim of this review is to present the evidence on safety and efficacy of smoking cessation interventions for those with schizophrenia spectrum illness. Smokers with schizophrenia spectrum disorders should receive varenicline or bupropion with or without nicotine replacement therapy in combination with behavioral treatment. Clinical practice guidelines now recommend duration of treatment be 12 weeks, but evidence indicates that maintenance pharmacotherapy for 1 year improves sustained abstinence rates. Controlled trials have found no evidence that in patients with serious mental illness, the use of pharmacotherapeutic cessation aids worsens psychiatric symptoms or increases the rate of psychiatric adverse events. Converging evidence indicates that a majority of smokers with serious mental illness want to quit smoking and that available pharmacotherapeutic cessation aids combined with behavioral support are both effective for, and well tolerated by, these smokers.

International Review of Neurobiology, Volume 124
ISSN 0074-7742
http://dx.doi.org/10.1016/bs.irn.2015.08.001

Addiction to tobacco-derived nicotine is highly prevalent among those with schizophrenia. Fifty years after the first United States Surgeon General's report of an association between smoking and cancer, adult smoking in the United States has declined by 55% in the general population, to 18% (Agaku, King, & Dube, 2014, USDHHS, 2004). Despite availability of smoking cessation treatments, specifically effective in this population, comparable decreases in smoking rates have not been observed among smokers with schizophrenia (Cook et al., 2014). Smoking prevalence among adults with serious mental illness (SMI) in the United States is higher today, at 53%, than it was in the general population in 1964 (Dickerson et al., 2013; McClave, McKnight-Eily, Davis, & Dube, 2010). The 7.3 million adults with SMI in the United States who smoke tobacco (NIMH, 2013; SAMHSA, 2014) comprise 2% of the US population but 13% of smokers in the United States (SAMHSA, 2014). Recent estimates indicate that 64–79% of those with schizophrenia spectrum disorders smoke tobacco regularly (Dickerson et al., 2013; Hartz, Pato, & Medeiros, 2014) and most state that they would like to quit (Siru, Hulse, & Tait, 2009; Tidey & Rohsenow, 2009a, 2009b).

Tobacco smoking is associated with significant morbidity and premature mortality in those with SMI. People with SMI in the United States suffer physical diseases at younger ages and die approximately 25 years earlier than those without mental illness, primarily from diseases directly attributable to tobacco smoking (Colton & Manderscheid, 2006; Parks, Svendsen, Singer, & Foti, 2006). Premature mortality in those with SMI is the largest lifespan disparity in the United States (Everett, Mahler, Biblin, Ganguli, & Mauer, 2008; Lutterman et al., 2003). While tobacco smoking was responsible for over 18% of all deaths in the United States in 2000 (Danaei et al., 2009; Mokdad, Marks, Stroup, & Gerberding, 2004), a recent report found that half of all deaths in those who had been hospitalized for schizophrenia, bipolar disorder, or major depressive disorder in the state of California were due to one of 19 diseases identified by the Centers for Disease Control and Prevention (CDC) as being causally linked to tobacco use (Callaghan et al., 2014). Age-adjusted mortality from smoking-related pulmonary and cardiovascular disease is two to six times higher among individuals with schizophrenia than in age-matched samples (Brown, Inskip, & Barraclough, 2000; Goff et al., 2005).

Those with schizophrenia are likely to be heavy smokers (de Leon, Diaz, Rogers, Browne, & Dinsmore, 2002) and extract more nicotine from each cigarette (Olincy, Young, & Freedman, 1997; Williams et al., 2005). Despite this, smokers with schizophrenia can be motivated and persistent in attempts to quit smoking (Addington, el-Guebaly, Addington, & Hodgins, 1997; Siru et al., 2009). The benefits of smoking cessation are dramatic for all age groups (Jha et al., 2013; Pirie, Peto, Reeves, Green, & Beral, 2013). Those who quit smoking between ages 25 and 34 live 10 years longer than those who do not quit; those who quit between ages 35 and 44 gain 9 years of life, those who quit between ages 45 and 54 gain 6 years, and those who quit between ages 55 and 64 gain 4 years compared to those who do not quit smoking (Jha et al., 2013). Smoking cessation in those with schizophrenia with multiple cardiovascular risk factors has been reported to drive reduction in Framingham cardiovascular risk despite abstinence associated weight gain (Thorndike, Achytes, Cather, Pachas, & Achtyes, In press).

Despite the very high prevalence of smoking and of smoking-related illness among smokers with SMI, these individuals have been excluded from most major clinical trials of treatments for nicotine dependence (Evins, 2013; Schroeder & Morris, 2010). However, there is growing evidence for safety and efficacy of specific cessation aids for specific psychiatric disorders. We summarize this evidence, focusing on evidence from randomized controlled trials (RCTs) with biochemically validated abstinence outcomes that compared smoking cessation interventions in adult smokers with a schizophrenia spectrum disorder. Clinical practice guidelines recommend use of combination of medication and counseling in all individuals making a smoking cessation attempt (Fiore et al., 2008), so the review is focused on studies of combination of pharmacotherapy and behavioral treatment. We used verified prolonged or continuous abstinence as the outcome of interest where available. First-line pharmacologic treatments in the general population include varenicline, bupropion, and nicotine replacement therapy (NRT) (Cahill, Stevens, & Lancaster, 2014). Bupropion and NRT increase cessation rates over placebo, while varenicline and combination NRT (transdermal nicotine patch combined with short-acting NRT (nicotine polacrilex gum, lozenge, inhaler, or nasal spray)) are associated with superior abstinence rates (Cahill et al., 2014).

There is impressive clinical trial evidence for efficacy and tolerability of the first-line pharmacotherapies, bupropion, either alone or in combination with NRT, and varenicline, for smoking cessation specifically for those with schizophrenia (Cather, Dyer, et al., 2013; Evins et al., 2007, 2014, 2001; Evins, Cather, et al., 2005; George et al., 2008, 2002; Tsoi, Porwal, & Webster, 2013; Weiner et al., 2012, 2011; Williams et al., 2012), confirmed in independent, systematic reviews and meta-analyses (Evins, Cather, &

Laffer, 2015; Tidey, In press; Tsoi et al., 2013; Weiner et al., 2012) and incorporated into clinical practice guidelines that recommend that providers prescribe pharmacotherapy for smoking cessation with behavioral support for all smokers with schizophrenia spectrum disorders who wish to try to quit (Buchanan et al., 2010).

1. EFFICACY OF BUPROPION, AS MONOTHERAPY OR IN COMBINATION WITH NRT

A meta-analysis of six placebo-controlled RCTs in 304 smokers with schizophrenia spectrum disorder supports the use of bupropion with or without NRT in smokers with schizophrenia (Evins et al., 2007, 2001; Evins, Cather, et al., 2005; George et al., 2008, 2002; Tsoi et al., 2013; Weiner et al., 2012; Williams et al., 2012). Treatment with behavioral therapy and bupropion with or without NRT tripled 7-day point-prevalence abstinence rates over placebo and behavioral treatment, risk ratio (RR) = 3.03, 95% confidence interval (CI) 1.35–4.88, with an average end of treatment cessation rate of 26% for those assigned to bupropion and 7% for those on placebo bupropion with or without NRT and behavioral therapy. In four studies of bupropion versus placebo monotherapy added to behavioral treatment, in 150 subjects with schizophrenia spectrum disorder, abstinence rates at the end of treatment ranged from 10% to 38% for those receiving bupropion and behavioral treatment and 0–9% for those receiving placebo and behavioral treatment (Evins, Cather, et al., 2005; Evins et al., 2001; George et al., 2002; Weiner et al., 2012). The very low cessation rate with behavioral treatment added to placebo in these trials is notable, as the behavioral treatment was of high intensity (e.g., weekly 1 h cognitive behavioral therapy (CBT) group therapy) in some trials, and differs from the significant, dose-dependent benefit for behavioral treatment for smoking cessation reported for smokers in the general population (Fiore et al., 2008; Hall, Humfleet, Reus, Munoz, & Cullen, 2004).

Though relapse rates were quite high after treatment discontinuation, in the 5 RCT's conducted in 214 subjects that evaluated abstinence rates at the 6-month follow-up, 3 months after treatment discontinuation, there remained a significant effect of bupropion with or without NRT over placebo with or without NRT on abstinence, 13.2% versus 3.7%, respectively, RR = 2.79, 95% CI 1.02–7.58 (Evins et al., 2007, 2001; Evins, Cather, et al., 2005; George et al., 2008, 2002; Tsoi et al., 2013). Importantly, there is no evidence for worsening of positive, negative, or depressive symptoms in those assigned to receive bupropion with or without NRT as compared to placebo in any of these trials individually or in aggregate (Evins et al., 2007; Evins, Cather, et al., 2005; George et al., 2008, 2002; Tsoi et al., 2013).

In 2010, the Schizophrenia Patient Outcomes Research Team (PORT) Guidelines for the first time made an evidence-based recommendation for pharmacologic smoking cessation treatment of smokers with schizophrenia. This recommendation states that those who want to quit or reduce smoking should be advised to use bupropion 150 mg twice per day for 10–12 weeks with or without NRT and a smoking cessation education or support group (Buchanan et al., 2010). Trials of bupropion in smokers with schizophrenia have initiated bupropion treatment 1–4 weeks prior to the quit date using standard dosing of 150 mg per day for 4–7 days then 150 mg twice daily for 11 weeks (Evins et al., 2007, 2001; Evins, Cather, et al., 2005; George et al., 2008, 2002; Weiner et al., 2012).

Thus, consistent clinical trial efficacy for bupropion alone and in combination with NRT in placebo-controlled RCTs for those with schizophrenia who wish to quit smoking (Cather, Friedman-Yakoobian, et al., 2013; Evins, Deckersbach, et al., 2005; Evins et al., 2007, 2001; George et al., 2008, 2002) has been confirmed in independent meta-analyses (Tsoi et al., 2013; Weiner et al., 2012), and integrated in the 2010, National Institute of Mental Health sponsored Schizophrenia PORT Guidelines for evidence-based treatment of smokers with schizophrenia (Buchanan et al., 2010).

While NRT monotherapy (Fiore et al., 2008; West, McNeill, & Raw, 2000), and combination of short- and long-acting NRT products in particular (Cahill et al., 2014), is effective first-line pharmacotherapy for smokers in the general population, there are no published placebo-controlled trials demonstrating efficacy of NRT monotherapy in outpatients with schizophrenia who expressed motivation to try to quit smoking. Extended duration treatment with NRT has demonstrated preliminary promise for maintenance of abstinence among smokers with schizophrenia who attained initial abstinence with NRT in one open-label study (Horst, Klein, Williams, & Werder, 2005). Controlled trials of NRT are needed in this population.

2. EFFICACY OF VARENICLINE

Varenicline has been well tolerated and effective for abstinence at the end of treatment in two RCT's of a 12-week course of varenicline or placebo added to brief individual behavioral support. In combined data from the two trials in 137 smokers with schizophrenia spectrum disorder, end of treatment abstinence rates were 21% on varenicline and 4% on placebo plus behavioral support, RR = 4.74, 95% CI 1.34–16.71 (Tsoi et al., 2013; Weiner et al., 2011; Williams et al., 2012). In one trial that followed subjects for 6 months, 3 months after treatment discontinuation, the effect of the 3-month treatment course did not remain at 6 months, with 12% and 2% meeting abstinence criteria in the varenicline and placebo groups, respectively (Williams et al., 2012). Maintenance pharmacotherapy with varenicline in those who attained initial abstinence has shown promise and is discussed below (Evins et al., 2014).

Additional strong evidence for efficacy for varenicline among those with SMI include abstinence rates of 42% with 12 weeks open-label varenicline among 203 smokers with SMI (Evins et al., 2014), and among those who attained abstinence in this trial, those who received maintenance varenicline were sixfold more likely to be abstinent at 52 weeks than those who switched from varenicline to placebo after 12 weeks (Evins et al., 2014). In smokers with remitted major depressive disorder, varenicline more than doubled continuous abstinence rates over placebo at the end of 12 weeks treatment and 6-month follow-up (Anthenelli et al., 2013). In the first adequately powered smoking cessation trial in those with bipolar disorder, 60 treated adult outpatients with bipolar disorder were randomly assigned to double-blind treatment with behavioral smoking cessation counseling and either varenicline or placebo for 12 weeks (Chengappa et al., 2014). At the end of treatment, 36% of those assigned to varenicline and 10% of those on placebo attained 4-week continuous abstinence at end of treatment [odds ratio (OR) = 4.8; 95% CI, 1.02–25.13; $p = 0.03$] (48% of those randomized to varenicline and 10% of those on placebo attained biochemically confirmed, 7-day point-prevalence abstinence at the end of treatment). Varenicline has been shown to be statistically significantly superior to bupropion or NRT for smoking cessation in the general population, and increasing evidence suggests that it is also superior to bupropion or NRT for smokers with SMI, though head to head comparisons are needed.

Contrary to anecdotal reports much trumpeted in the press, no controlled trial to date of smokers with SMI or schizophrenia in particular has shown a signal for increased neuropsychiatric adverse events other than sleep disturbance with varenicline compared to placebo, bupropion, or NRT (Anthenelli et al., 2013; Chengappa et al., 2014; Evins et al., 2014; Gibbons & Mann, 2013; Pachas et al., 2012; Tsoi et al., 2013; Weiner et al., 2011; Williams et al., 2012; Wu et al., 2012). This evidence has now been added as supplemental data in the FDA prescribing information.

3. DURATION OF TREATMENT

Because of the high rates and rapidity of relapse observed after discontinuation of pharmacotherapeutic cessation aids in those with schizophrenia, as well as the high smoking prevalence rates and greater severity of dependence than that observed in the general population, it has been postulated that smokers with schizophrenia spectrum disorders may need maintenance pharmacotherapeutic treatment to achieve the sustained tobacco abstinence needed to realize the substantial health benefits of smoking cessation (Chengappa et al., 2014; Evins et al., 2007, 2014; Schroeder & Morris, 2010). Maintenance pharmacotherapeutic treatment has been reported to reduce relapse rates over maintenance behavioral treatment alone among smokers with psychiatric illness in open- (Cather, Dyer, et al., 2013), single-blind (Horst et al., 2005), and double-blind trials (Evins et al., 2014). In a double-blind, placebo-controlled maintenance treatment trial in 203 smokers with schizophrenia, schizoaffective, or bipolar disorder, 87 (43%) attained at least 2 weeks abstinence at the end of a 12-week open-label treatment phase with varenicline and group CBT and were randomized to continue CBT plus double-blind varenicline or identical placebo for an additional 40 weeks. At week 52, the end of treatment, point-prevalence abstinence rates were 60% for those treated with varenicline and 19% for those treated with placebo. At week 64 and week 76 follow-up visits, 45% and 30%, respectively, of those treated with varenicline were continuously abstinent versus 15% and 11% of those on placebo (Evins et al., 2014), suggesting that maintenance smoking cessation pharmacotherapy may be highly beneficial in those with SMI.

4. BEHAVIORAL TREATMENT

Like smokers in the general population, efficacy of pharmacotherapy for smoking cessation in those with schizophrenia is enhanced by behavioral treatment, but unlike the general population, behavioral treatment alone is associated with a very low average cessation rate of 4% (Tsoi et al., 2013). Abnormalities in brain nicotine receptors in those with schizophrenia spectrum disorders (Greenwood et al., 2011; Leonard et al., 2000; Stephens et al., 2012), that may confer a particular cognitive benefit from nicotine

(Barr et al., 2008; Jubelt et al., 2008; Sacco, Bannon, & George, 2004; Wing, Sacco, & George, 2011), and medications that substitute the actions of nicotine in the brain, such as bupropion (Culhane et al., 2008; Evins, Deckersbach, et al., 2005), NRT (Barr et al., 2008; Jubelt et al., 2008), and varenicline (Hong et al., 2011; Roh et al., 2014; Shim et al., 2012) may improve cognitive impairments associated with schizophrenia and maintain abstinence (Dolan et al., 2004; Roh et al., 2014; Sacco et al., 2005; Weinberger et al., 2007).

Clinical practice guidelines recommend treatment for all smokers with pharmacotherapeutic cessation aids combined with psychosocial or behavioral therapy (Fiore et al., 2008), and all of the studies reviewed here tested pharmacotherapy given in the context of a standard psychosocial treatment. These studies were not designed to directly assess whether a psychosocial intervention is required, nor do they provide information on the most effective psychosocial interventions to treat smoking cessation in those with SMI. Unfortunately, there are no studies that have compared use of pharmacotherapeutic cessation aids with and without a psychosocial intervention. The few studies that have examined the efficacy of psychosocial interventions for smoking cessation in those with SMI (Baker et al., 2006; Buchanan et al., 2010; George et al., 2000) support their benefit when combined with psychopharmacologic treatment but do not provide sufficient data to delineate the key components of the interventions.

Behavioral smoking cessation interventions for those with SMI have employed a range of treatment strategies including: educational (e.g., health risks of smoking, effectiveness of different treatment approaches, relationship between smoking and psychiatric symptoms), motivational enhancement (e.g., identifying personally relevant benefits of quitting, and costs of continuing to smoke), and cognitive-behavioral elements (e.g., identifying smoking triggers, developing coping strategies for these triggers, planning for high risk situations, relapse prevention). The optimal frequency, duration, and format (individual vs. group) of behavioral treatment and the active ingredients of these multicomponent interventions remain unclear.

There has been a call for more work on use of incentives to reduce smoking and improve other health behaviors in people with SMI (Tidey, 2012). Contingency reinforcement for reduced cotinine but not for group attendance was associated with reduction in urine cotinine and expired carbon monoxide (CO) in one study in smokers with schizophrenia (Tidey, Rohsenow, Kaplan, Swift, & Reid, 2011). Some evidence suggests NRT combined with monetary reinforcement for lower expired CO concentration is superior to placebo + contingency reinforcement for smoking reduction among smokers with schizophrenia while contingencies are in place (Tsoi et al., 2013).

5. TREATMENT ADHERENCE

Treatment adherence has been quite high (e.g., over 80% attendance at CBT group sessions) in treatment studies in samples of smokers with schizophrenia and bipolar disorder (Chengappa et al., 2014; Evins et al., 2007, 2014). The majority of those who discontinue prematurely have not attained significant periods of abstinence (Evins et al., 2007, 2014; Evins, Cather, et al., 2005).

6. SAFETY ISSUES

For smokers with schizophrenia, both bupropion and varenicline appear to be effective for smoking cessation and neither bupropion nor varenicline appears to increase risk of neuropsychiatric adverse effects over control treatments during a smoking cessation attempt, though the clinical trials performed to date were powered to detect efficacy and not safety with respect to rare adverse events (Evins et al., 2014; George et al., 2008; Tsoi et al., 2013; Weiner et al., 2011; Williams et al., 2012).

In addition to concerns regarding the safety of specific medications, providers and patients may be concerned about whether smoking cessation worsens psychiatric outcomes or stability. To date, no study has identified abstinence as negatively impacting positive, negative, or depressive symptoms in people with schizophrenia (Evins et al., 2007; Evins, Cather, et al., 2005; George et al., 2008, 2002; Williams et al., 2012). Although several studies reported no effect of pharmacotherapeutic cessation aids or tobacco abstinence on extrapyramidal symptoms (Evins et al., 2007; Evins, Cather, et al., 2005; George et al., 2002, 2000), one study in smokers with schizophrenia spectrum disorders reported increased muscle stiffness with bupropion plus NRT (George et al., 2008). Because tobacco smoking is associated with increased hepatic clearance of many psychotropic drugs, particularly those metabolized by cytochrome P450 1A2 and 2E1 (Schrenk, Brockmeier, Morike, Bock, & Eichelbaum, 1998; Zevin & Benowitz, 1999), it is recommended that clinicians monitor patients who reduce or quit smoking for evidence of reduced clearance of psychotropic medications metabolized by these enzymes and consider dose adjustment accordingly.

In clinical trials of pharmacotherapeutic cessation aids in people with psychiatric illness, incidence of adverse events is often high, but not different in those assigned to active pharmacotherapy versus placebo, as seen in trials of varenicline in those with schizophrenia (Williams et al., 2012) and depression (Anthenelli et al., 2013). Gibbons and Mann conducted a meta-analysis examining neuropsychiatric adverse events across 17 randomized clinical trials of varenicline with 1004 people with psychiatric disorders and 7023 people without psychiatric disorders (Gibbons & Mann, 2013). In these trials, the rate of suicidal thoughts and behavior, depression, or aggression/agitation was no higher for people assigned to varenicline than those assigned to placebo either during treatment or shortly after treatment discontinuation. A retrospective cohort study, commissioned by the FDA, conducted with Department of Defense clinical records, reported in the same paper, compared 30- and 60-day rates of neuropsychiatric events charted in approximately 20,000 people who used varenicline and 16,000 people who used the nicotine patch. Rates of neuropsychiatric disorders and neuropsychiatric adverse events were significantly lower overall in those prescribed varenicline than for those who had taken NRT (Gibbons & Mann, 2013). Importantly, this chart review was conducted in records prior to the first FDA safety warning about varenicline (August 1, 2006–August 31, 2007), reducing the risk for selection bias and stimulated reporting of neuropsychiatric adverse events. Additionally, the study used propensity score matching to minimize the bias related to the differential selection effects for the two treatments.

7. CONCLUSION

Low rates of guideline-concordant smoking cessation treatment to those with schizophrenia in the community contribute to the abysmally low rates of smoking cessation and the subsequent mortality disparity in this population (Druss, Bradford, Rosenheck, Radford, & Krumholz, 2001; Goff et al., 2005; Heffner, Strawn, DelBello, Strakowski, & Anthenelli, 2011; Himelhoch & Daumit, 2003; Schroeder & Morris, 2010; Ziedonis et al., 2008). Population-wide smoking cessation rates among adults with SMI are less than half those in the general population (Lasser et al., 2000). However, 20–48% abstinence rates in individuals with SMI are reported in clinical trials of standard pharmacotherapies and behavioral treatment

(Anthenelli et al., 2013; Chengappa et al., 2014; Evins et al., 2014; Tsoi et al., 2013; Williams et al., 2012). Achievement of the abstinence rates in those with SMI observed in clinical trials in the community through improved delivery of effective, evidence-based, smoking cessation treatments would result in the elimination of tobacco use among 1.4–3 million people. This achievement is possible with tools available to us today and would dramatically impact the 25-year reduction in lifespan observed in those with SMI (Thun et al., 2013), largely due to smoking-related cardiovascular disease (Parks et al., 2006).

REFERENCES

Addington, J., el-Guebaly, N., Addington, D., & Hodgins, D. (1997). Readiness to stop smoking in schizophrenia. *Canadian Journal of Psychiatry*, *42*, 49–52.

Agaku, I. T., King, B. A., & Dube, S. R. (2014). Current cigarette smoking among adults—United States, 2005–2012. *MMWR. Morbidity and Mortality Weekly Report*, *63*, 29–34.

Anthenelli, R. M., Morris, C., Ramey, T. S., Dubrava, S. J., Tsilkos, K., Russ, C., et al. (2013). Effects of varenicline on smoking cessation in adults with stably treated current or past major depression: A randomized trial. *Annals of Internal Medicine*, *159*, 390–400.

Baker, A., Richmond, R., Haile, M., Lewin, T. J., Carr, V. J., Taylor, R. L., et al. (2006). A randomized controlled trial of a smoking cessation intervention among people with a psychotic disorder. *The American Journal of Psychiatry*, *163*, 1934–1942.

Barr, R. S., Culhane, M. A., Jubelt, L. E., Mufti, R. S., Dyer, M. A., Weiss, A. P., et al. (2008). The effects of transdermal nicotine on cognition in nonsmokers with schizophrenia and nonpsychiatric controls. *Neuropsychopharmacology*, *33*, 480–490.

Brown, S., Inskip, H., & Barraclough, B. (2000). Causes of the excess mortality of schizophrenia. *The British Journal of Psychiatry*, *177*, 212–217.

Buchanan, R. W., Kreyenbuhl, J., Kelly, D. L., Noel, J. M., Boggs, D. L., Fischer, B. A., et al. (2010). The 2009 schizophrenia PORT psychopharmacological treatment recommendations and summary statements. *Schizophrenia Bulletin*, *36*, 71–93.

Cahill, K., Stevens, S., & Lancaster, T. (2014). Pharmacological treatments for smoking cessation. *JAMA*, *311*, 193–194.

Callaghan, R. C., Veldhuizen, S., Jeysingh, T., Orlan, C., Graham, C., Kakouris, G., et al. (2014). Patterns of tobacco-related mortality among individuals diagnosed with schizophrenia, bipolar disorder, or depression. *Journal of Psychiatric Research*, *48*, 102–110.

Cather, C., Dyer, M. A., Burrell, H. A., Hoeppner, B., Goff, D. C., & Evins, A. E. (2013). An open trial of relapse prevention therapy for smokers with schizophrenia. *Journal of Dual Diagnosis*, *9*, 87–93.

Cather, C., Friedman-Yakoobian, M., Gottlieb, J. D., Park, E., Goff, D., & Evins, A. E. (2013). *A randomized controlled trial of motivational interviewing compared to psychoeducation for smoking in pre-contemplators with severe mental illness*. Baltimore: Society for Research on Nicotine and Tobacco.

Chengappa, K. N. R., Perkins, K. A., Brar, J. S., Schlicht, P. J., Turkin, S. R., Hetrick, M. L., et al. (2014). Varenicline for smoking cessation in bipolar disorder: A randomized, double-blind, placebo-controlled study. *The Journal of Clinical Psychiatry*, *75*, 765–772.

Colton, C. W., & Manderscheid, R. W. (2006 Apr). Congruencies in increased mortality rates, years of potential life lost, and causes of death among public mental health clients in eight states. *Preventing Chronic Disease* [serial online] [date cited: 08/31/15]. Available from: http://www.cdc.gov/pcd/issues/2006/apr/05_0180.htm.

Cook, B., Wayne, G., Kafali, E., Liu, Z., Shu, C., & Flores, M. (2014). Trends in smoking among adults with mental illness and association between mental health treatment and smoking cessation. *JAMA*, *311*, 172–182.
Culhane, M. A., Schoenfeld, D. A., Barr, R. S., Cather, C., Deckersbach, T., Freudenreich, O., et al. (2008). Predictors of early abstinence in smokers with schizophrenia. *The Journal of Clinical Psychiatry*, *69*, 1743–1750.
Danaei, G., Ding, E. L., Mozaffarian, D., Taylor, B., Rehm, J., et al. (2009). The preventable causes of death in the United States: Comparative risk assessment of dietary, lifestyle, and metabolic risk factors. *PLoS Medicine*, *6*(4), e1000058. http://dx.doi.org/10.1371/journal.pmed.1000058.
de Leon, J., Diaz, F. J., Rogers, T., Browne, D., & Dinsmore, L. (2002). Initiation of daily smoking and nicotine dependence in schizophrenia and mood disorders. *Schizophrenia Research*, *56*, 47–54.
Dickerson, F., Stallings, C. R., Origoni, A. E., Vaughan, C., Khushalani, S., Schroeder, J., et al. (2013). Cigarette smoking among persons with schizophrenia or bipolar disorder in routine clinical settings, 1999–2011. *Psychiatric Services*, *64*, 44–50.
Dolan, S. L., Sacco, K. A., Termine, A., Seyal, A. A., Dudas, M. M., Vessicchio, J. C., et al. (2004). Neuropsychological deficits are associated with smoking cessation treatment failure in patients with schizophrenia. *Schizophrenia Research*, *70*, 263–275.
Druss, B. G., Bradford, W. D., Rosenheck, R. A., Radford, M. J., & Krumholz, H. M. (2001). Quality of medical care and excess mortality in older patients with mental disorders. *Archives of General Psychiatry*, *58*, 565–572.
Everett, A., Mahler, J., Biblin, J., Ganguli, R., & Mauer, B. (2008). Improving the health of mental health consumers: Effective policies and practices. *International Journal of Mental Health*, *37*, 8–48.
Evins, A. E. (2013). Reassessing the safety of varenicline. *American Journal of Psychiatry*, *170*, 1385–1387.
Evins, A. E., Cather, C., Culhane, M. A., Birnbaum, A., Horowitz, J., Hsieh, E., et al. (2007). A 12-week double-blind, placebo-controlled study of bupropion SR added to high-dose dual nicotine replacement therapy for smoking cessation or reduction in schizophrenia. *Journal of Clinical Psychopharmacology*, *27*, 380–386.
Evins, A. E., Cather, C., Deckersbach, T., Freudenreich, O., Culhane, M. A., Olm-Shipman, C. M., et al. (2005). A double-blind placebo-controlled trial of bupropion sustained-release for smoking cessation in schizophrenia. *Journal of Clinical Psychopharmacology*, *25*, 218–225.
Evins, A. E., Cather, C., & Laffer, A. (2015). Treatment of tobacco use disorders in smokers with serious mental illness: Toward clinical best practices. *Harvard Review of Psychiatry*, *23*, 90–98.
Evins, A. E., Cather, C., Pratt, S. A., Pachas, G. N., Hoeppner, S. S., Goff, D. C., et al. (2014). Maintenance treatment with varenicline for smoking cessation in patients with schizophrenia and bipolar disorder: A randomized clinical trial. *JAMA*, *311*, 145–154.
Evins, A. E., Deckersbach, T., Cather, C., Freudenreich, O., Culhane, M. A., Henderson, D. C., et al. (2005). Independent effects of tobacco abstinence and bupropion on cognitive function in schizophrenia. *The Journal of Clinical Psychiatry*, *66*, 1184–1190.
Evins, A. E., Mays, V. K., Rigotti, N. A., Tisdale, T., Cather, C., & Goff, D. C. (2001). A pilot trial of bupropion added to cognitive behavioral therapy for smoking cessation in schizophrenia. *Nicotine & Tobacco Research*, *3*, 397–403.
Fiore, M., Jaen, C., Baker, T., Bailey, W., Benowitz, N., & Curry, S. (2008). Treating tobacco use and dependence: 2008 update. In UDoHaH Services (Ed.), *Clinical practice guideline. Executive summary*. Rockville, Maryland: Public Health Service.

George, T. P., Vessicchio, J. C., Sacco, K. A., Weinberger, A. H., Dudas, M. M., Allen, T. M., et al. (2008). A placebo-controlled trial of bupropion combined with nicotine patch for smoking cessation in schizophrenia. *Biological Psychiatry*, *63*, 1092–1096.

George, T. P., Vessicchio, J. C., Termine, A., Bregartner, T. A., Feingold, A., Rounsaville, B. J., et al. (2002). A placebo controlled trial of bupropion for smoking cessation in schizophrenia. *Biological Psychiatry*, *52*, 53–61.

George, T. P., Ziedonis, D. M., Feingold, A., Pepper, W. T., Satterburg, C. A., Winkel, J., et al. (2000). Nicotine transdermal patch and atypical antipsychotic medications for smoking cessation in schizophrenia. *The American Journal of Psychiatry*, *157*, 1835–1842.

Gibbons, R. D., & Mann, J. J. (2013). Varenicline, smoking cessation, and neuropsychiatric adverse events. *American Journal of Psychiatry*, *170*, 1460–1467.

Goff, D. C., Cather, C., Evins, A. E., Henderson, D. C., Freudenreich, O., Copeland, P. M., et al. (2005). Medical morbidity and mortality in schizophrenia: Guidelines for psychiatrists. *Journal of Clinical Psychiatry*, *66*, 183–194. quiz 47, 273–4.

Greenwood, T. A., Lazzeroni, L. C., Murray, S. S., Cadenhead, K. S., Calkins, M. E., Dobie, D. J., et al. (2011). Analysis of 94 candidate genes and 12 endophenotypes for schizophrenia from the Consortium on the Genetics of Schizophrenia. *The American Journal of Psychiatry*, *168*, 930–946.

Hall, S. M., Humfleet, G. L., Reus, V. I., Munoz, R. F., & Cullen, J. (2004). Extended nortriptyline and psychological treatment for cigarette smoking. *The American Journal of Psychiatry*, *161*, 2100–2107.

Hartz, S., Pato, C., & Medeiros, H. (2014). Comorbidity of severe psychotic disorders with measures of substance use. *JAMA Psychiatry*, *71*, 248–254.

Heffner, J. L., Strawn, J. R., DelBello, M. P., Strakowski, S. M., & Anthenelli, R. M. (2011). The co-occurrence of cigarette smoking and bipolar disorder: Phenomenology and treatment considerations. *Bipolar Disorders*, *13*, 439–453.

Himelhoch, S., & Daumit, G. (2003). To whom do psychiatrists offer smoking-cessation counseling? *The American Journal of Psychiatry*, *160*, 2228–2230.

Hong, L. E., Thaker, G. K., McMahon, R. P., Summerfelt, A., Rachbeisel, J., Fuller, R. L., et al. (2011). Effects of moderate-dose treatment with varenicline on neurobiological and cognitive biomarkers in smokers and nonsmokers with schizophrenia or schizoaffective disorder. *Archives of General Psychiatry*, *68*, 1195–1206.

Horst, W. D., Klein, M. W., Williams, D., & Werder, S. F. (2005). Extended use of nicotine replacement therapy to maintain smoking cessation in persons with schizophrenia. *Neuropsychiatric Disease and Treatment*, *1*, 349–355.

Jha, P., Ramasundarahettige, C., Landsman, V., Rostron, B., Thun, M., Anderson, R. N., et al. (2013). 21st-century hazards of smoking and benefits of cessation in the United States. *The New England Journal of Medicine*, *368*, 341–350.

Jubelt, L. E., Barr, R. S., Goff, D. C., Logvinenko, T., Weiss, A. P., & Evins, A. E. (2008). Effects of transdermal nicotine on episodic memory in non-smokers with and without schizophrenia. *Psychopharmacology*, *199*, 89–98.

Lasser, K., Boyd, J., Woolhandler, S., Himmelstein, D., McCormick, D., & Bor, D. (2000). Smoking and mental illness: A population-based prevalence study. *JAMA*, *284*, 2606–2610.

Leonard, S., Breese, C., Adams, C., Benhammou, K., Gault, J., Stevens, K., et al. (2000). Smoking and schizophrenia: Abnormal nicotinic receptor expression. *European Journal of Pharmacology*, *393*, 237–242.

Lutterman, T., Ganju, V., Schacht, L., Shaw, R., Monihan, K., & Huddle, M. (2003). *Sixteen state study on mental health performance measures*. Rockville: Center for Mental Health Services, Substance Abuse and Mental Health Services Administration.

McClave, A. K., McKnight-Eily, L. R., Davis, S. P., & Dube, S. R. (2010). Smoking characteristics of adults with selected lifetime mental illnesses: Results from the 2007 national health interview survey. *American Journal of Public Health*, *100*, 2464–2472.

Mokdad, A. H., Marks, J. S., Stroup, D. F., & Gerberding, J. L. (2004). Actual causes of death in the United States, 2000. *JAMA*, *291*, 1238–1245.

Any Mental Illness Among Adults. (2013) (Accessed March 5, 2013, at http://www.nimh.nih.gov/statistics/1ANYDIS_ADULT.shtml).

Olincy, A., Young, D. A., & Freedman, R. (1997). Increased levels of the nicotine metabolite cotinine in schizophrenic smokers compared to other smokers. *Biological Psychiatry*, *42*, 1–5.

Pachas, G. N., Cather, C., Pratt, S. A., Hoeppner, B., Nino, J., Carlini, S. V., et al. (2012). Varenicline for smoking cessation in schizophrenia: Safety and effectiveness in a 12-week, open-label trial. *Journal of Dual Diagnosis*, *8*, 117–125.

Parks, J., Svendsen, D., Singer, P., & Foti, M. (2006). *Morbidity and mortality in people with serious mental illness*. Alexandria, VA: National Association of State Mental Health Program Directors (NASMHPD) Medical Directors Council.

Pirie, K., Peto, R., Reeves, G. K., Green, J., & Beral, V. (2013). The 21st century hazards of smoking and benefits of stopping: A prospective study of one million women in the UK. *Lancet*, *381*, 133–141.

Roh, S., Hoeppner, S. S., Schoenfeld, D., Fullerton, C. A., Stoeckel, L. E., & Evins, A. E. (2014). Acute effects of mecamylamine and varenicline on cognitive performance in non-smokers with and without schizophrenia. *Psychopharmacology*, *231*, 765–775.

Sacco, K. A., Bannon, K. L., & George, T. P. (2004). Nicotinic receptor mechanisms and cognition in normal states and neuropsychiatric disorders. *Journal of Psychopharmacology*, *18*, 457–474.

Sacco, K. A., Termine, A., Seyal, A., Dudas, M. M., Vessicchio, J. C., Krishnan-Sarin, S., et al. (2005). Effects of cigarette smoking on spatial working memory and attentional deficits in schizophrenia: Involvement of nicotinic receptor mechanisms. *Archives of General Psychiatry*, *62*, 649–659.

Substance Abuse and Mental Health Service Administration. (2014) (Accessed July 28, 2014, at http://www.samhsa.gov/).

Schrenk, D., Brockmeier, D., Morike, K., Bock, K. W., & Eichelbaum, M. (1998). A distribution study of CYP1A2 phenotypes among smokers and non-smokers in a cohort of healthy Caucasian volunteers. *European Journal of Clinical Pharmacology*, *53*, 361–367.

Schroeder, S. A., & Morris, C. D. (2010). Confronting a neglected epidemic: Tobacco cessation for persons with mental illnesses and substance abuse problems. *Annual Review of Public Health*, *31*, 297–314.

Shim, J. C., Jung, D. U., Jung, S. S., Seo, Y. S., Cho, D. M., Lee, J. H., et al. (2012). Adjunctive varenicline treatment with antipsychotic medications for cognitive impairments in people with schizophrenia: A randomized double-blind placebo-controlled trial. *Neuropsychopharmacology*, *37*, 660–668.

Siru, R., Hulse, G. K., & Tait, R. J. (2009). Assessing motivation to quit smoking in people with mental illness: A review. *Addiction*, *104*, 719–733.

Stephens, S. H., Franks, A., Berger, R., Palionyte, M., Fingerlin, T. E., Wagner, B., et al. (2012). Multiple genes in the 15q13-q14 chromosomal region are associated with schizophrenia. *Psychiatric Genetics*, *22*, 1–14.

Thorndike, A. N., Hoeppner, S. S., Cather, C. Pachas, G. N., Achtyes, E. D., & Evins, A. E. (in press). Weight, cardiovascular risk, and tobacco abstinence in smokers with serious mental illness: A subgroup analysis of a randomized trial. *Journal of Clinical Psychiatry*.

Thun, M. J., Carter, B. D., Feskanich, D., Freedman, N. D., Prentice, R., Lopez, A. D., et al. (2013). 50-year trends in smoking-related mortality in the United States. *The New England Journal of Medicine*, *368*, 351–364.

Tidey, J. W. (2012). Using incentives to reduce substance use and other health risk behaviors among people with serious mental illness. *Preventive Medicine*, *55*(Suppl.), S54–S60.

Tidey, J., & Miller, M. (in press). Smoking cessation and reduction in people with chronic mental illness. *BMJ*.

Tidey, J. W., & Rohsenow, D. J. (2009a). Intention to quit moderates the effect of bupropion on smoking urge. *Nicotine & Tobacco Research*, *11*, 308–312.

Tidey, J. W., & Rohsenow, D. J. (2009b). Smoking expectancies and intention to quit in smokers with schizophrenia, schizoaffective disorder and non-psychiatric controls. *Schizophrenia Research*, *115*, 310–316.

Tidey, J. W., Rohsenow, D. J., Kaplan, G. B., Swift, R. M., & Reid, N. (2011). Effects of contingency management and bupropion on cigarette smoking in smokers with schizophrenia. *Psychopharmacology (Berl)*, *217*, 279–287.

Tsoi, D. T., Porwal, M., & Webster, A. C. (2013). Interventions for smoking cessation and reduction in individuals with schizophrenia. *Cochrane Database of Systematic Reviews*, *2*.

USDHHS. (2004). *The health consequences of smoking: A report of the surgeon genera*. U.S. Department of Health and Human Services. Department of Health and Human Services, Centers for Disease Control and Prevention, National Center for Chronic Disease Prevention and Health Promotion, Office on Smoking and Health.

Weinberger, A. H., Sacco, K. A., Creeden, C. L., Vessicchio, J. C., Jatlow, P. I., & George, T. P. (2007). Effects of acute abstinence, reinstatement, and mecamylamine on biochemical and behavioral measures of cigarette smoking in schizophrenia. *Schizophrenia Research*, *91*, 217–225.

Weiner, E., Ball, M. P., Buchholz, A. S., Gold, J. M., Evins, A. E., McMahon, R. P., et al. (2012). Bupropion sustained release added to group support for smoking cessation in schizophrenia: A new randomized trial and a meta-analysis. *The Journal of Clinical Psychiatry*, *73*, 95–102.

Weiner, E., Buchholz, A., Coffay, A., Liu, F., McMahon, R. P., Buchanan, R. W., et al. (2011). Varenicline for smoking cessation in people with schizophrenia: A double blind randomized pilot study. *Schizophrenia Research*, *129*, 94–95.

West, R., McNeill, A., & Raw, M. (2000). Smoking cessation guidelines for health professionals: An update. Health education authority. *Thorax*, *55*, 987–999.

Williams, J. M., Anthenelli, R. M., Morris, C. D., Treadow, J., Thompson, J. R., Yunis, C., et al. (2012). A randomized, double-blind, placebo-controlled study evaluating the safety and efficacy of varenicline for smoking cessation in patients with schizophrenia or schizoaffective disorder. *The Journal of Clinical Psychiatry*, *73*, 654–660.

Williams, J. M., Ziedonis, D. M., Abanyie, F., Steinberg, M. L., Foulds, J., & Benowitz, N. L. (2005). Increased nicotine and cotinine levels in smokers with schizophrenia and schizoaffective disorder is not a metabolic effect. *Schizophrenia Research*, *79*, 323–335.

Wing, V. C., Sacco, K. A., & George, T. P. (2011). Spatial working memory impairments induced by cigarette smoking abstinence are correlated with plasma nicotine levels in schizophrenia. *Schizophrenia Research*, *128*, 171–172.

Wu, B. S., Weinberger, A. H., Mancuso, E., Wing, V. C., Haji-Khamneh, B., Levinson, A. J., et al. (2012). A preliminary feasibility study of varenicline for smoking cessation in bipolar disorder. *Journal of Dual Diagnosis*, *8*, 131–132.

Zevin, S., & Benowitz, N. L. (1999). Drug interactions with tobacco smoking. An update. *Clinical Pharmacokinetics*, *36*, 425–438.

Ziedonis, D., Hitsman, B., Beckham, J. C., Zvolensky, M., Adler, L. E., Audrain-McGovern, J., et al. (2008). Tobacco use and cessation in psychiatric disorders: National Institute of Mental Health report. *Nicotine & Tobacco Research*, *10*, 1691–1715.

PART 3

Mood disorders

CHAPTER SIX

Role of the Brain's Reward Circuitry in Depression: Transcriptional Mechanisms

Eric J. Nestler[1]
Fishberg Department of Neuroscience and Friedman Brain Institute, Icahn School of Medicine at Mount Sinai, New York, USA
[1]Corresponding author: e-mail address: eric.nestler@mssm.edu

Contents

Abstract

Increasing evidence supports an important role for the brain's reward circuitry in controlling mood under normal conditions and contributing importantly to the pathophysiology and symptomatology of a range of mood disorders, such as depression. Here we focus on the nucleus accumbens (NAc), a critical component of the brain's reward circuitry, in depression and other stress-related disorders. The prominence of anhedonia, reduced motivation, and decreased energy level in most individuals with depression supports the involvement of the NAc in these conditions. We concentrate on several transcription factors (CREB, ΔFosB, SRF, NFκB, and β-catenin), which are altered in the NAc in rodent depression models—and in some cases in the NAc of depressed humans, and which produce robust depression- or antidepressant-like effects when manipulated in the NAc in animal models. These studies of the NAc have established novel approaches toward modeling key symptoms of depression in animals and could enable the development of antidepressant medications with fundamentally new mechanisms of action.

International Review of Neurobiology, Volume 124
ISSN 0074-7742
http://dx.doi.org/10.1016/bs.irn.2015.07.003

1. INTRODUCTION

Depression and related mood disorders are among the world's greatest public health problems. While there are many effective treatments of depression, roughly half of affected individuals are inadequately treated by available medications and psychotherapeutic approaches (Trivedi & Daly, 2008). In addition, virtually all existing antidepressant medications, which act initially on the brain's serotonergic or noradrenergic systems, are based on serendipitous discoveries made more than a half-century ago (Berton & Nestler, 2006). Despite tremendous effort, the field has not yet succeeded in developing fundamentally new antidepressants with distinct mechanisms of action. One reason for this lack of progress is that we still lack a comprehensive understanding of the neural circuitry that malfunctions in depression as well as of the molecular pathology that drives abnormal circuit function. A stark reminder of this fact is that, if we had an opportunity to biopsy the brains of patients with depression, it is not at all clear which brain regions should be biopsied. The likelihood that depression comprises numerous, distinct disease states, and the lack of any clear distinction between depression and other stress-related disorders such as posttraumatic stress disorder and anxiety disorders, also have contributed to our lack of progress. In fact, this heterogeneity of clinical syndromes raises the possibility that different subtypes of depression and related conditions may be mediated by molecular pathology localized to different brain areas, which might be responsive to very different types of treatments.

Earlier work on depression focused on the hippocampus, frontal regions of cerebral cortex, and amygdala, among other regions. This impressive body of work is reviewed elsewhere (Duman & Monteggia, 2006; Miller & Hen, 2014; Nestler et al., 2002; Turner, Watson, & Akil, 2012). In more recent years, we and other groups have been interested in a role for the brain's reward regions in also contributing to depression and antidepressant treatment. Studies from the drug addiction field have identified the nucleus accumbens (NAc), part of the ventral striatum, and its dopaminergic inputs from the ventral tegmental area (VTA) of the midbrain, as the one of the most important anatomical substrates for drug reward as well as for natural rewards, such as food, sex, and social interactions (Koob & Le Moal, 2001; Wise, 1998). A striking observation related to brain reward is the extent to which abnormalities in reward and motivation are seen in depression and related disorders. For example, most depressed patients prominently exhibit a reduced ability to experience pleasure (anhedonia) and loss of motivation, as well as abnormalities in several neurovegetative functions such as appetite, sleep, energy level, and circadian rhythms (American Psychiatric Association, 2013).

Of course, these various brain areas cannot be thought of as distinct, since they function in highly overlapping and interacting circuits (Fig. 1). For example, the VTA and NAc receive strong glutamatergic inputs from several frontal cortical regions, hippocampus, and amygdala (see Everitt & Wolf, 2002; Hyman, Malenka, & Nestler, 2006; Kalivas, 2004). All of these regions, in turn, receive innervation from VTA dopamine neurons, where dopaminergic transmission has been shown to profoundly affect the functioning of these regions in electrophysiological and behavioral paradigms (e.g., see Arnsten, Wang, & Paspalas, 2012; Pezze & Feldon, 2004; Wittmann et al., 2005).

The objective of this review is to summarize the growing evidence for a role of the NAc in the regulation of mood and motivation under normal conditions, and in mediating many of the prominent behavioral

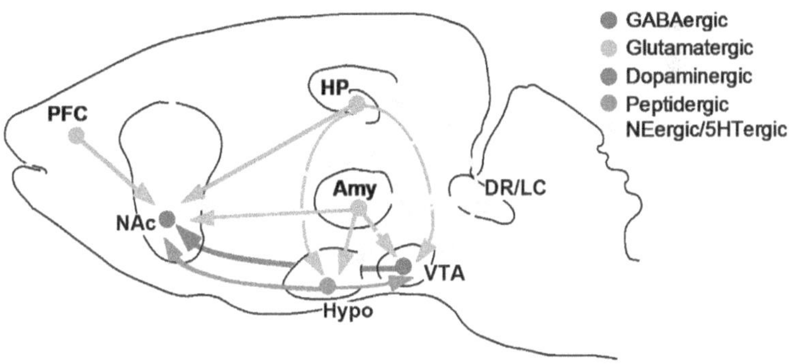

Figure 1 *The neural circuitry of mood.* The figure shows a highly simplified summary of a series of neural circuits in the brain that are believed to contribute to the regulation of mood. While most research in the depression field until recently has focused on hippocampus (HP) and cerebral cortex (e.g., prefrontal cortex or PFC), there is the increasing realization that several subcortical structures implicated in reward, fear, and motivation are also critically involved. These include the nucleus accumbens (NAc), amygdala (Amy), and hypothalamus (Hypo). The figure shows only a subset of the many known interconnections among these various brain regions. The figure also shows the innervation of several of these brain regions by monoaminergic neurons. The ventral tegmental area (VTA) provides dopaminergic input to the NAc; inputs to most of the other brain areas are not shown in the figure. Norepinephrine (NE, from the locus coeruleus or LC) and serotonin (5HT from the dorsal raphe and other raphe nuclei) innervate all of the regions shown in the figure. In addition, strong connections between the hypothalamus and VTA–NAc pathway have been established in recent years. *From Nestler and Carlezon (2006) with permission.*

abnormalities seen in depression and other mood disorders, with a particular focus on transcriptional mechanisms that operate in this brain region and have been implicated in depression and its treatment. We highlight how such studies of the NAc may provide novel targets for the development of new antidepressant treatments.

2. THE NAc REWARD CIRCUIT IN MOOD REGULATION

The involvement of the NAc in mood regulation and depression is supported by an increasing literature. The notion that midbrain dopamine systems contribute to depression-like behaviors was first proposed on the basis of studies with dopamine receptor antagonists (Wise, 1982). The ensuing several decades saw only sporadic publications reporting an association between the two (reviewed by Nestler & Carlezon, 2006). These early studies showed that stress, in the context of animal models of depression, potently activates VTA dopamine neurons and stimulates dopaminergic transmission to its limbic targets including the NAc. There were also reports that antidepressant treatments alter dopaminergic activity in the VTA or its targets, and that experimental manipulation of dopaminergic transmission in the VTA–NAc pathway can regulate depression-like behavior in acute stress assays in rodents.

The pace of progress has increased dramatically over the past decade. First, as will be seen below, the manipulation of a host of genes within the NAc was demonstrated to exert dramatic effects in more sophisticated animal models of depression. Second, in parallel to these preclinical studies, brain imaging investigations of humans documented abnormal functioning of the VTA–NAc in depression (Pizzagalli, 2014; Silbersweig, 2013; Tremblay et al., 2005). As just one example, depressed patients show reduced activation of the NAc as measured by functional magnetic resonance imaging in response to rewarding stimuli, direct evidence of a "reward deficit" in this syndrome. Third, advances in neural circuitry studies, enabled by track tracing tools and optogenetic approaches, have begun to define subsets of VTA and NAc neurons that play different roles in reward and stress responses. Distinct subsets of VTA dopamine neurons, for example, display very different responses to stressful stimuli: while some neurons are activated, others are suppressed. And there is increasing evidence that these subsets may be "wired" differently in the brain, with distinct afferent inputs and efferent projections (Chaudhury et al., 2013; Lammel, Lim, & Malenka, 2014; Tye et al., 2013). Similarly, the two major subtypes of medium spiny neurons (MSNs) in the NAc, those predominantly express D1 dopamine receptors versus those that express D2 receptors, which together comprise ~95% of all neurons in this region, differently control stress responsiveness (Francis et al., 2014; Lammel et al., 2014).

3. TRANSCRIPTIONAL AND EPIGENETIC MECHANISMS IN THE NAc IN MOOD REGULATION

There is now robust evidence for a role of several specific molecular pathways in the NAc, most of which were first implicated in regulating drug and natural reward in this circuit, in animal models of depression and antidepressant action. Prominent among such pathways are proteins important in the control of gene transcription and the regulation of chromatin. Representative examples are discussed here.

3.1 CREB-Mediated Transcription in the NAc in Mood Regulation

The transcription factor CREB (cAMP response element-binding protein) is stimulated in the NAc by exposure to several types of drugs of abuse or stress. This effect seems to be shared by both D1- and D2-type MSNs, and numerous studies have established that CREB activity in the NAc has a profound effect on an animal's responsiveness to emotional stimuli (Blendy, 2006; Carlezon, Duman, & Nestler, 2005; Conti & Blendy, 2004). CREB function in the NAc is normally regulated by glutamatergic and dopaminergic inputs (Dudman et al., 2003), suggesting that—by determining the set point of NAc neurons (Dong et al., 2006)—it represents an emotional gate for behavioral responsivity. This view is now supported by a large body of data. Elevations of CREB activity within the NAc, achieved via viral-mediated gene transfer or in inducible transgenic mice, reduce the rewarding effects of cocaine, morphine, and sucrose (Barrot et al., 2002; Carlezon et al., 1998; McClung & Nestler, 2003; Pliakas et al., 2001), which indicates that a sustained elevation of CREB activity in the NAc produces anhedonia-like signs. In fact, this CREB phenotype appears to reflect a generalized numbing of behavioral responses to emotional stimuli, since animals with increased CREB function in the NAc also show reduced responses to a wide range of aversive conditions (Barrot et al., 2002). Elevations of CREB activity produce pro-depression-like symptoms in several acute and subchronic stress models (Newton et al., 2002; Pliakas et al., 2001).

Conversely, reductions in CREB activity in the rat NAc, through viral- or transgenic-mediated expression of a dominant negative CREB mutant (mCREB) or through a local knockout of CREB from the NAc of floxed CREB mice, increases the rewarding effects of cocaine, morphine, and sucrose (Barrot et al., 2002; Carlezon et al., 1998; DiNieri et al., 2009) and produces antidepressant-like effects in several stress assays (Conti, Cryan, Dalvi, Lucki, & Blendy, 2002; Newton et al., 2002; Pliakas et al., 2001), including chronic social defeat stress (Covington et al., 2011)—an ethologically validated model of depression (Berton et al., 2006; Krishnan et al., 2007).

CREB function in the NAc also has a profound effect on anxiety-like behavior in rodents. Disruption of CREB function within the NAc, achieved by viral overexpression of mCREB, produces anxiety-like effects, whereas increased CREB function causes opposite changes (Barrot et al., 2002, 2005; Wallace et al., 2009). On the other hand, global knockdown of CREB reduces anxiety-like behavior (Valverde et al., 2004), perhaps owing to CREB actions outside the NAc. The notion that elevated CREB function in NAc causes certain depression-like symptoms, while reduced CREB function in this region causes anxiety-like behavior, may seem paradoxical, but can be understood within the hypothesized role CREB plays in this reward circuit under normal conditions. Our hypothesis is that CREB in the NAc is a key regulator of the reactivity of brain reward circuits and thereby regulates individual sensitivity to emotional stimuli (Carlezon et al., 2005). Short-term increases in CREB activity in NAc, induced by normal rewarding or aversive stimuli, dampen responses to subsequent stimuli and facilitate the ability to actively deal with the situation at hand (e.g., consumption of reward and escape from danger). Under more pathological conditions, however, larger and more sustained increases in CREB activity, induced by drugs of abuse or excessive stress, would lead to an excessive dampening of emotional reactivity and to the behavioral phenotype outlined above. Conversely, sustained reductions in CREB activity, which are seen under conditions of prolonged social isolation (Barrot et al., 2005; Wallace et al., 2009), heighten emotional reactivity, and in the extreme induce a state of anxiety. This work highlights the notion that extreme increases or decreases in CREB function in NAc may be detrimental and contribute to the symptomology of different mood disorders (Carlezon et al., 2005).

This role for CREB in the NAc in depression models is in stark contrast to CREB's activity in the hippocampus and other regions in many of the same behavioral models. In hippocampus, for example, CREB is an important mediator of antidepressant effects (Duman & Duman, 2005). These findings underscore the importance of identifying the region-specific target genes through which CREB exerts these various effects, as such targets could be mined in antidepressant drug discovery efforts.

3.2 ΔFosB and SRF in the NAc Pathway in Mood Regulation

A large literature has established that ΔFosB, a Fos family transcription factor, is induced in the NAc uniquely in response to chronic exposure to virtually any drug of abuse (Nestler, 2008). Moreover, once induced, the accumulated ΔFosB persists in the NAc for several weeks during a period of drug withdrawal due to the unusual stability of the protein. Such drug induction of ΔFosB in NAc is selective for D1-type MSNs with the one exception of opiate drugs of abuse which induce the protein equally in D1- and D2-type neurons (Lobo et al., 2013). A wealth of evidence now supports the notion that induction of ΔFosB in D1 MSNs promotes natural reward and enhances behavioral responses, including self-administration, of drugs of abuse, while inhibition of ΔFosB function in this cell type exerts the opposite effects (Nestler, 2008).

A role for ΔFosB in stress responses was first implicated in 2004, when chronic exposure to restraint stress was found to induce ΔFosB in the NAc (Perrotti et al., 2004). However, it was not until 2010 when information became available about the functional significance of this phenomenon. Chronic social defeat stress was found to induce ΔFosB predominantly in mice resistant to the deleterious effects of the stress, so-called resilient mice (Vialou, Robison, et al., 2010). Such induction is specific to D1-type MSNs (Lobo et al., 2013), consistent with our knowledge that ΔFosB in this cell type promotes reward and motivation. By contrast, a lower level of induction of ΔFosB in the NAc of susceptible mice (those that succumb to the deleterious effects of chronic stress) is seen in D2-type MSNs only (Lobo et al., 2013). While the functional consequences of ΔFosB induction in D2 MSNs remain poorly understood, these data suggest it might reduce hedonic responses.

Moreover, selective overexpression of ΔFosB in D1 MSNs of NAc was shown to promote resilience in the social defeat paradigm and exert antidepressant-like responses in mice susceptible to the stress (Donahue, Muschamp, Russo, Nestler, & Carlezon, 2014; Muschamp, Nemeth, Robison, Nestler, & Carlezon, 2012; Vialou, Robison, et al., 2010). In contrast, antagonism of ΔFosB activity, achieved via overexpression of a dominant negative mutant, made mice more susceptible. Interestingly, another type of chronic stress, prolonged social isolation in adulthood, decreases ΔFosB levels in the NAc and overexpression of ΔFosB in this region of isolated mice reverses the depression-like behavioral abnormalities exhibited by the animals (Vialou, Robison, et al., 2010). Depressed humans, like isolated mice, display lower levels of ΔFosB in the NAc, lending important clinical validation to the importance of ΔFosB in regulating depression-related behavioral abnormalities.

These behavioral effects of ΔFosB in the NAc, like those of CREB, show regional specificity. In the medial prefrontal cortex (mPFC), ΔFosB is induced selectively in susceptible mice after chronic social defeat stress, and in this region, ΔFosB overexpression promotes susceptibility while blocking ΔFosB function exerts the opposite effect (Vialou et al., 2014). Interestingly, we found that ΔFosB's prosusceptible effect in the mPFC is mediated in part by controlling the output of this region to the NAc. These region-specific effects of ΔFosB once again point to the importance of identifying target genes of this transcription factor that predominate in one region or another.

3.3 Other Transcription Factors in the NAc in Mood Regulation

Several other transcription factors in the NAc have been implicated in depression. Studies of ΔFosB have led to the discovery that another transcription factor, serum response factor (SRF), also exerts proresilience effects through actions in the NAc. ΔFosB induction in the NAc of resilient mice, after a period of chronic social defeat stress, is mediated via SRF, which binds to the *FosB* promoter and increases transcription of the gene (Vialou, Maze, et al., 2010). Interestingly, CREB, which plays a role in mediating ΔFosB induction in response to chronic cocaine, has no role in stress induction of ΔFosB (Vialou et al., 2012). Further research is needed to understand why CREB—which is induced in D1 NAc MSNs in response to chronic stress—does not control FosB transcription in the context of stress, while it does so in the context of cocaine exposure. Further research is also needed to identify additional targets of SRF, which like ΔFosB might contribute to behavioral resilience. The importance of this effort is underscored by the observation that SRF levels are reduced in the NAc of depressed humans (Vialou, Maze, et al., 2010).

Another transcription factor implicated in depression is β-catenin, which we demonstrated recently displays reduced activity in the NAc of susceptible mice after chronic social defeat stress, with resilient mice exhibiting a robust induction of β-catenin activity (Fig. 2) (Dias et al., 2014; Wilkinson et al., 2011). This regulation predominates in D2 MSNs. Importantly, depressed humans, like stressed mice, show reduced levels of β-catenin activity in the NAc. Viral-mediated expression of a dominant negative β-catenin mutant, or local knockdown of β-catenin from the NAc of floxed β-catenin mice, induces heightened susceptibility to defeat stress, whereas overexpression of β-catenin—in D2 MSNs selectively—exerts a proresilient effect (Dias et al., 2014). As will be discussed below, genome-wide analysis of β-catenin targets in the NAc has revealed numerous target genes through which the transcription factor mediates this interesting behavioral phenotype.

NFκB (nuclear factor κB) is a transcription factor that is best studied in the immune system, although it has been shown to regulate neural function in more recent years. Interest in NFκB function in the NAc was sparked by the finding that it is a target gene for ΔFosB and is induced in this brain region by chronic cocaine administration (Ang et al., 2001). Years later,

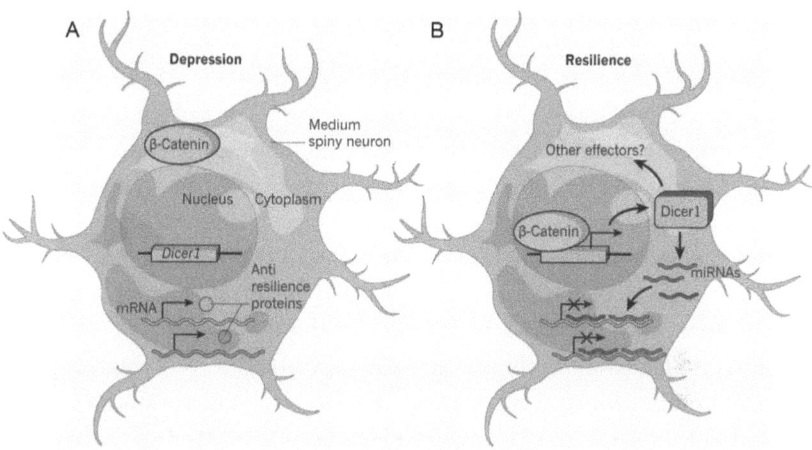

Figure 2 *β-Catenin mediates stress resilience through Dicer1/miRNA regulation.* (A) D2-type MSNs are less activated in chronically stressed mice. As a consequence, β-catenin protein remains in the cytoplasm in these cells, unable to enter the nucleus, and the *Dicer1* gene is thus inactive. Antiresilience (or prosusceptible) proteins may therefore be produced from mRNAs that would otherwise have been inhibited by miRNAs generated by the DICER1. (B) In resilient mice, β-catenin enters the nucleus of activated D2 MSNs, thereby turning on *Dicer1* transcription. Elevated levels of DICER1 increase production of miRNAs and possibly other effectors of resilience. This might, in turn, inhibit the production of antiresilience proteins, because of binding and inhibition of mRNA by miRNAs. *From Schratt (2014) with permission.*

it was found that viral-mediated expression of NFκB in the NAc promotes behavioral responses to cocaine, whereas expression of a dominant negative mutant exerts the opposite effect (Russo et al., 2009). It was surprising then to find that NFκB induction in the NAc occurs selectively in susceptible mice after a course of chronic social defeat stress and that NFκB expression in this region promotes susceptibility while inhibition of NFκB activity promotes resilience (Christoffel et al., 2011). On the other hand, NFκB activity in the NAc appears to exert opposite functional effects in the NAc of female mice (LaPlant et al., 2009), something not examined by Christoffel et al. (2011). The cell type in NAc where NFκB exerts these various addiction- and depression-related phenotypes remains unknown and is an active topic of current research.

3.4 Epigenetic Mechanisms in the NAc in Mood Regulation

Advances in chromatin biology have made it possible for the first time to investigate the epigenetic changes that occur in concert with transcriptional regulation in the context of depression models (Peña, Bagot, Labonté, & Nestler, 2014; Sun, Kennedy, & Nestler, 2013; Vialou, Feng, Robison, & Nestler, 2013). Thus, a large literature has demonstrated the important role of histone acetylation and methylation and DNA methylation in controlling gene expression, and each of these chromatin regulatory mechanisms has been implicated in depression and its treatment (Fig. 3). Most studied to date are HDACs (histone deacetylases), which generally inhibit gene expression. Sustained inhibition of HDACs selectively in the NAc exerts a potent antidepressant-like effect in several acute and chronic stress assays, which is stronger and more consistent than that elicited by standard antidepressants such as fluoxetine (Covington et al., 2009). Genome-wide analysis of gene expression changes induced by such HDAC inhibition identified a set of genes, only a portion of which were similarly regulated by chronic fluoxetine administration, which reveal paths for the generation of novel antidepressants. The HDAC subtype mediating this effect is not known, since knockout of HDAC5, albeit from all brain regions, makes mice more susceptible to social defeat (Renthal et al., 2007). Less is known about how chronic stress influences HDACs in the NAc. Expression of HDAC2 and HDAC5 is reduced in the NAc of mice subjected to chronic social defeat stress, with similar effects seen in depressed humans (Covington et al., 2009; Renthal et al., 2007). In contrast, total HDAC activity is reportedly increased in the NAc of mice subjected to

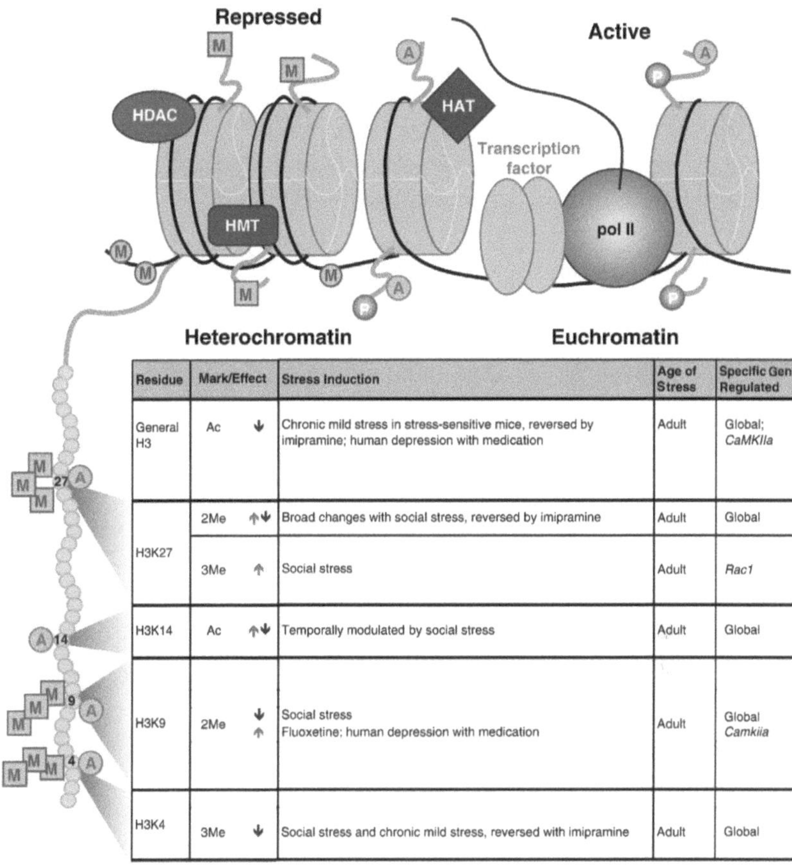

Residue	Mark/Effect		Stress Induction	Age of Stress	Specific Genes Regulated
General H3	Ac	↓	Chronic mild stress in stress-sensitive mice, reversed by imipramine; human depression with medication	Adult	Global; *CaMKIIa*
H3K27	2Me	↑↓	Broad changes with social stress, reversed by imipramine	Adult	Global
	3Me	↑	Social stress	Adult	*Rac1*
H3K14	Ac	↑↓	Temporally modulated by social stress	Adult	Global
H3K9	2Me	↓ ↑	Social stress Fluoxetine; human depression with medication	Adult	Global *Camkiia*
H3K4	3Me	↓	Social stress and chronic mild stress, reversed with imipramine	Adult	Global

Figure 3 *Examples of chromatin modifications regulated in the NAc by stress or antidepressant treatment.* Illustration (top) indicates histone octamers (pink (light gray in the print version)) in heterochromatin (left) and euchromatin (right), along with associated proteins and histone tail/DNA modifications. Table (bottom) lists histone tail modifications of specific residues—depicted on the expanded histone tail illustration (left)—that are regulated by various stress paradigms or antidepressant treatments within the NAc. Arrows indicate an increase (green (light gray in the print version)) or decrease (blue (dark gray in the print version)) in specific modifications. Abbreviations: A, acetylation; P, phosphorylation; M (in a square), histone methylation; M (in a circle), DNA methylation; HAT, histone acetyltransferase; HDAC, histone deacetylase; HMT, histone methyltransferase; pol II, RNA polymerase II. See Peña et al. (2014) for references. *Modified from Peña et al. (2014) with permission.*

early life stress, effects reversed by antidepressant administration, although levels of expression of various HDAC isoforms was not examined in this study (Réus et al., 2013). Future work is needed to resolve these discrepant findings.

Histone methylation in the NAc is also implicated in depression. Chronic social defeat stress decreases expression of G9a and GLP (G9a-like protein), two histone methyltransferases that catalyze the dimethylation of Lys9 of histone H3 (H3K9me2) (Covington et al., 2011), a mark associated with gene repression. This effect contributes to depression-related behavioral abnormalities, since local knockdown of G9a from the NAc of floxed G9a mice increases susceptibility to social defeat stress, while viral-mediated G9a overexpression in this region promotes resilience. An earlier genome-wide study of changes in H3K9me2 in NAc of susceptible and resilient mice identified numerous genes that might mediate these behavioral effects (Wilkinson et al., 2009). Several additional histone methyltransferases are altered in the NAc upon exposure to chronic stress; however, their influence on stress vulnerability has not yet been investigated (see Peña et al., 2014).

Several forms of chronic stress, including chronic social defeat stress and early life stress, have been shown to increase the expression of DNMTs (DNA methyltransferases) in the NAc (Anier et al., 2014; LaPlant et al., 2010). Importantly, overexpression of one of the regulated DNMTs, DNMT3a, in the NAc makes mice more susceptible to social defeat stress, while intra-NAc administration of a DNMT inhibitor (RG108) exerts antidepressant-like effects (LaPlant et al., 2010).

A recent study has demonstrated that another type of epigenetic mechanism—namely, microRNAs (miRNAs), also act in the NAc to control depression-related behavior. miRNAs are small (~20 nt) noncoding RNAs which exert powerful control over the translation of mRNAs into proteins. Our interest in miRNAs was sparked by the finding, alluded to above, that genome-wide characterization of β-catenin target genes in the NAc identified *Dicer1*—the protein product of which catalyzes a crucial step in the biogenesis of miRNAs—as a prominent target that is induced in this brain region of resilient mice (Dias et al., 2014). Indeed, viral-mediated DICER1 expression in NAc makes mice more resilient. This led to a genome-wide study of all miRNAs that are regulated in the NAc of susceptible versus resilient mice and whose regulation is mediated by β-catenin. The identified miRNAs now become interesting targets for future antidepressant drug discovery research (Dias et al., 2014).

While the unusually potent pro- or antidepressant effects of these various epigenetic mechanisms, acting in the NAc, have stoked interest in mining these discoveries in the development of new treatments, a major caveat is that all of these mechanisms are ubiquitous. Consequently, it is likely that direct manipulations of HDACs, histone methyltransferases, or DNMTs would produce side effects that render them unsuitable as antidepressants. This once again emphasizes the importance of identifying the target genes that mediated these actions. Still, it is possible that the particularly potent behavioral actions of these mechanisms are precisely because they alter numerous genes at once. This points to the potential unique utility of miRNAs, the manipulation of which would regulate multiple mRNAs but a relatively small number, and thus might be amenable to drug discovery efforts.

3.5 Identification and Characterization of Target Genes for Specific Transcription Factors in the NAc

The discussion above highlights the importance of going beyond transcriptional and epigenetic mechanisms *per se* in the quest to develop improved antidepressant treatments based on our increasing appreciation of the role of the NAc in depression. In general, two approaches have been used in the field: a candidate gene approach whereby investigators have focused on a small number of genes based on *a priori* hypotheses of their role in depression models and unbiased genome-wide efforts to search for novel target genes.

Candidate gene approaches have been fruitful. Based on the knowledge that the opioid peptide dynorphin—which activates κ-opioid receptors—is a CREB target gene in other systems, we and others have established that dynorphin is one CREB target that is induced in the NAc by chronic stress and contributes to depression-related behavioral abnormalities (Fig. 4) (Bruchas, Land, & Chavkin, 2010; Van't Veer & Carlezon, 2013). This has led to interest in κ-opioid antagonists for the treatment of depression. In contrast, there is evidence that ΔFosB suppresses dynorphin expression in the NAc (Zachariou et al., 2006), which might contribute to its proresilience effects. BDNF (brain-derived neurotrophic factor) and its signaling pathways are another example. BDNF, like dynorphin, is a CREB target and local synthesis of BDNF likely contributes to stress susceptibility (Berton et al., 2006; Krishnan et al., 2007). Moreover, genes encoding proteins that mediate the actions of BDNF show reduced binding of H3K9me2, leading to induction of those genes, enhanced BDNF signaling, and prosusceptible effects (Covington et al., 2011). Finally, the GluA2 AMPA glutamate receptor subunit is induced in the NAc of resilient mice only, an action mediated by ΔFosB, and direct manipulation of this subunit within the NAc confirms that its upregulation promotes behavioral resilience (Vialou, Robison, et al., 2010).

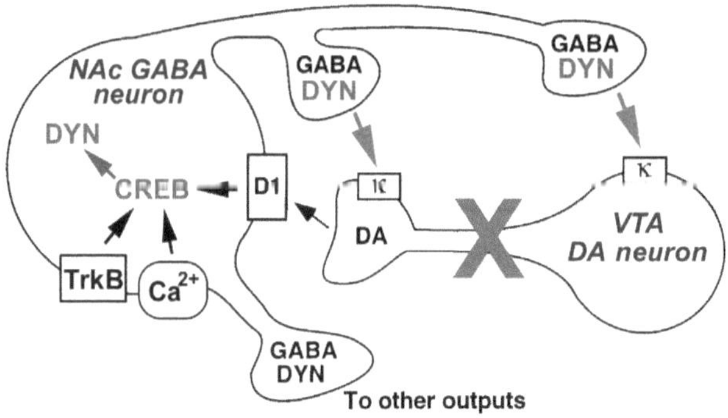

Figure 4 *CREB and dynorphin in the NAc in depression.* The figure shows a simplified hypothetical scheme by which CREB induction of dynorphin (DYN) in the NAc contributes to certain symptoms of depression. CREB is activated by D1 dopamine receptors (through activation of the cAMP pathway) or by Ca^{2+}- or TrkB-regulated signal transduction pathways, which leads to increased expression of DYN. DYN feeds back on κ-opioid receptors located on the terminals and cell bodies/dendrites of VTA dopamine (DA) neurons. Stimulation of these κ receptors inhibits the VTA neurons, which may contribute to anhedonia and related symptoms of depression. Antagonists of κ receptors may thus block the consequences of CREB-induced increases in DYN activity, and exert antidepressant activity in some individuals. *From Nestler and Carlezon (2006) with permission.*

The limitation of candidate gene approaches is that we still know relatively little about depression and hence are limited in our choice of specific genes to investigate. The discovery of DICER1 and miRNAs as downstream mediators of the proresilience effects of β-catenin is a good illustration of the power of unbiased, open-ended approaches. ChIP-seq (chromatin immunoprecipitation followed by deep sequencing) is the method used to identify targets of a transcription factor or chromatin modification genome wide and was used to characterize β-catenin targets (Dias et al., 2014). An earlier variant of ChIP-seq, termed ChIP-chip (ChIP followed by analysis on promoter chips) was used to study CREB and H3K9me2 targets in the NAc in the chronic social defeat stress and social isolation paradigms (Wilkinson et al., 2009). However, ChIP-chip is far less quantitative than ChIP-seq and covers only promoter regions of genes, in contrast to ChIP-seq which provides a truly genome-wide measure. A major need of future research, therefore, is to carry out ChIP-seq for the increasing number of transcription factors and chromatin regulatory mechanisms that are being shown to control depression-related behavior via actions in the NAc. As such data are generated, advanced bioinformatic tools will be used to identify the molecular pathways that are most highly regulated in a state of susceptibility versus resilience, or in depressed human

NAc, findings which can then be mined for fundamentally novel approaches to antidepressant drug discovery efforts (Maze et al., 2014).

4. FUTURE DIRECTIONS

A major need of future research is to better define how the numerous and diverse molecular pathways discussed above influence the several cell types in the NAc and how that altered function controls mood and depression-related behavioral abnormalities. Early steps of this work have been achieved for CREB (Dong et al., 2006; Huang et al., 2008) and ΔFosB (Grueter, Robison, Neve, Nestler, & Malenka, 2013): CREB controls the intrinsic excitability and NMDA responses of both D1 and D2 MSNs, while ΔFosB exerts selective effects on AMPA glutamate responses in D1 MSNs only. Likewise, optogenetic tools have been used to demonstrate opposite effects of D1 versus D2 MSN activation in controlling stress responses (Francis et al., 2014). A goal of current research is to define the various outputs of D1 and D2 NAc MSNs in mediating these effects.

A surprising finding to date is the complex relationship between the brain's reward circuitry and mood disorders. One might have assumed that "more reward" equals "less depression": for example, that a protein acting in a certain brain structure that increase cocaine's rewarding effects would exert an antidepressant-like effect based on the assumption that the protein also boosts natural reward (Russo & Nestler, 2013). To put it another way: one might expect that the influence of a protein in depression models is predictable based on its effects in addiction models and vice versa. However, this is clearly not the case: there is no predictable relationship between the effects of a given protein in the NAc in depression models versus drug addiction models (Table 1); this is despite the considerable comorbidity between depression and addiction syndromes. One complicating factor is that addiction likely involves adaptations that impair brain reward and others that promote reward-related memories (Russo & Nestler, 2013). The heterogeneity of cell types within the NAc is another likely source of different actions observed for a protein in stress versus drug models. However, cell type is not the only explanation. For instance, CREB appears to have very different target genes in the NAc in depression versus addiction models, even within the same cell types (Covington et al., 2011; McClung & Nestler, 2003; Wallace et al., 2009; Wilkinson et al., 2009). This suggests that the stimulus itself (i.e., stress or drugs) can engage different intracellular pathways, thereby regulating chromatin structure and gene expression in

Table 1 Effect of Molecular Mediators in the NAc on Depression- Versus Addiction-Like Behavior

Protein	Effect in Depression Models	Effect in Addiction Models
CREB	↑ Susceptibility	↓ Drug reward
ΔFosB	↓ Susceptibility	↑ Drug reward
SRF	↓ Susceptibility	0 Drug reward
NFκB	↑ Susceptibility	↑ Drug reward
β-Catenin	↓ Susceptibility	N/A
HDAC inhibition	↓ Susceptibility	↑ Drug reward
G9a inhibition	↑ Susceptibility	↑ Drug reward
DNMT inhibition	↓ Susceptibility	↑ Drug reward
BDNF-TrkB	↑ Susceptibility	↑ Cocaine reward ↓ Morphine reward
GluA2	↓ Susceptibility	↑ Drug reward
Dynorphin	↑ Susceptibility	↓ Drug reward

The table illustrates examples from our laboratory which indicate that there is not a clear relationship between an effect of a protein in the NAc on stress susceptibility and on drug reward. One complicating factor may be differential effects of a protein in D1-type versus D2-type NAc MSNs, which have only recently begun to be parsed. Another possible explanation is complex combinatorial effects unique to a given condition. For example, we know that a given transcription factor (e.g., CREB) has very different target genes in NAc under stress or cocaine conditions, even after different types of stress (Wilkinson et al., 2009), presumably due to a host of other factors induced uniquely in each situation.
Adapted from Russo and Nestler (2013).

their own unique ways. Future studies using cell-selective molecular profiling and viral gene transfer approaches will be crucial for shedding light on these complex stimulus-specific effects on reward-related behavior and depression.

ACKNOWLEDGMENT

This work was supported by grants from the National Institute of Mental Health (P50MH096890 and R01MH051399).

REFERENCES

American Psychiatric Association. (2013). *Diagnostic and statistical manual of mental disorders* (5th ed.). Washington, DC: American Psychiatric Press.

Ang, E., Chen, J. S., Zagouras, P., Magna, H., Holland, J., Schaeffer, E., et al. (2001). Induction of NFκB in nucleus accumbens by chronic cocaine administration. *Journal of Neurochemistry*, *79*, 221–224.

Anier, K., Malinovskaja, K., Pruus, K., Aonurm-Helm, A., Zharkovsky, A., & Kalda, A. (2014). Maternal separation is associated with DNA methylation and behavioural changes in adult rats. *European Neuropsychopharmacology, 24*, 459–468.

Arnsten, A. F., Wang, M. J., & Paspalas, C. D. (2012). Neuromodulation of thought: Flexibilities and vulnerabilities in prefrontal cortical network synapses. *Neuron, 76*, 223–239.

Barrot, M., Olivier, J. D. A., Perrotti, L. I., Impey, S., Storm, D. R., Neve, R. L., et al. (2002). CREB activity in the nucleus accumbens shell controls gating of behavioral responses to emotional stimuli. *Proceedings of the National Academy of Sciences of the United States of America, 99*, 11435–11440.

Barrot, M., Wallace-Black, D., Bolanos, C. A., Graham, D., Perrotti, L. I., Neve, R. L., et al. (2005). Regulation of anxiety and initiation of sexual anxiety by CREB in the nucleus accumbens. *Proceedings of the National Academy of Sciences of the United States of America, 102*, 8357–8362.

Berton, O., McClung, C. A., DiLeone, R. J., Krishnan, V., Russo, S., Graham, D., et al. (2006). Essential role of BDNF in the mesolimbic dopamine pathway in social defeat stress. *Science, 311*, 864–868.

Berton, O., & Nestler, E. J. (2006). New approaches to antidepressant drug discovery: Beyond monoamines. *Nature Reviews. Neuroscience, 7*, 137–151.

Blendy, J. A. (2006). The role of CREB in depression and antidepressant treatment. *Biological Psychiatry, 59*, 1144–1150.

Bruchas, M. R., Land, B. B., & Chavkin, C. (2010). The dynorphin/kappa opioid system as a modulator of stress-induced and pro-addictive behaviors. *Brain Research, 1314*, 44–55.

Carlezon, W. A., Jr., Duman, R. S., & Nestler, E. J. (2005). The many faces of CREB. *Trends in Neurosciences, 28*, 436–445.

Carlezon, W. A., Jr., Thome, J., Olson, V. G., Lane-Ladd, S. B., Brodkin, E. S., Hiroi, N., et al. (1998). Regulation of cocaine reward by CREB. *Science, 282*, 2272–2275.

Chaudhury, D., Walsh, J. J., Friedman, A. K., Juarez, B., Ku, S. M., Koo, J. W., et al. (2013). Rapid regulation of depression-related behaviours by control of midbrain dopamine neurons. *Nature, 493*, 532–536.

Christoffel, D. J., Golden, S. A., Dumitriu, D., Robison, A. J., Janssen, W. G., Ahn, H. F., et al. (2011). IκB kinase regulates social defeat stress-induced synaptic and behavioral plasticity. *Journal of Neuroscience, 31*, 314–321.

Conti, A. C., & Blendy, J. A. (2004). Regulation of antidepressant activity by cAMP response element binding proteins. *Molecular Neurobiology, 30*, 143–155.

Conti, A. C., Cryan, J. F., Dalvi, A., Lucki, L., & Blendy, J. A. (2002). cAMP response element-binding protein is essential for the upregulation of brain-derived neurotrophic factor transcription, but not the behavioral or endocrine responses to antidepressant drugs. *Journal of Neuroscience, 22*, 3262–3268.

Covington, H. E., III, Maze, I., LaPlant, Q. C., Vialou, V. F., Yoshinori, O. N., Berton, O., et al. (2009). Antidepressant actions of HDAC inhibitors. *Journal of Neuroscience, 29*, 11451–11460.

Covington, H. E., III, Maze, I., Sun, H. S., Wu, E. Y., Dietz, D., Lobo, M. K., et al. (2011). A role for repressive histone methylation in cocaine-induced vulnerability to stress. *Neuron, 71*, 656–670.

Dias, C., Feng, J., Sun, H. S., Shao, N. Y., Mazei-Robison, M. S., Damez-Werno, D., et al. (2014). β-catenin mediates stress resilience through Dicer1/microRNA regulation. *Nature, 516*, 51–55.

DiNieri, J. A., Nemeth, C., Parsegian, A., Carle, T., Gurevich, V. V., Gurevich, E., et al. (2009). Altered sensitivity to rewarding and aversive drugs in mice with inducible disruption of cAMP response element-binding protein function within the nucleus accumbens. *Journal of Neuroscience, 29*, 1855–1859.

Donahue, R. J., Muschamp, J. W., Russo, S. J., Nestler, E. J., & Carlezon, W. A., Jr. (2014). Effects of striatal ΔFosB overexpression and ketamine on social defeat stress-induced anhedonia in mice. *Biological Psychiatry*, *76*, 550–558.

Dong, Y., Green, T., Saal, D., Marie, H., Neve, R., Nestler, E. J., et al. (2006). CREB modulates excitability of nucleus accumbens neurons. *Nature Neuroscience*, *9*, 475–477.

Dudman, J. T., Eaton, M. E., Rajadhyaksha, A., Macías, W., Taher, M., Barczak, A., et al. (2003). Dopamine D1 receptors mediate CREB phosphorylation via phosphorylation of the NMDA receptor at Ser897-NR1. *Journal of Neurochemistry*, *87*, 922–934.

Duman, C. H., & Duman, R. S. (2005). Neurobiology and treatment of anxiety: Signal transduction and neural plasticity. *Handbook of Experimental Pharmacology*, 305–334.

Duman, R. S., & Monteggia, L. M. (2006). A neurotrophic model for stress-related mood disorders. *Biological Psychiatry*, *59*, 1116–1127.

Everitt, B. J., & Wolf, M. E. (2002). Psychomotor stimulant addiction: A neural systems perspective. *Journal of Neuroscience*, *22*, 3312–3320.

Francis, T. C., Chandra, R., Friend, D. M., Finkel, E., Dayrit, G., Miranda, J., et al. (2014). Nucleus accumbens medium spiny neuron subtypes mediate depression-related outcomes to social defeat stress. *Biological Psychiatry*, *77*, 212–222.

Grueter, B. A., Robison, A. J., Neve, R. L., Nestler, E. J., & Malenka, R. C. (2013). ΔFosB differentially modulates nucleus accumbens direct and indirect pathway function. *Proceedings of the National Academy of Sciences of the United States of America*, *110*, 1923–1928.

Huang, Y. H., Lin, Y., Brown, T. E., Han, M. H., Saal, D. B., Neve, R. L., et al. (2008). CREB modulates the functional output of nucleus accumbens neurons: A critical role of N-methyl-D-aspartate glutamate receptor (NMDAR) receptors. *Journal of Biological Chemistry*, *283*, 2751–2760.

Hyman, S. E., Malenka, R. C., & Nestler, E. J. (2006). Neural mechanisms of addiction: The role of reward-related learning and memory. *Annual Review of Neuroscience*, *29*, 565–598.

Kalivas, P. W. (2004). Glutamate systems in cocaine addiction. *Current Opinion in Pharmacology*, *4*, 23–29.

Koob, G. F., & Le Moal, M. (2001). Drug addiction, dysregulation of reward, and allostasis. *Neuropsychopharmacology*, *24*, 97–129.

Krishnan, V., Han, M.-H., Graham, D. L., Berton, O., Renthal, W., Russo, S. J., et al. (2007). Molecular adaptations underlying susceptibility and resistance to social defeat in brain reward regions. *Cell*, *131*, 391–404.

Lammel, S., Lim, B. K., & Malenka, R. C. (2014). Reward and aversion in a heterogeneous midbrain dopamine system. *Neuropharmacology*, *76*(Pt. B), 351–359.

LaPlant, Q., Chakravarty, S., Vialou, V., Mukherjee, S., Koo, J.-W., Kalahasti, G., et al. (2009). Role of NFκB in ovarian hormone mediated stress hypersensitivity in female mice. *Biological Psychiatry*, *65*, 874–880.

LaPlant, Q., Vialou, V., Covington, H. E., Dumitriu, D., Feng, J., Warren, B., et al. (2010). DNMT3a regulates emotional behavior and spine plasticity in the nucleus accumbens. *Nature Neuroscience*, *13*, 1137–1143.

Lobo, M. K., Zaman, S., Damez-Werno, D. M., Koo, J. W., Bagot, R. C., DiNieri, J. A., et al. (2013). ΔFosB induction in striatal medium spiny neuron subtypes in response to chronic pharmacological, emotional, and optogenetic stimuli. *Journal of Neuroscience*, *33*, 18381–18395.

Maze, I., Shen, L., Zhang, B., Garcia, B. A., Shao, N. Y., Mitchell, A., et al. (2014). Analytical tools and current challenges in the modern era of neuroepigenomics. *Nature Neuroscience*, *17*, 1476–1490.

McClung, C. A., & Nestler, E. J. (2003). Regulation of gene expression and cocaine reward by CREB and ΔFosB. *Nature Neuroscience*, *11*, 1208–1215.

Miller, B. R., & Hen, R. (2014). The current state of the neurogenic theory of depression and anxiety. *Current Opinion in Neurobiology*, *30C*, 51–58.

Muschamp, J. W., Nemeth, C. L., Robison, A. J., Nestler, E. J., & Carlezon, W. A., Jr. (2012). ΔFosB enhances the rewarding effects of cocaine while reducing the pro-depressive effects of the kappa-opioid receptor agonist U50488. *Biological Psychiatry*, *71*, 44–50.

Nestler, E. J. (2008). Transcriptional mechanisms of addiction: Role of deltaFosB. *Philosophical Transactions of the Royal Society of London. Series B: Biological Sciences*, *363*, 3245–3255.

Nestler, E. J., Barrot, M., DiLeone, R. J., Eisch, A. J., Gold, S. J., & Monteggia, L. M. (2002). Neurobiology of depression. *Neuron*, *34*, 13–25.

Nestler, E. J., & Carlezon, W. A., Jr. (2006). The mesolimbic dopamine reward circuit in depression. *Biological Psychiatry*, *59*, 1151–1159.

Newton, S. S., Thome, J., Wallace, T., Shirayama, Y., Dow, A., Schlesinger, L., et al. (2002). Inhibition of CREB or dynorphin in the nucleus accumbens produces an antidepressant-like effect. *Journal of Neuroscience*, *22*, 10883–10890.

Peña, C. J., Bagot, R. C., Labonté, B., & Nestler, E. J. (2014). Epigenetic signaling in psychiatric disorders. *Journal of Molecular Biology*, *426*, 3389–3412.

Perrotti, L. I., Hadeishi, Y., Ulery, P., Barrot, M., Monteggia, L., Duman, R. S., et al. (2004). Induction of deltaFosB in reward-related brain structures after chronic stress. *Journal of Neuroscience*, *24*, 10594–10602.

Pezze, M. A., & Feldon, J. (2004). Mesolimbic dopaminergic pathways in fear conditioning. *Progress in Neurobiology*, *74*, 301–320.

Pizzagalli, D. A. (2014). Depression, stress, and anhedonia: Toward a synthesis and integrated model. *Annual Review of Clinical Psychology*, *10*, 393–423.

Pliakas, A. M., Carlson, R. R., Neve, R. L., Konradi, C., Nestler, E. J., & Carlezon, W. A., Jr. (2001). Altered responsiveness to cocaine and increased immobility in the forced swim test associated with elevated CREB expression in the nucleus accumbens. *Journal of Neuroscience*, *21*, 7397–7403.

Renthal, W., Maze, I., Krishnan, V., Covington, H. E., Xiao, G. H., Kumar, A., et al. (2007). Histone deacetylase 5 epigenetically controls behavioral adaptations to chronic emotional stimuli. *Neuron*, *56*, 517–529.

Réus, G. Z., Abelaira, H. M., dos Santos, M. A., Carlessi, A. S., Tomaz, D. B., Neotti, M. V., et al. (2013). Ketamine and imipramine in the nucleus accumbens regulate histone deacetylation induced by maternal deprivation and are critical for associated behaviors. *Behavioural Brain Research*, *256*, 451–456.

Russo, S. J., & Nestler, E. J. (2013). The brain reward circuitry in mood disorders. *Nature Reviews. Neuroscience*, *14*, 609–625.

Russo, S. J., Wilkinson, M. B., Mazei-Robison, M. S., Dietz, D. M., Maze, I., Krishnan, V., et al. (2009). Nuclear factor κB signaling regulates neuronal morphology and cocaine reward. *Journal of Neuroscience*, *29*, 3529–3537.

Schratt, G. (2014). A molecular knife to dice depression. *Nature*, *516*, 45–46.

Silbersweig, D. (2013). Default mode subnetworks, connectivity, depression and its treatment: Toward brain-based biomarker development. *Biological Psychiatry*, *74*, 5–6.

Sun, H., Kennedy, P. J., & Nestler, E. J. (2013). Epigenetics of the depressed brain: Role of histone acetylation and methylation. *Neuropsychopharmacology*, *38*, 124–137.

Tremblay, L. K., Naranjo, C. A., Graham, S. J., Herrmann, N., Mayberg, H. S., Hevenor, S., et al. (2005). Functional neuroanatomical substrates of altered reward processing in major depressive disorder revealed by a dopaminergic probe. *Archives of General Psychiatry*, *62*, 1228–1236.

Trivedi, M. H., & Daly, E. J. (2008). Treatment strategies to improve and sustain remission in major depressive disorder. *Dialogues in Clinical Neuroscience*, *10*, 377–384.

Turner, C. A., Watson, S. J., & Akil, H. (2012). The fibroblast growth factor family: Neuromodulation of affective behavior. *Neuron*, *76*, 160–174.

Tye, K. M., Mirzabekov, J. J., Warden, M. R., Ferenczi, E., Tsai, H. C., Finkelstein, J., et al. (2013). Dopamine neurons modulate neural encoding and expression of depression-related behaviour. *Nature*, *493*, 537–541.

Valverde, O., Mantamadiotis, T., Torrecilla, M., Ugedo, L., Pineda, J., Bleckmann, S., et al. (2004). Modulation of anxiety-like behavior and morphine dependence in CREB-deficient mice. *Neuropsychopharmacology*, *29*, 1122–1133.

Van't Veer, A., & Carlezon, W. A., Jr. (2013). Role of kappa-opioid receptors in stress and anxiety-related behavior. *Psychopharmacology*, *229*, 435–452.

Vialou, V., Bagot, R. C., Cahill, M. E., Ferguson, D., Robison, A. J., Dietz, D. M., et al. (2014). Prefrontal cortical circuit for depression- and anxiety-related behaviors mediated by cholecystokinin: Role of ΔFosB. *Journal of Neuroscience*, *34*, 3878–3887.

Vialou, V., Feng, J., Robison, A. J., Ku, S. M., Ferguson, D., Scobie, K. N., et al. (2012). Serum response factor and cAMP response element binding protein are both required for cocaine induction of ΔFosB. *Journal of Neuroscience*, *32*, 7577–7584.

Vialou, V., Feng, J., Robison, A. J., & Nestler, E. J. (2013). Epigenetic mechanisms of depression and antidepressant action. *Annual Review of Pharmacology and Toxicology*, *53*, 59–87.

Vialou, V., Maze, I., Renthal, W., LaPlant, Q. C., Watts, E. L., Mouzon, E., et al. (2010). Serum response factor promotes resilience to chronic social stress through the induction of DeltaFosB. *Journal of Neuroscience*, *30*, 14585–14592.

Vialou, V., Robison, A. J., LaPlant, Q. C., Covington, H. E., III, Dietz, D. M., Ohnishi, Y. N., et al. (2010). DeltaFosB in brain reward circuits mediates resilience to stress and antidepressant responses. *Nature Neuroscience*, *13*, 745–752.

Wallace, D. L., Han, M.-H., Graham, D. L., Green, T. A., Vialou, V., Iniguez, S. D., et al. (2009). CREB regulation of nucleus accumbens excitability mediates social isolation-induced behavioral deficits. *Nature Neuroscience*, *12*, 200–209.

Wilkinson, M. B., Dias, C., Magida, J., Mazei-Robison, M., Lobo, M. K., Kennedy, P., et al. (2011). A novel role of the WNT-dishevelled-GSK3β signaling cascade in the mouse nucleus accumbens in a social defeat model of depression. *Journal of Neuroscience*, *31*, 9084–9092.

Wilkinson, M. B., Xiao, G. H., Kumar, A., LaPlant, Q., Renthal, W., Sikder, D., et al. (2009). Imipramine treatment and resiliency exhibit similar chromatin regulation in a key brain reward region. *Journal of Neuroscience*, *29*, 7820–7832.

Wise, R. A. (1982). Neuroleptics and operant behavior: The anhedonia hypothesis. *Behavioral and Brain Sciences*, *5*, 39–87.

Wise, R. A. (1998). Drug-activation of brain reward pathways. *Drug and Alcohol Dependence*, *51*, 13–22.

Wittmann, B. C., Schott, B. H., Guderian, S., Frey, J. U., Heinze, H. J., & Duzel, E. (2005). Reward-related fMRI activation of dopaminergic midbrain is associated with enhanced hippocampus-dependent long-term memory formation. *Neuron*, *45*, 459–467.

Zachariou, V., Bolanos, C. A., Selley, D. E., Theobald, D., Cassidy, M. P., Kelz, M. B., et al. (2006). ΔFosB: An essential role for ΔFosB in the nucleus accumbens in morphine action. *Nature Neuroscience*, *9*, 205–211.

CHAPTER SEVEN

Nicotine Addiction and Psychiatric Disorders

Munir Gunes Kutlu, Vinay Parikh, Thomas J. Gould[1]
Temple University, Philadelphia, Pennsylvania, USA
[1]Corresponding author: e-mail address: tgould@temple.edu

Contents

Abstract

Even though smoking rates have long been on the decline, nicotine addiction still affects 20% of the US population today. Moreover, nicotine dependence shows high comorbidity with many mental illnesses including, but are not limited to, attention deficit hyperactivity disorder, anxiety disorders, and depression. The reason for the high rates of smoking in patients with mental illnesses may relate to attempts to self-medicate with nicotine. While nicotine may alleviate the symptoms of mental disorders, nicotine abstinence has been shown to worsen the symptoms of these disorders. In this chapter, we review the studies from animal and human research examining the bidirectional relationship between nicotine and attention deficit hyperactivity disorder, anxiety disorders, and depression as well as studies examining the roles of specific subunits of nicotinic acetylcholine receptors (nAChRs) in the interaction between nicotine and these mental illnesses. The results of these studies suggest that activation, desensitization, and upregulation of nAChRs modulate the effects of nicotine on mental illnesses.

International Review of Neurobiology, Volume 124
ISSN 0074-7742
http://dx.doi.org/10.1016/bs.irn.2015.08.004

1. INTRODUCTION

Since 1965, tobacco use has declined in the United States from 42% to 20% in 2004 (CDC, 2008, http://www.cdc.gov/tobacco/data_statistics/tables/trends/cig_smoking). However, use rates have remained relatively steady since 2004. While many reasons may contribute to absence of further decline in smoking, one possibility is that this population of smokers contains individuals that have increased vulnerability to nicotine addiction. Increasing evidence suggests individuals with mental illness and/or cognitive impairments may be at increased risk of smoking. In support, the rate of smoking in individuals that reported mental illness in the past month was 41% (Lasser et al., 2000). This is a doubling of what is reported by the CDC for the general population. In this chapter, we will examine the relationships between smoking/nicotine and psychiatric disorders of cognition and affect such as attention deficit hyperactivity disorder (ADHD), anxiety disorders, and depression. In addition, the nicotinic acetylcholinergic receptor (nAChR) subtypes associated with these relationships will be examined.

As discussed in other chapters of this book, nAChRs are pentameric, ionotropic receptors that gate Na+ and Ca++ and can be homomeric, consisting of all α subunits, or heteromeric, consisting of α and β subunits, in the central nervous system (Decker, Brioni, Bannon, & Arneric, 1995; Hogg, Raggenbass, & Bertrand, 2003; Jones, Sudweeks, & Yakel, 1999; McGehee, 1999). The predominant nAChRs in the central nervous system are the α7 and the α4β2 nAChRs (Marks & Collins, 1982; Whiteaker et al., 1999). α7 nAChRs are functionally different showing decreased affinity for agonists and increased sensitivity to desensitization as opposed to the α4β2 nAChRs (Alkondon & Albuquerque, 2004; Gotti et al., 2009; Marks, Burch, & Collins, 1983; Mihailescu & Drucker-Colin, 2000; Olale, Gerzanich, Kuryatov, Wang, & Lindstrom, 1997; Picciotto, Caldarone, King, & Zachariou, 2000; Schwartz & Kellar, 1983). Moreover, the addition of different types of β subunits or changes in stoichiometry of α4 and β2 subunits can change the functional properties of α4β2 nAChRs (Kuryatov, Luo, Cooper, & Lindstrom, 2005; Nelson, Kuryatov, Choi, Zhou, & Lindstrom, 2003; Salminen et al., 2007, 2004; Zwart & Vijverberg, 1998). Thus, understanding the contribution of different nAChR subtypes to the behavioral and neurochemical effects of nicotine in individuals with ADHD, anxiety disorders, and depression may provide insights into higher prevalence of tobacco smoking in people with these psychiatric conditions.

2. NICOTINE'S EFFECTS ON ADHD

ADHD may be one of the most common childhood disorders. The key symptoms of ADHD are inattention, hyperactivity, and impulsivity (Gehricke et al., 2007). It is estimated that ADHD affects approximately 6.5–8.4% of children (Barbaresi et al., 2002, 2004) and between 1.9% and 6% of adults (Kessler et al., 2006; Weiss & Murray, 2003). There are numerous risk factors and changes in brain function associated with ADHD. For instance, fMRI analysis suggests that ADHD may be associated with a decrease in connectivity between the dorsal anterior cingulate cortex and the posterior cingulate cortex and precuneus (Castellanos et al., 2008); brain regions associated with higher cognitive function including working memory (Hampson, Driesen, Skudlarski, Gore, & Constable, 2006). In addition to changes in brain regions involved in cognition, ADHD may also involve changes in neurotransmitter systems associated with cognition and attention such as acetylcholine (for review, see Beane & Marrocco, 2004). Furthermore, cognitive deficits associated with ADHD are similar to nicotine withdrawal-associated changes in cognition seen in smokers such as deficits in sustained attention, response inhibition, and working memory (Beane & Marrocco, 2004; Dovis, der Oord, Huizenga, Wiers, & Prins, 2014; Hughes, Keenan, & Yellin, 1989; Jacobsen et al., 2005; Mendrek et al., 2006; for review, see Ashare, Falcone, & Lerman, 2014).

The similarities between ADHD symptoms and nicotine withdrawal symptoms and the potential involvement of the cholinergic system in ADHD could suggest that individuals with ADHD may be an at-risk group for smoking; this is supported by data. Forty-two percent of males with ADHD were smokers and 38% of females with ADHD were smokers; this compares to 28.1% and 23.5% smokers for males and females without ADHD (Pomerleau, Downey, Stelson, & Pomerleau, 1995). Furthermore, the same study found that the quit ratio was substantially lower in individuals with ADHD compared to the rest of the non-mentally ill population, 29% versus 48.5%, respectively. These findings have been replicated by other scientists, for example, Lambert and Hartsough (1998) found lifetime tobacco dependence was 40% in individuals with ADHD compared to 19%. However, a remaining important question is whether smoking is a result of ADHD or whether ADHD is a result of smoking.

2.1 Relationship Between Nicotine Exposure and ADHD in Humans

Evidence suggests a complex relationship exists between ADHD and smoking with ADHD contributing to smoking but smoking also contributing to the development of attention deficits. For example, ADHD predicted future smoking and the transition into associated nicotine addiction (Fuemmeler, Kollins, & McClernon, 2007) and adolescents with ADHD were more likely to experiment with cigarettes and become smokers (Tercyak, Lerman, & Audrain, 2002). Therefore, smoking may be an attempt to self-medicate for symptoms of ADHD. Acute nicotine administered via the patch improved attention in adults and young adults with ADHD (Levin et al., 1996; Potter & Newhouse, 2008). In addition, nicotine improved behavioral inhibition in highly impulsive people (Potter, Bucci, & Newhouse, 2012). However, individuals with ADHD may also be more susceptible to the negative effects of smoking. In a study of twins, a greater increase in attention deficits across years was seen for smokers versus never-smoking twin cohorts (Treur et al., in press). This study suggests that smoking can worsen attention problems. In addition, in a study that specifically examined individuals with ADHD, inattention symptoms during childhood, but not hyperactivity, was associated with greater nicotine withdrawal symptoms during adulthood (Ameringer & Leventhal, 2012). This relationship between ADHD and greater nicotine withdrawal symptoms is supported by other studies. Female, but not male, smokers with ADHD had greater withdrawal symptoms (McClernon et al., 2011), and in other studies, smokers with ADHD showed great withdrawal symptoms and willingness to work harder for cigarette puffs (Kollins et al., 2013; Pomerleau et al., 2003). Furthermore, another study found that both male and female smokers with ADHD had a greater level of nicotine dependence than smokers without ADHD (Wilens et al., 2008). The emergence of increased ADHD symptoms during periods of abstinence was associated with increased risk of relapse (Rukstalis, Jepson, Patterson, & Lerman, 2005), which demonstrates that the increased withdrawal symptoms impact successful quit attempts and thus health.

In order to improve treatment for both ADHD and nicotine dependence, an understanding of what factors underlie the increased risk of nicotine dependence in individuals with ADHD is necessary. While undoubtedly many factors contribute to this relationship, increasing evidence suggests that differences in attention processes and ADHD symptoms may be related to genetic variants in genes encoding nicotinic receptor subunits. Lee, Fuemmeler, McClernon, Ashley-Koch, and Kollins (2013) found a significant interaction between single nucleotide polymorphisms of CHRNB3 and ADHD where the AA variant and the ADHD symptom of inattention were associated with greater increase in the level of cigarette smoking across adolescence. The study also reported differences in cigarette smoking associated with variants of CHRNA6. Another study found that two alleles (rs578776 and rs3743078) in CHRNA3 were associated with an increased risk of smoking, but only among individuals with ADHD (Polina et al., 2014). Moreover, Viñals et al. (2012) showed that transgenic mice that overexpress $\alpha3/\alpha5/\beta4$ nAChRs exhibited less impulsive-like behavior than the wild-type controls. Also, Cohen et al. (2012) found that mutant mice with hypersensitive $\alpha6$ nAChR subunits showed spontaneous hyperactivity in their home cages. In addition, changes in other genes associated with cholinergic function may also contribute to ADHD as a study that examined polymorphisms in a gene encoding the high-affinity choline transporter (CHT; *SLC5A7*) found a two- to threefold increase in Val89 allele, which is associated with reduced choline transport function, in individuals with ADHD (English et al., 2009). These studies suggest that differences in nAChR function may contribute to increased vulnerability to nicotine addiction for individuals with ADHD and that genetics plays a role in this vulnerability.

The studies reviewed so far demonstrate a clear relationship between smoking and ADHD and suggest treating nicotine addiction in individuals with ADHD may be more challenging. Several studies, however, suggest that treating ADHD symptoms may reduce smoking. Methylphenidate is a stimulant commonly used to treat ADHD, and in adolescents, methylphenidate was also shown to reduce smoking (Hammerness et al., 2013; Schoenfelder, Faraone, & Kollins, 2014). In addition, another study found that osmotic controlled-release delivery of methylphenidate for 11 weeks to adult smokers with severe ADHD promoted abstinence to smoking in these individuals in part by improving ADHD symptoms (Nunes et al., 2013). In contrast, individuals with lower ADHD scores exhibited lower abstinence rates as opposed to the placebo-treated group. However, a study by Vansickel, Stoops, Glaser, Poole, and Rush (2011) reported that acutely administered methylphenidate increased smoking in adult ADHD subjects. It is possible that methylphenidate's effectiveness in reducing smoking in ADHD subjects may depend upon the duration of treatment and symptom severity. Further research is required to delineate behavioral and neural mechanisms that underlie the observed relation between stimulant medications and smoking in ADHD subjects.

2.2 Effects of Nicotine on Animal Models of ADHD

Studies in rats and mice have led to advances in understanding the neural substrates of attention, the effects of nicotine on attention and cognitive control, and the development of models of ADHD that allow further examining of the effects of nicotine on ADHD. Cholinergic deafferentation produced by infusion of the cholinotoxin 192-IgG saporin into the basal forebrain produced robust attentional impairments in rats (McGaughy, Kaiser, & Sarter, 1996). Moreover, studies employing microdialysis and electrochemical recordings to measure ACh release from animals performing attentional tasks show performance-associated increases in cortical cholinergic transmission (Arnold, Burk, Hodgson, Sarter, & Bruno, 2002; Howe et al., 2013; Parikh, Kozak, Martinez, & Sarter, 2007). Another study employing transgenic mice with a heterozygous deletion of the CHT gene reported attentional deficits in a signal detection task and reduced ability to sustain acetylcholine release (Parikh, Peters, Blakely, & Sarter, 2013). Together, these results provide support for the hypothesis that cholinergic function is required for normal attentional processes and that modulation of cholinergic signaling alters attention. As discussed earlier, polymorphism in the CHT gene was found in individuals exhibiting ADHD symptoms. Therefore, attentional impairments observed in ADHD subjects might occur as a consequence of disruption in cholinergic signaling.

Multiple studies have shown that nicotine alters behavioral processes that impact cognitive and attentional control in rodents. In rats performing a perceptual attentional set-shifting task, acute nicotine administration in rats enhanced extradimensional set-shifting that requires switching between the two perceptual dimensions of a compound stimulus (Allison & Shoaib, 2013). The same study also reported that intradimensional set-shifting that entails switching within the same dimension was also facilitated by acute nicotine. However in a recent study conducted in our laboratory, chronic nicotine treatment for 4 weeks did not alter strategy set-shifting in an operant-based cognitive flexibility task in mice (Ortega, Tracy, Gould, & Parikh, 2013). Rather, it impaired reversal learning and these cognitive deficits were associated with increased perseverative responding to the previously rewarded stimulus indicating deficits in response inhibition. Similarly, a study that examined inhibitory control found that in male C57BL/6J mice, acute nicotine enhanced inhibitory control and reduced impulsivity but tolerance developed to these effects with chronic nicotine treatment (Leach, Cordero, & Gould, 2013). Together, these studies suggest that initially nicotine may have positive effects on cognitive/attentional control and response inhibition but with prolonged used, the positive effects disappear and cognitive functions may worsen in ADHD.

The 5-choice serial reaction time task (5CSRTT), developed to assess visual attentional processes in rodents (Bari, Dalley, & Robbins, 2008), is sensitive to the effects of nicotine. This paradigm requires animals to detect brief flashes of light presented in a pseudorandom order in one out of multiple (five or nine) spatial locations over a large number of trials. Visual cues are detected by responding in the appropriate aperture within a certain amount of time. A correct choice is rewarded with a food pellet. If the animals fail to respond, respond in the wrong aperture or at an inappropriate time, a short period of darkness (time-out) is presented as punishment and no reward is delivered. A number of behavioral measures including correct responses, premature responses, perseverative responses, omissions, and response latencies are recorded to assess attentional control functions. Male hooded Lister rats administered nicotine prior to each 5CSRTT session initially had increased omissions but this effect dissipated between weeks 1 and 2 and thereafter increased correct responses and anticipatory response and decreased omissions were seen (Hahn & Stolerman, 2002). Half the rats received daily injections of nicotine but this did not alter the effects. Because the half-life of nicotine is 45 min in rats compared to 2 h in humans (Matta et al., 2007), it is difficult to determine whether this nicotine administration protocol reflects acute, subchronic, or chronic administration. However, given that nicotine initially enhanced attentional set-shifting and response inhibition but those cognitive effects disappeared with chronic administration in the aforementioned studies, the effects of nicotine on attentional measures observed on 5CSRTT task may reflect acute effects only.

In a similar study, nicotine administered prior to testing in the 5CSRTT improved accuracy while reducing omissions and reaction time in male hooded Lister rats (Hahn, Shoaib, & Stolerman, 2002). In addition, nicotine was able to reverse attentional deficits induced by the addition of auditory distractor cues. Another study, however, suggests that the effects of nicotine on 5CSRTT may be mediated by additional factors. Mirza and Bright (2001) found that nicotine administered prior to testing dose-dependently increased correct responses in male Sprague–Dawley rats but had no effect in male Lister hooded rats. Since similar doses of nicotine were used in the Mirza and Bright (2001) and Hahn et al. (2002) studies, which use Lister hooded rats, unidentified environmental factors must have contributed to the difference in the effects of nicotine on the 5CSRTT in the Lister hooded rats.

Just as seen in human studies, nicotine withdrawal disrupts attention in rodent models. Lister Hooded rats withdrawn from chronic nicotine treatment had deficits in 5CSRTT that were related to increased omissions (Shoaib & Bizarro, 2005). Also, the α4β2 nAChR antagonist dihydro-β-erythroidine (DhβE) precipitated similar withdrawal deficits in rats treated with chronic nicotine but the α7 nAChR antagonist methyllycaconitine (MLA) produced no deficits. Similar to the Shoaib and Bizarro (2005) study, Semenova, Stolerman, and Markou (2007) found that nicotine withdrawal increased omissions and decreased correct response in Wistar rats. This study also found that acute nicotine increased correct responses and decreased omissions, while chronic nicotine increased premature response but also increased correct responses.

As found in the Shoaib and Bizarro (2005) study, multiple studies suggest that cholinergic signaling via α4β2 nAChRs is critically involved in the effects of nicotine on attention. For example, nicotine and the α4β2 nAChR agonist SIB 1765F increased correct responding and decreased response latency for the 5CSRTT in male Lister hooded rats but also increased premature responding. The α7 nAChR agonist AR-R 17779 was without effect, though only one dose was tested (Grottick & Higgins, 2000). In an operant sustained attention task, neither nicotine nor an α4β2 nAChR agonist, S38232, was effective but when a distractor cue was added, S38232 enhanced attention (Howe et al., 2010). Interestingly, if nicotine was paired with the α7 nAChR antagonist MLA, enhancement of attention under distracting conditions was seen; this suggests agonism of α4β2 nAChR has pro-attention effects but antagonism of α7 nAChRs improves attention. Similar effect was seen in C57BL/6N mice; nicotine and an α4β2 nAChR agonist, ABT-418, enhanced attention assessed in 5-choice continuous performance tests, whereas an α7 nAChR agonist, PNU 282987, was without effect (Young, Meves, & Geyer, 2013).

A series of experiments further demonstrated a role of α4β2 nAChRs in attention. Dizocilpine-induced deficits in a signal detection task in female Sprague–Dawley rats were ameliorated with an α4β2 nAChR agonist, AZD3480; donepezil, an acetylcholinesterase inhibitor; and sazetidine-A, which has high affinity for α4β2 nAChRs and after brief activation produces a long-lasting desensitization of the receptors (Rezvani, Cauley, Johnson, Gatto, & Levin, 2012; Rezvani et al., 2011; Rezvani, Cauley, Xiao, Kellar, & Levin, 2013). In addition, spontaneously hypertensive rats have been used as a model of ADHD and these rats have been shown to have lower levels of brain α4β2 nAChR binding but no changes in α7 nAChR binding (Wigestrand et al., 2011).

While overwhelming evidence suggests α4β2 nAChRs are involved in attention, other nAChR subunits may also be involved. In a study with α5 KO mice, the α5 KO decreased nicotinic currents in layer VI pyramidal neurons in prefrontal cortex and increased accuracy in the 5-CSRTT when task parameters were made more difficult (Bailey, De Biasi, Fletcher, & Lambe, 2010). This suggests that α4α5β2 nAChRs may also play a key role in attentional processes. Finally, while the pharmacology studies suggest α7 nAChRs are not critically involved in attention, gene knockout studies suggest complete absence of α7 nAChR may alter attention. Young et al. (2004) found that α7 KO mice had increased omissions in the 5CSRTT. Similarly, Hoyle, Genn, Fernandes, and Stolerman (2006) reported that α7 KO mice have decreased correct responses but increased anticipatory responses. Thus, even though α7 agonists were without effect on attention (Grottick & Higgins, 2000; Young et al., 2013), the complete absence of α7 nAChRs could alter attention, possibly through changes during development or other compensatory changes that affect attention in adulthood. Clearly, this issue requires further examination.

2.3 Developmental Nicotine Exposure and Its Effects on ADHD

While genetic factors such as differences in nAChR expression and function could contribute to ADHD, environmental factors such as developmental exposure are also critical factors in ADHD. Individuals prenatally exposed to constituents of cigarette smoke have higher rates of ADHD but it is difficult to determine if this is directly related to nicotine or if other factors contribute to this relationship. For example, individuals with ADHD may smoke in an attempt to self-medicate (Gehricke et al., 2007), and thus, mothers that smoke during pregnancy may be smoking in an attempt to self-medicate ADHD or subclinical ADHD symptoms. Therefore, the offspring of these mothers could express ADHD because of an inherited risk factor. Studies of laboratory animals can address these issues.

In a study of prenatal cigarette smoke, male B6C3F1 mice, but not female mice, had increased hyperactivity, decreased striatal and cortical dopamine and serotonin, and reduced BDNF mRNA and protein (Yochum et al., 2014). These effects may be due to the nicotine in the cigarette smoke as Zhu et al. (2012) found that prenatal nicotine exposure increased hyperactivity and reduced cingulate cortical volume and dopamine turnover in male and female young adult C57BL/6J mice. Treatment with methylphenidate decreased hyperactivity in the mice, as in patients with ADHD, and increased dopamine turnover, suggesting that this model has strong external validity. Similarly, adult male and female Lister hooded rats prenatally exposed to nicotine had decreased correct responding, increased omission and increased anticipatory responses in the 5CSRTT, and increased expression of the D5 dopamine gene in the striatum (Schneider et al., 2011).

Prenatal nicotine exposure may also alter nAChR function. Adult mice prenatally exposed to nicotine had increased dendritic branching of medial prefrontal layer VI pyramidal neurons but also decreased nAChR signaling indicated by reduced nAChR responses to 1 mM acetylcholine (Bailey, Tian, Kang, O'Reilly, & Lambe, 2014). These effects were mediated by α5 nAChR subunits as they were reversed in α5 KO mice. In another study, prenatal nicotine exposure increased α4β2 nAChRs in the frontal cortex, hippocampus, caudate, and brainstem and increased both α4β2 and α7 nAChRs in the occipital cortex of rhesus monkeys (Slotkin et al., 2005). Changes in cholinergic signaling in frontal cortex and hippocampus could impact ADHD symptoms such as inattention and impulsivity.

The effects of prenatal nicotine on development of ADHD are greatly concerning but the effects of prenatal exposure may have an even greater effect on mental health and health care than original expected. Prenatal nicotine exposure may produce epigenetic changes that result in transgenerational inheritance of ADHD symptoms (Zhu, Lee, Spencer, Biederman, & Bhide, 2014). C57BL/6J female mice received nicotine in the drinking water starting 3 weeks prior to mating and continued throughout pregnancy. Locomotor activity was significantly increased in F2 and F3 generation male and female mice, even though the mice were not exposed to nicotine; however, inheritance occurred only through the maternal line. These results suggest that prenatal exposure to nicotine produces a transgenerational inheritance of the ADHD symptom of hyperactivity along the maternal line. Further work is needed but these findings could have potentially dramatic impact of public policy and health-care practices related to nicotine exposure.

The developmental effects of nicotine exposure may not be limited to prenatal exposure as adolescent nicotine exposure, but not adult nicotine exposure, produced long-lasting deficits in visuospatial divided and sustained attention, affecting accuracy and premature responses in rats (Counotte et al., 2011). In addition, mGluR2 protein levels and function on presynaptic prefrontal cortical terminals of glutamate synapses were reduced. However, they found local infusion of a group II mGluR agonist into the medial prefrontal cortex reversed deficits in attention. Collectively, these studies suggest that prenatal and adolescent nicotine exposure contribute to the development of ADHD, and thus, limiting nicotine exposure during development is of critical importance.

Likewise, considerable evidence from human studies indicates that developmental exposure to nicotine and/or smoking may increase the odds of developing ADHD. In a study examining males 6–17 years old, the odds ratio for ADHD when mothers smoked during pregnancy was 2.7 (Milberger, Biederman, Faraone, Chen, & Jones, 1996). The study also found that the odds ratio for ADHD when the mother had ADHD was 2.2 and was 1.8 when the father had ADHD, suggesting maternal ADHD and smoking have similar genetic contributions to offspring ADHD. In a follow-up study that contained both males and females (though heavily skewed toward male) with an average age of 13, the odds ratios for ADHD were even higher: 4.4 for ADHD when mothers smoked during pregnancy and 5.4 for when mothers had ADHD, but were lower for when the father had ADHD: 1.2 (Milberger, Biederman, Faraone, & Jones, 1998). Similarly, in a study conducted in male and female teenagers with an average age of 14 years, teachers, mothers, and fathers reported significantly higher scores for ADHD symptoms in offspring whose mothers smoked during pregnancy (Indredavik, Brubakk, Romundstad, & Vik, 2007). Finally, in a study of children aged 6–12 and diagnosed with ADHD, children exposed prenatally to maternal cigarette smoke had more severe ADHD symptoms and neurocognitive deficits as compared to unexposed children (Thakur et al., 2013). These studies strongly suggest that developmental exposure to constituents of cigarette smoke can exacerbate ADHD symptoms and may even contribute to the development of ADHD. Important unresolved issues include the underlying neurobiological substrates and whether nicotine is sufficient to increase susceptibility to ADHD. The latter issue is of increasing importance as pregnant women may be prescribed nicotine replacement therapy and the emergence of e-cigarettes as a nicotine free-base delivery system. Laboratory animal models provide a means to address these issues.

In summary, the relationship between nicotine and ADHD is complex. Individuals with ADHD may smoke in an attempt to self-medicate, but over time positive effects may dissipate and worsening of symptoms may develop. In addition, smoking and nicotine exposure may facilitate the development of ADHD; nowhere may this be more critical than in developmental exposure to nicotine. Both prenatal and adolescent nicotine exposure were associated with increased expression of ADHD symptoms. Furthermore, prenatal nicotine exposure may produce epigenetic changes that increase

ADHD symptoms in future generations that are not exposed to nicotine. The relationship between ADHD and nicotine/smoking may involve multiple mechanisms including signaling via α4β2 nAChRs and other nAChR subtypes. Further work is needed to fully understand the relationship between ADHD and nicotine and the underlying substrates in order to improve health care.

3. INVOLVEMENT OF nAChRs IN ANXIETY AND ANXIETY DISORDERS

Anxiety disorders including but not limited to panic disorders, phobias, generalized anxiety disorder, and posttraumatic stress disorder (PTSD) are a cluster of disorders that affects approximately 40 million Americans (18.1%; Kessler, 1997; Kessler, Chiu, Demler, & Walters, 2005) and costs more than $42 billion a year, almost one-third of the United States $148 billion total mental health costs (Greenberg et al., 1999). Disorders under the anxiety disorder category usually develop after a highly stressful traumatic event (Mineka & Zinbarg, 2006) and they are among the most frequently diagnosed psychological disorders (Breslau, Novak, & Kessler, 2004). In the following sections, we will review the literature on the relationship between anxiety disorders and nicotine dependence in humans. In addition, we will examine the effects of acute, chronic, and withdrawal from chronic nicotine and the involvement of specific nAChR subtypes on the animal models of fear and anxiety.

3.1 Relationship Between Nicotine Exposure and Anxiety Disorders in Humans

Numerous studies have identified a bidirectional link between nicotine dependence and anxiety disorders (Breslau, Davis, & Schultz, 2003; Breslau et al., 2004; Feldner, Babson, & Zvolensky, 2007; Fu et al., 2007; Koenen et al., 2005). Specifically, smoking rates have been shown to be significantly higher in the population with anxiety disorders than it is in the nonclinical population, 45.3% and 22.5%, respectively (Lasser et al., 2000; Ziedonis et al., 2008). On the other hand, anxiety disorders have also been shown to be significantly more prevalent in the people diagnosed with nicotine dependence (22%) than in the nondependent population (11.1%; Grant, Hasin, Chou, Stinson, & Dawson, 2004; Grant, Stinson, et al., 2004). Consistent with the high rates of nicotine dependence in patients with anxiety disorders, prior smoking has been found to be associated with increased susceptibility to develop PTSD in the event of a trauma (Koenen et al., 2005), an increased risk of panic attacks and development of panic disorders (Goodwin, Lewinsohn, & Seeley, 2005). In addition, following trauma, smoking initiation and daily smoking rates also increased (Breslau et al., 2003, 2004). PTSD patients also showed lower rates of quitting (Hapke et al., 2005; Lasser et al., 2000), suffered from worse nicotine withdrawal symptoms (Dedert et al., 2011), and as a result showed shorter times to first smoking lapse (Beckham, Calhoun, Dennis, Wilson, & Dedert, 2012) than non-PTSD population. Similarly, patients with social phobia have also demonstrated increased rates of smoking initiation (Sonntag, Wittchen, Höfler, Kessler, & Stein, 2000). Furthermore, several studies have shown that the presence of PTSD symptoms, such as hyperarousal and emotional numbing, is a predictor for nicotine dependence and these symptoms are reduced by nicotine intake (Beckham et al., 2005; Feldner et al., 2007; Greenberg et al., 2012; Thorndike, Wernicke, Pearlman, & Haaga, 2006). Therefore, it is possible that while nicotine dependence increases one's vulnerability to anxiety disorders, smoking may serve as a mean to alleviate symptoms associated with anxiety disorders, which, in turn, increases nicotine dependence among patients with anxiety disorders.

Investigating the direct relationship between nAChRs and PTSD by using the radiotracer [^{123}I]5-IA-85380 ([^{123}I]5-IA) and single-photon emission computed tomography, Czermak et al. (2008) found that PTSD patients who never smoked showed significantly higher β2 nAChR density in the mesiotemporal cortex including the amygdala and hippocampus compared to healthy individuals who never smoked. Furthermore, the same study found a significant correlation between β2 nAChR binding in the thalamus and the reexperiencing symptom among the PTSD patients. Both thalamus and mesiotemporal cortex dysfunction have been functionally linked to the pathogenesis of PTSD (Lanius et al., 2001; Shin, Rauch, & Pitman, 2006). These results suggest that β2-containing nAChRs may play an important role in the epidemiology of PTSD. As discussed earlier, nicotine binds and modulates a variety of nAChR subunits. Therefore, it is possible that the modulation of β2-containing nAChRs by nicotine intake may also directly modulate PTSD symptomatology.

3.2 Effects of Nicotine on Animal Models of Anxiety Disorders

Fear conditioning, a behavioral paradigm in which the subject learns an association between a neutral stimulus and an aversive unconditioned stimulus akin to the associations formed during trauma, has been widely utilized as a transitional animal model to study the traumatic experience common for all anxiety disorders (Briscione, Jovanovic, & Norrholm, 2014). Previous studies have identified two types of fear memories: (1) hippocampus dependent (contextual and trace fear conditioning) and (2) hippocampus independent (cued fear conditioning; Clark & Squire, 1998; Logue, Paylor, & Wehner, 1997; McEchron, Bouwmeester, Tseng, Weiss, & Disterhoft, 1998; Phillips & LeDoux, 1992; Solomon, Vander Schaaf, Thompson, & Weisz, 1986). The effects of nicotine and other nAChR agonists/antagonists on fear conditioning have been extensively studied (see Gould & Leach, 2014 for a review). For example, there is ample evidence suggesting that acute nicotine enhances hippocampus-dependent contextual (Davis, Porter, & Gould, 2006; Gould, 2003; Gould, Feiro, & Moore, 2004; Gould & Higgins, 2003; Gould & Wehner, 1999; Wehner et al., 2004) and trace (Davis & Gould, 2007; Gould et al., 2004) fear conditioning, whereas it has no effect on hippocampus-independent cued fear conditioning (e.g., Gould & Higgins, 2003; Gould & Wehner, 1999). Furthermore, there is evidence suggesting that systemic administration of a high-affinity nAChR antagonist, DhβE, reverses the effects of nicotine, while a low-affinity α7 nAChR antagonist, MLA, has no effect (Davis, Kenney, & Gould, 2007). Similarly, several studies using knockout (KO) mice have shown that animals lacking β2 nAChRs did not show enhancement of contextual (Davis & Gould, 2007) or trace fear conditioning (Davis & Gould, 2007; Lotfipour et al., 2013) by nicotine. These results suggest that nicotine enhances hippocampal-dependent fear memories through activation of β2-containing nAChRs.

While the enhancing effects of acute nicotine are well documented, several studies have shown that while chronic nicotine has no effect on fear conditioning, withdrawal from chronic nicotine impairs contextual and trace fear conditioning (André, Gulick, Portugal, & Gould, 2008; Davis, James, Siegel, & Gould, 2005; Portugal & Gould, 2009; Portugal, Wilkinson, Kenney, Sullivan, & Gould, 2012; Raybuck & Gould, 2009). There is also evidence suggesting that during chronic nicotine administration, hippocampal nAChRs desensitize and upregulate and the resulting hypersensitive cholinergic system may be responsible for the effects of nicotine withdrawal on hippocampus-dependent learning (Dani & Heinemann, 1996; Gould et al., 2012; Marks, Grady, & Collins, 1993; Wilkinson & Gould, 2013). In support, Gould et al. (2012) found that chronic nicotine increased nAChR binding in the hippocampus and the duration of nAChR upregulation paralleled the duration of withdrawal deficits in hippocampus-dependent learning. Also supporting the role of hypersensitive nAChRs in the withdrawal effects, Wilkinson and Gould (2013) found that reintroducing acute nicotine into the system during nicotine withdrawal lead to an even greater enhancement of the contextual fear conditioning compared to the effects of acute nicotine in previously nicotine naïve mice. However, while upregulation seems to be necessary for the behavioral effects of nicotine withdrawal, there is evidence showing that nicotine withdrawal and tolerance are dissociable processes as tolerance was shown to occur before withdrawal and in the absence of nAChR upregulation (Gould, Wilkinson, Yildirim, Blendy, & Adoff, 2014). Therefore, this suggests that the rapidly developing nAChR desensitization may be responsible for the tolerance effects, whereas upregulation of nAChRs, which requires a longer period of time to develop, is necessary for the withdrawal effects.

Similar to the KO studies suggesting a central role of the β2 nAChRs in the acute effects of nicotine on hippocampal fear learning, Portugal, Kenney, and Gould (2008) found that β2 nAChR KO animals also did not show withdrawal deficits in contextual fear learning. In addition, infusions of the high-affinity nAChR antagonist DhβE into the dorsal hippocampus precipitated withdrawal deficits in both contextual (Davis & Gould, 2009) and trace fear learning (Raybuck & Gould, 2009). Overall, in line with the human studies linking hippocampal β2 nAChRs with PTSD (Czermak et al., 2008), results from the studies using acute, chronic, and withdrawal from chronic nicotine suggest that the effects of nicotine on hippocampus-dependent fear memories require the activation and upregulation of the high-affinity β2-containing nAChRs.

In addition to the effects of nicotine on acquisition of fear conditioning, several studies have also investigated the effects of nicotine on fear extinction, a form of inhibitory learning that occurs when the conditioned stimulus is repeatedly presented in the absence of the aversive unconditioned stimulus. Fear extinction has been widely utilized as a translational animal model for the exposure therapies used for anxiety disorders (Briscione et al., 2014; Myers & Davis, 2006; Quirk & Mueller, 2007). In humans, several studies have demonstrated that PTSD patients show impaired fear extinction (Blechert, Michael, Vriends, Margraf, & Wilhelm, 2007; Michael, Blechert, Vriends, Margraf, & Wilhelm, 2007; Milad et al., 2009) as well as difficulty learning safety discrimination, another form of inhibitory learning where the subjects are trained to differentiate between a safe versus dangerous cue or context (Jovanovic, Kazama,

Bachevalier, & Davis, 2012; Jovanovic et al., 2010, 2009; Lissek et al., 2005; for a review, see Christianson et al., 2012). Therefore, according to some accounts, PTSD may be attributed to impaired safety learning (Davis, Falls, & Gewirtz, 2000; Lissek et al., 2005). The effects of nicotine exposure on safety learning and extinction have been investigated using the animal models of fear extinction and contextual safety discrimination (Elias, Gulick, Wilkinson, & Gould, 2010; Kutlu & Gould, 2014; Kutlu, Oliver, & Gould, 2014; Tian et al., 2008). For example, Elias et al. (2010) found that acute nicotine enhanced extinction and impaired renewal of cued fear memories when administered during the extinction phase, whereas nicotine administration during both training and extinction phases impaired extinction and enhanced renewal of cued fear. This suggests that acute nicotine may enhance recovery and extinction of cued fear conditioning by potentially strengthening encoding and facilitating new inhibitory learning during acquisition and extinction, respectively. On the other hand, Kutlu and Gould (2014) showed that acute nicotine impaired contextual fear extinction, which, unlike cued extinction, requires direct involvement of the hippocampus (Tronson et al., 2009). Also recently, Kutlu et al. (2014) examined the role of nicotine on contextual safety discrimination and found that acute nicotine impaired this form of safety learning. As mentioned above, safety learning is already compromised in the patients with PTSD, and therefore, these results suggest that nicotine may make treatment of PTSD even harder by further impairing this type of learning. Finally, Tian and colleagues (2008) tested the effects of prior chronic nicotine exposure on subsequent extinction and showed that prior nicotine impaired cued extinction while having no effect of contextual fear. In summary, these studies suggest that nicotine may cause further disruption of the already impaired safety learning in PTSD patients.

Overall, numerous studies that have investigated the relationship between nicotine and fear learning suggest that acute nicotine enhances acquisition and impairs extinction of hippocampus-dependent fear memories. In addition, multiple studies have also shown that withdrawal from chronic nicotine results in impaired hippocampus-dependent fear learning. Interestingly, as described above, acute nicotine administration during nicotine withdrawal results in even greater enhancement of fear learning than acute administration alone. This suggests that during withdrawal from chronic nicotine administration, the cholinergic system becomes hypersensitive to the enhancing effects of acute nicotine. Therefore, these results indicate that abstinence from nicotine may worsen the conditions of the

PTSD patients. This is because while PTSD symptoms are reduced during nicotine intake (Beckham et al., 2005; Feldner et al., 2007; Greenberg et al., 2012; Thorndike et al., 2006), they resurface during abstinence (Dedert et al., 2011). Therefore, most PTSD patients reinitiate smoking to alleviate their symptoms (Beckham et al., 2012). However, while reinitiating smoking may help patients reduce their symptoms, Wilkinson and Gould's (2013) results showing increased sensitivity to the effects of reintroduction of nicotine during nicotine withdrawal suggest that it may also enhance fear memories associated with new trauma and could prolong the course of the disorder.

Animal models of anxiety have been widely utilized for the neurobiological and pharmacological investigation of the anxiety disorders (Rodgers, 1997). These behavioral paradigms take advantage of the innate explanatory behaviors of rodents and natural fears; these tasks include the open field (OF; decreased time spent in the central part of the OF arena), elevated plus maze (EPM; decreased time spent in open arms), and marble burying test (increased duration of burying an object, e.g., glass marbles) as well as the social anxiety test (decreased time spent by pairs of male rats in social interaction; Lister, 1990). The effects of nicotine and nicotinic agents on anxiety-related paradigms are somewhat contradicting (Table 1). For example, Irvine et al. (2001) found that subcutaneous injections of acute nicotine (0.1 mg/kg) increased anxiety-like behavior in the EPM 30 min after injection. Similarly, Ouagazzal et al. (1999) showed that higher doses of acute nicotine (0.5 and 1 mg/kg) administered intraperitoneally (i.p.) also produced increased anxiety in the EPM. However, Irvine et al. (2001) also found that injections 5 min before the EPM test had an anxiolytic effect after 7 days of repeated administration. Finally, rats tested 24 h after the last nicotine injection showed increased anxiety during withdrawal and this effect was reversed by dorsal hippocampal infusions of nicotine (Irvine et al., 2001). The anxiogenic effects of nicotine withdrawal have also been shown to be reversed by acute nicotine challenge (Ericson et al., 2000). In contrast to the results of the Irvine et al. (2001) study on anxiety in the EPM, Irvine et al. (1999) previously found an opposite pattern for the effects of nicotine injection timing on anxiety-like behavior in a social interaction test. Specifically, Irvine et al. (1999) found that nicotine injections 5 min before the behavioral test decreased social interaction, an anxiogenic effect, while injections 30 min before the task increased social interaction, an anxiolytic effect. This shows that acute nicotine may have differential effects on different anxiety paradigms. Also in contrast to the studies reporting the

Table 1 Effects of Acute Nicotine on Anxiety

Study	Strain/ Species	Task	Nicotine Dose	Injection Time	Effect
Irvine, Cheeta, and File (2001)	Hooded Lister rats	EPM	0.1 mg/kg s.c.	30 min	Increased anxiety
				5 min	Decreased anxiety
Ouagazzal, Kenny, and File (1999)	Hooded Lister rats	EPM	0.5 and 1 mg/kg i.p.	5 min	Increased anxiety
O'Neill and Brioni (1994)	Wistar rats	EPM	0.3 mg/kg i.p.	30 min	Decreased anxiety
Ericson, Olausson, Engel, and Söderpalm (2000)	CD1 mice	EPM	0.35 mg/kg s.c.	Immediate	Decreased anxiety
Irvine, Cheeta, and File (1999)	Hooded Lister rats	Social interaction	0.1 mg/kg s.c.	30 min	Decreased anxiety
				5 min	Increased anxiety

anxiogenic effects of acute nicotine, O'Neill and Brioni (1994) found that i. p. injections of nicotine decreased anxiety as measured in the EPM paradigm. Moreover, Ericson et al. (2000) found that subcutaneous acute nicotine administration (0.35 mg/kg) significantly increased open arm duration in rats. Finally, File, Kenny, and Ouagazzal (1998) found that lower doses of acute nicotine (0.01 and 0.1 mg/kg) also had anxiolytic effects, while higher doses (0.5 and 1.0 mg/kg) had anxiogenic effects in a social interaction test. The contradicting effects of nicotine reported by different studies may also be explained by the different species and strains used in these experiments. For example, both Irvine et al. (1999, 2001) and Ouagazzal et al. (1999) studies, which found that nicotine has an anxiogenic effect, used male hooded Lister rats, whereas other studies showing anxiolytic effects of nicotine (Ericson et al., 2000; O'Neill & Brioni, 1994) used male Wistar rats or CD1 mice.

Several nAChR subtypes have been identified to play modulatory roles in the anxiety-like behavior. As in the effects of nicotine on fear conditioning, there is evidence that β2-containing nAChRs mediate nicotine's effects on anxiety. For example, an α4β2 nAChR agonist, ABT-418, was found to increase open arm time in the EPM and both ABT-418 and an α4β2 nAChR partial agonist, ABT-089, reversed the anxiogenic effects of nicotine withdrawal (Brioni et al., 1994; Decker et al., 1994; Yohn, Turner, & Blendy, 2014). Brioni et al. (1994) also showed that the anxiolytic effects of ABT-418 were reversed by the nAChR antagonist mecamylamine, which suggests that sustained nAChR activity is required for the anxiolytic effect of ABT-418. Similarly, McGranahan, Patzlaff, Grady, Heinemann, and Booker (2011) found that selective elimination of α4β2 nAChRs located on dopaminergic neurons reversed the anxiolytic effects of nicotine. There is also evidence that while genetically modified mice that lack β2 subtype of nAChRs showed normal levels of anxiety in the EPM (Picciotto et al., 1998), α4 KO mice showed increased anxiety in the same paradigm (Labarca et al., 2001; Ross et al., 2000). Also, Paylor et al. (1998) showed that α7 nAChR-lacking KO mice show decreased levels of anxiety in the EPM paradigm, while an α7-selective agonist, PNU-282987, was shown to increase anxiety in the OF paradigm (Pandya & Yakel, 2013). Similarly, desensitization of α7 nAChRs by using an α7 partial agonist, ABT-107, was found to reverse the anxiogenic effects of nicotine withdrawal (Yohn et al., 2014). Overall, these results suggest that in animals, nicotine has differential effects on anxiety in different strains/species and anxiety-related animal models. Results from the studies using genetically modified mice and pharmacological nAChR agonist/antagonists also showed that while elimination of β2-containing nAChRs can prevent the anxiolytic effects of nicotine, activation of these receptors can reverse the anxiogenic effects of nicotine withdrawal. Conversely, while activation of α7 nAChRs has an anxiogenic effect, inactivation of these receptors via using a partial antagonist α7 reverses the anxiogenic effects of nicotine withdrawal. Additionally, there is also evidence showing that other nAChR subtypes may play a role in modulating anxiety as well. For example, Gangitano, Salas, Teng, Perez, and De Biasi (2009) demonstrated that α5 KO animals showed altered hypothalamus–pituitary–adrenal axis function as they were shown to have lower basal corticosterone levels. In addition, while α5 KO animals showed no behavioral changes in the OF and light–dark box paradigms, female KO mice, but not male mice, showed reduced anxiety-like behavior in the EPM. Another nAChR subtype that might be important for anxiety modulation is β4 nAChR. Salas, Pieri, Fung, Dani, and De Biasi (2003) showed that β4 KO mice showed reduced anxiety in the EPM compared to WT littermates as well as reduced heart rate in this paradigm. However, β4 KO animals showed higher heart rate in another anxiety paradigm, social isolation. These studies suggest that in addition to the involvement of the major nAChR subtypes such as β2 and α7, anxiety phenotypes are controlled by a variety of different nAChRs.

In summary, the studies cited above reveal a strong bidirectional relationship between nicotine, anxiety, and fear learning. While human studies suggest that smoking may alleviate the symptoms associated with anxiety disorders (Beckham et al., 2005; Feldner et al., 2007; Greenberg et al., 2012; Thorndike et al., 2006), nicotine abstinence worsens those symptoms (Dedert et al., 2011) and results in higher rates of reinitiation of smoking (Beckham et al., 2012; Hapke et al., 2005; Lasser et al., 2000). In line with these results, the animal studies using fear conditioning suggest that initial acute nicotine results in enhancement of hippocampus-dependent fear learning and impairment of fear extinction, whereas acute nicotine during nicotine withdrawal further enhances hippocampus-dependent fear memories, which may prolong the course of the disorder. Evidence from the animal studies also suggests that the effects of nicotine on hippocampus-dependent fear memories require the activation and upregulation of the high-affinity β2-containing nAChRs (Davis & Gould, 2006, 2007, 2009; Davis et al., 2007; Kenney, Raybuck, & Gould, 2012; Portugal et al., 2008; Raybuck & Gould, 2009). However, the results from the studies using animal models of anxiety are less conclusive as the effects of nicotine on anxiety-related behavior in animals vary based on the strains/species and the task used in the study. This suggests that anxiety measured in different animal models may actually have different underlying neurobiological and genetic mechanisms. Nevertheless, results from these studies converge on the importance of α7/α4-containing nAChRs in the effects of nicotine on anxiety. Further investigation of the relationship between nicotine, specific nAChR subtypes, and anxiety as measured in animal models is still required to understand the underlying mechanisms of nicotine's effects on anxiety.

4. EFFECTS OF NICOTINE DEPENDENCE, WITHDRAWAL, AND nAChR REGULATION ON DEPRESSION

Depression is a common mental disorder that manifests itself with symptoms such as depressed mood, lowered interest in pleasure, fatigue, and psychomotor agitation or retardation (American Psychiatric Association, 2013). It has a lifetime prevalence of 10–25% in women and

6–10% in men (Moore & Bona, 2001), and affects approximately 20 million adults in the United States (CDC, 2010). Depression is especially common among the individuals with chronic health problems such as obesity, alcoholism, and smoking (Strine et al., 2008). In this section, we will review results suggesting a bidirectional relationship between depression and nicotine.

4.1 The Relationship Between Nicotine Dependence and Depression in Humans

An in-depth review of the relationship between depression and nicotine dependence in humans is provided in another chapter of this book. Similar to the link between anxiety disorders and smoking, depression and nicotine dependence have also a reciprocal relationship (Ischaki & Gratziou, 2009; John, Meyer, Rumpf, & Hapke, 2004). Previous studies have shown that depression symptoms are important determinants of smoking initiation, maintenance, and cessation, whereas nicotine dependence is associated with vulnerability for depression (Breslau, Kilbey, & Andreski, 1991; Fergusson, Goodwin, & Horwood, 2003; Glassman et al., 1990; Hall, Muñoz, Reus, & Sees, 1993; Morrell & Cohen, 2006; Paperwalla, Levin, Weiner, & Saravay, 2004). For example, nicotine dependence, as defined in Diagnostic and Statistical Manual of Mental Disorders, Revised Third Edition, predicted a threefold risk of developing major depression (Breslau & Johnson, 2000), and this relationship was maintained across different age groups such as adolescents (Killen et al., 2004), adults (Fergusson et al., 2003), and adults over age 60 (Glassman, Covey, Stetner, & Rivelli, 2001). There is also evidence suggesting that smoking during adolescence results in a fourfold increase in the likelihood of developing depressive symptoms later in life (Brook, Schuster, & Zhang, 2004; Choi, Patten, Gillin, Kaplan, & Pierce, 1997; Goodman & Capitman, 2000; Wu & Anthony, 1999). Another study found that the risk of developing depression was four times higher in heavy smokers than nonsmokers and increased time of smoking dependency was correlated with increased risk of depression (Klungsøyr, Nygård, Sørensen, & Sandanger, 2006), which suggests that the vulnerability for depression increases with higher rates of smoking.

Just as smokers show higher rates of depression, patients with major depression have been shown to have higher levels of smoking than the nonclinical population (Covey, Glassman, & Stetner, 1997, 1998; Fergusson et al., 2003). According to the self-medication hypothesis of nicotine dependence (Carmody, 1992; Markou, Kosten, & Koob, 1998; Pomerleau &

Pomerleau, 1985), this might be because nicotine reduces negative affect and works as an antidepressant. In support, patients with major depression increased smoking when they experienced depressive symptoms (Schleicher, Harris, Catley, & Nazir, 2009). In line with the self-medication hypothesis, there are also several studies showing that a majority of smokers enrolled in smoking cessation programs have a history of depressive episodes (Dalack, Glassman, Rivelli, Covey, & Stetner, 1995; Glassman et al., 1988; Lerman et al., 1996), and those with a history of depression failed to quit smoking twice as often as smokers without a prior depressive episode (Glassman et al., 1990). Similarly, it has also been indicated that smoking cessation increases the severity of depression symptoms (Covey et al., 1997; Swan, Ward, & Jack, 1996; West, Hajek, & Belcher, 1989). Furthermore, smokers with a history of depression have been shown to be more likely to experience another depressive episode 6 months after smoking cessation (Glassman et al., 2001). Consequently, these results suggest that the increased severity of depression and withdrawal symptoms upon smoking cessation might contribute to the lower levels of successful smoking cessation in people with depression.

4.2 Effects of Nicotine on Animal Models of Depression

Depression is a multifaceted mental disease with a distinct symptomatology, some of which can be captured in several different animal models (Matthews, Christmas, Swan, & Sorrell, 2005). Animal models of depression can be categorized as acute depression models (e.g., the forced swim test and tail-suspension test) and chronic depression models (e.g., learned helplessness and prolonged subordination stress; Stone & Lin, 2011). Consistent with the reports from human studies suggesting that nicotine decreases depressive symptoms, numerous studies have linked nicotine and nAChR agents with the modulation of depression-like behaviors in animals (see Mineur & Picciotto, 2010 for a review). In line with the above-described self-medication hypothesis of nicotine in depression, previous studies have shown that nicotine had an antidepressant effect in the forced swim test (Andreasen & Redrobe, 2009; Suemaru et al., 2006; Tizabi, Getachew, Rezvani, Hauser, & Overstreet, 2009, Tizabi et al., 1999; Vázquez-Palacios, Bonilla-Jaime, & Velázquez-Moctezuma, 2005), the chronic mild stress sucrose test (Andreasen, Henningsen, Bate, Christiansen, & Wiborg, 2011), and learned helplessness (Semba, Mataki, Yamada, Nankai, & Toru, 1998). Multiple studies have suggested that β2-containing nAChRs may play an important role in modulation of depression-like behavior. For example, KO mice that lack the β2 subunit of nAChRs show decreased amount of baseline immobility in both the FST and tail-suspension test, indication of reduced depression-like phenotype, compared to wild-type animals (Caldarone et al., 2004). Furthermore, it has also been shown that mecamylamine, a nonselective nAChR antagonist, has similar antidepressant-like effects in the FST and tail-suspension test in wild-type but not in β2 KO animals (Andreasen, Nielsen, & Redrobe, 2009; Caldarone et al., 2004; Rabenstein, Caldarone, & Picciotto, 2006). Also consistent with the modulatory role of β2-containing high-affinity nAChRs in the depression-like behavior, an α4β2 selective high-affinity nAChR antagonist, DHβE, has been found to decrease immobility in the FST and tail-suspension test (Andreasen, Olsen, Wiborg, & Redrobe, 2009; Rabenstein et al., 2006). Apart from the β2-containing nAChRs, there is also evidence showing that α7 subtype of nAChRs may also play a modulatory role in the depression-like behavior in animals. For example, Andreasen, Olsen, et al. (2009) found that MLA (α7-nAChR selective antagonist) also reduced immobility in the FST, while another study found that mecamylamine had antidepressant effects in the FST for wild-type but not on α7-nAChR KO mice. In addition to the β2 and α7 nAChRs, α5 may also be involved in the modulation of depressive-like behavior. For example, using a tail-suspension test and α5 nAChR-null mice, Gangitano et al. (2009) found that female but not male α5 KO mice showed reduced depressive-like behavioral phenotype. These studies suggest that inactivation of β2-containing nAChRs along with α7 nAChR activation results in reduction of depression-like behavior in the animal models of depression. Nevertheless, other nAChR subtypes, such as α5, may also be involved in the depression-related behavioral phenotype.

Interestingly, like nAChR antagonists mecamylamine, DhβE, and MLA, varenicline, a partial agonist of the α4β2 nAChRs and a full α7 agonist developed to aid smoking cessation in humans (Mihalak, Carroll, & Luetje, 2006), has also been found to have antidepressant effects in the FST (Rollema et al., 2009). Being a partial agonist of the α4β2 nAChRs, varenicline increases nAChR activity and maintains it at a submaximal level, and therefore, it prevents binding of the endogenous acetylcholine and effectively desensitizes nAChRs (Mineur & Picciotto, 2010; Papke & Heinemann, 1994). Similarly, other nicotinic partial agonists such as sazetidine-A and ispronicline (TC-1734 or AZD3480) also have antidepressant effects (Gatto et al., 2004; Xiao et al., 2006). However, Turner, Castellano, and Blendy (2010) reported that sazetidine-A, but not varenicline, decreased immobility in the FST and tail-suspension test measures of depression. The discrepancy between Rollema et al. (2009) and Turner et al. (2010) results is likely to be due to the different mouse strains used by these two studies, C57BL/6J mice and 129SvJ-C57BL/6J F1 hybrid mice, respectively. In support, previous research showed that the parental lines of the 129SvJ-C57BL/6J F1 hybrid mice show differences in motor function, anxiety-related behavior, and sensorimotor responsivity, which may contribute to the differential effects of partial nAChR agonists on depression-like behavior (Tarantino, Gould, Druhan, & Bucan, 2000). Overall, the results from the studies using partial nAChR agonist and direct nAChR antagonists converge on the importance of the nAChR desensitization in the antidepressant effects of nicotine.

In summary, studies indicate that depression is strongly affected by nicotine exposure. The human studies cited suggest that initially smoking may alleviate symptoms of depression and works as a mean to self-medicate, whereas smoking cessation results in the exacerbation of the depressive symptoms. In line with the human studies, the studies utilizing animal models of depression also showed that nicotine reduced depressive-like behavior in animals in several different models of depression (e.g., FST, chronic mild stress sucrose test, and learned helplessness). Evidence from the pharmacological inactivation studies and studies using global KO animals suggest that nicotine's antidepressant effect seems to be modulated by the β2-containing and α7 nAChRs. Interestingly, partial agonists of the α4β2 nAChRs such as sazetidine-A, and varenicline, which was developed for smoking cessation, seem to also help with the alleviation of the depressive-like behavior. Future studies that clarify the role of nAChR subtypes in depression will help develop better nAChR-based pharmacological agents for depression treatment.

5. CONCLUSION

The significantly higher rates of smoking are seen in the population with mental disorders such as ADHD, anxiety disorders, and depression, in comparison to the nonpatient population, suggesting a strong relationship between nicotine and symptomatology of these disorders. The studies reviewed in this chapter show that often acute nicotine produces effects that result in the short-term reduction of the symptoms associated with the mental illness. Consequently, patients with these disorders usually transition into chronic use of nicotine for the self-medication purposes. Nevertheless, there is also strong evidence suggesting that patients with mental disorders usually have difficulty quitting smoking. This is because the symptoms of the disorders usually worsen during the period of withdrawal, which results in shorter period of abstinence and eventually reinitiation of smoking. Several subunits of nAChRs, such as α4, β2, and α7, have been shown to directly modulate the severity of the symptoms of mental disorders and the effects of nicotine on these symptoms. Therefore, future studies examining the roles of specific nAChR subunits in mental illness may help to develop better treatments for mental disorders.

ACKNOWLEDGMENT

This work was funded with grant support from the National Institute on Drug Abuse (T.J.G., DA017949).

REFERENCES

Alkondon, M., & Albuquerque, E. X. (2004). The nicotinic acetylcholine receptor subtypes and their function in the hippocampus and cerebral cortex. *Progress in Brain Research*, *145*, 109–120.

Allison, C., & Shoaib, M. (2013). Nicotine improves performance in an attentional set shifting task in rats. *Neuropharmacology*, *64*, 314–320.

American Psychiatric Association. (2013). *DSM 5*. Washington, DC: American Psychological Association.

Ameringer, K. J., & Leventhal, A. M. (2012). Symptom dimensions of attention deficit hyperactivity disorder and nicotine withdrawal symptoms. *Journal of Addictive Diseases*, *31*(4), 363–375.

André, J. M., Gulick, D., Portugal, G. S., & Gould, T. J. (2008). Nicotine withdrawal disrupts both foreground and background contextual fear conditioning but not pre-pulse inhibition of the acoustic startle response in C57BL/6 mice. *Behavioural Brain Research*, *190*(2), 174–181.

Andreasen, J. T., Henningsen, K., Bate, S., Christiansen, S., & Wiborg, O. (2011). Nicotine reverses anhedonic-like response and cognitive impairment in the rat chronic mild stress model of depression: Comparison with sertraline. *Journal of Psychopharmacology*, *25*(8), 1134–1141.

Andreasen, J. T., Nielsen, E. Ø., & Redrobe, J. P. (2009). Chronic oral nicotine increases brain [3H] epibatidine binding and responsiveness to antidepressant drugs, but not nicotine, in the mouse forced swim test. *Psychopharmacology*, *205*(3), 517–528.

Andreasen, J. T., Olsen, G. M., Wiborg, O., & Redrobe, J. P. (2009). Antidepressant-like effects of nicotinic acetylcholine receptor antagonists, but not agonists, in the mouse forced swim and mouse tail suspension tests. *Journal of Psychopharmacology*, *23*(7), 797–804.

Andreasen, J. T., & Redrobe, J. P. (2009). Antidepressant-like effects of nicotine and mecamylamine in the mouse forced swim and tail suspension tests: Role of strain, test and sex. *Behavioural Pharmacology*, *20*(3), 286–295.

Arnold, H. M., Burk, J. A., Hodgson, E. M., Sarter, M., & Bruno, J. P. (2002). Differential acetylcholine release in rats performing a sustained attention task versus behavioral control tasks that do not explicitly tax attention. *Neuroscience*, *114*(2), 451–460.

Ashare, R. L., Falcone, M., & Lerman, C. (2014). Cognitive function during nicotine withdrawal: Implications for nicotine dependence treatment. *Neuropharmacology*, *76*, 581–591.

Bailey, C. D., De Biasi, M., Fletcher, P. J., & Lambe, E. K. (2010). The nicotinic acetylcholine receptor α5 subunit plays a key role in attention circuitry and accuracy. *The Journal of Neuroscience*, *30*(27), 9241–9252.

Bailey, C. D., Tian, M. K., Kang, L., O'Reilly, R., & Lambe, E. K. (2014). Chrna5 genotype determines the long-lasting effects of developmental in vivo nicotine exposure on prefrontal attention circuitry. *Neuropharmacology*, 77, 145–155.

Barbaresi, W. J., Katusic, S. K., Colligan, R. C., Pankratz, V. S., Weaver, A. L., Weber, K. J., et al. (2002). How common is attention-deficit/hyperactivity disorder? Incidence in a population-based birth cohort in Rochester, Minnesota. *Archives of Pediatrics & Adolescent Medicine*, *156*(3), 217–224.

Barbaresi, W., Katusic, S., Colligan, R., Weaver, A., Pankratz, V., Mrazek, D., et al. (2004). How common is attention-deficit/hyperactivity disorder? Towards resolution of the controversy: Results from a population-based study. *Acta Paediatrica*, *93*(s445), 55–59.

Bari, A., Dalley, J. W., & Robbins, T. W. (2008). The application of the 5-choice serial reaction time task for the assessment of visual attentional processes and impulse control in rats. *Nature Protocols*, *3*(5), 759–767.

Beane, M., & Marrocco, R. T. (2004). Norepinephrine and acetylcholine mediation of the components of reflexive attention: Implications for attention deficit disorders. *Progress in Neurobiology*, *74*(3), 167–181.

Beckham, J. C., Calhoun, P. S., Dennis, M. F., Wilson, S. M., & Dedert, E. A. (2012). Predictors of lapse in first week of smoking abstinence in PTSD and non-PTSD smokers. *Nicotine & Tobacco Research*, *15*(6), 1122–1129.

Beckham, J. C., Feldman, M. E., Vrana, S. R., Mozley, S. L., Erkanli, A., Clancy, C. P., et al. (2005). Immediate antecedents of cigarette smoking in smokers with and without posttraumatic stress disorder: A preliminary study. *Experimental and Clinical Psychopharmacology*, *13*(3), 219.

Blechert, J., Michael, T., Vriends, N., Margraf, J., & Wilhelm, F. H. (2007). Fear conditioning in posttraumatic stress disorder: Evidence for delayed extinction of autonomic, experiential, and behavioural responses. *Behaviour Research and Therapy*, *45*(9), 2019–2033.

Breslau, N., Davis, G. C., & Schultz, L. R. (2003). Posttraumatic stress disorder and the incidence of nicotine, alcohol, and other drug disorders in persons who have experienced trauma. *Archives of General Psychiatry*, *60*(3), 289–294.

Breslau, N., & Johnson, E. O. (2000). Predicting smoking cessation and major depression in nicotine-dependent smokers. *American Journal of Public Health*, *90*(7), 1122.

Breslau, N., Kilbey, M., & Andreski, P. (1991). Nicotine dependence, major depression, and anxiety in young adults. *Archives of General Psychiatry*, *48*(12), 1069–1074.

Breslau, N., Novak, S. P., & Kessler, R. C. (2004). Psychiatric disorders and stages of smoking. *Biological Psychiatry*, *55*(1), 69–76.

Brioni, J. D., O'Neill, A. B., Kim, D. J., Buckley, M. J., Decker, M. W., & Arneric, S. P. (1994). Anxiolytic-like effects of the novel cholinergic channel activator ABT-418. *Journal of Pharmacology and Experimental Therapeutics*, *271*(1), 353–361.

Briscione, M. A., Jovanovic, T., & Norrholm, S. D. (2014). Conditioned fear associated phenotypes as robust, translational indices of trauma-, stressor-, and anxiety-related behaviors. *Frontiers in Psychiatry*, *5*, 88.

Brook, J. S., Schuster, E., & Zhang, C. (2004). Cigarette smoking and depressive symptoms: A longitudinal study of adolescents and young adults. *Psychological Reports*, *95*(1), 159–166.

Caldarone, B. J., Harrist, A., Cleary, M. A., Beech, R. D., King, S. L., & Picciotto, M. R. (2004). High-affinity nicotinic acetylcholine receptors are required for antidepressant effects of amitriptyline on behavior and hippocampal cell proliferation. *Biological Psychiatry, 56*(9), 657–664.

Carmody, T. P. (1992). Affect regulation, nicotine addiction, and smoking cessation. *Journal of Psychoactive Drugs, 24*(2), 111–122.

Castellanos, F. X., Margulies, D. S., Kelly, C., Uddin, L. Q., Ghaffari, M., Kirsch, A., et al. (2008). Cingulate-precuneus interactions: A new locus of dysfunction in adult attention-deficit/hyperactivity disorder. *Biological Psychiatry, 63*(3), 332–337.

Centers for Disease Control. (2008). *Trends in current cigarette smoking among high school students and adults, United States, 1965–2007.*

Centers for Disease Control and Prevention (CDC) (2010). Tobacco use among middle and high school students—United States, 2000-2009. *MMWR. Morbidity and Mortality Weekly Report, 59*(33), 1063.

Choi, W. S., Patten, C. A., Gillin, J. C., Kaplan, R. M., & Pierce, J. P. (1997). Cigarette smoking predicts development of depressive symptoms among US adolescents. *Annals of Behavioral Medicine, 19*(1), 42–50.

Christianson, J. P., Fernando, A. B., Kazama, A. M., Jovanovic, T., Ostroff, L. E., & Sangha, S. (2012). Inhibition of fear by learned safety signals: A mini-symposium review. *The Journal of Neuroscience, 32*(41), 14118–14124.

Clark, R. E., & Squire, L. R. (1998). Classical conditioning and brain systems: The role of awareness. *Science, 280*(5360), 77–81.

Cohen, B. N., Mackey, E. D., Grady, S. R., McKinney, S., Patzlaff, N. E., Wageman, C. R., et al. (2012). Nicotinic cholinergic mechanisms causing elevated dopamine release and abnormal locomotor behavior. *Neuroscience, 200*, 31–41.

Counotte, D. S., Goriounova, N. A., Li, K. W., Loos, M., Van Der Schors, R. C., Schetters, D., et al. (2011). Lasting synaptic changes underlie attention deficits caused by nicotine exposure during adolescence. *Nature Neuroscience, 14*(4), 417–419.

Covey, L. S., Glassman, A. H., & Stetner, F. (1997). Major depression following smoking cessation. *The American Journal of Psychiatry, 154*(2), 263.

Covey, L. S., Glassman, A. H., & Stetner, F. (1998). Cigarette smoking and major depression. *Journal of Addictive Diseases, 17*(1), 35–46.

Czermak, C., Staley, J. K., Kasserman, S., Bois, F., Young, T., Henry, S., et al. (2008). β2 Nicotinic acetylcholine receptor availability in post-traumatic stress disorder. *The International Journal of Neuropsychopharmacology, 11*(03), 419–424.

Dalack, G. W., Glassman, A. H., Rivelli, S., Covey, L., & Stetner, F. (1995). Mood, major depression, and fluoxetine response in cigarette smokers. *American Journal of Psychiatry, 152*, 398–403.

Dani, J. A., & Heinemann, S. (1996). Molecular and cellular aspects of nicotine abuse. *Neuron, 16*(5), 905–908.

Davis, M., Falls, W. A., & Gewirtz, J. (2000). Neural systems involved in fear inhibition: Extinction and conditioned inhibition. In M. S. Myslobodsky & I. Weiner (Eds.), *Contemporary issues in modeling psychopathology* (pp. 113–141). Boston: Kluwer Academic Publishers.

Davis, J. A., & Gould, T. J. (2006). The effects of DHBE and MLA on nicotine-induced enhancement of contextual fear conditioning in C57BL/6 mice. *Psychopharmacology, 184*(3–4), 345–352.

Davis, J. A., & Gould, T. J. (2007). β2 subunit-containing nicotinic receptors mediate the enhancing effect of nicotine on trace cued fear conditioning in C57BL/6 mice. *Psychopharmacology, 190*(3), 343–352.

Davis, J. A., & Gould, T. J. (2009). Hippocampal nAChRs mediate nicotine withdrawal-related learning deficits. *European Neuropsychopharmacology, 19*(8), 551–561.

Davis, J. A., James, J. R., Siegel, S. J., & Gould, T. J. (2005). Withdrawal from chronic nicotine administration impairs contextual fear conditioning in C57BL/6 mice. *The Journal of Neuroscience*, *25*(38), 8708–8713.
Davis, J. A., Kenney, J. W., & Gould, T. J. (2007). Hippocampal α4β2 nicotinic acetylcholine receptor involvement in the enhancing effect of acute nicotine on contextual fear conditioning. *The Journal of Neuroscience*, *27*(40), 10870–10877.
Davis, J. A., Porter, J., & Gould, T. J. (2006). Nicotine enhances both foreground and background contextual fear conditioning. *Neuroscience Letters*, *394*(3), 202–205.
Decker, M. W., Brioni, J. D., Bannon, A. W., & Arneric, S. P. (1995). Diversity of neuronal nicotinic acetylcholine receptors: Lessons from behavior and implications for CNS therapeutics. *Life Sciences*, *56*(8), 545–570.
Decker, M. W., Brioni, J. D., Sullivan, J. P., Buckley, M. J., Radek, R. J., Raszkiewicz, J. L., et al. (1994). (S)-3-methyl-5-(1-methyl-2-pyrrolidinyl) isoxazole (ABT 418): A novel cholinergic ligand with cognition-enhancing and anxiolytic activities: II. In vivo characterization. *Journal of Pharmacology and Experimental Therapeutics*, *270*(1), 319–328.
Dedert, E. A., Calhoun, P. S., Harper, L. A., Dutton, C. E., McClernon, F. J., & Beckham, J. C. (2011). Smoking withdrawal in smokers with and without posttraumatic stress disorder. *Nicotine & Tobacco Research*, *14*(3), 372–376.
Dovis, S., Van der Oord, S., Huizenga, H. M., Wiers, R. W., & Prins, P. J. (2014). Prevalence and diagnostic validity of motivational impairments and deficits in visuospatial short-term memory and working memory in ADHD subtypes. *European Child & Adolescent psychiatry*, *24*(5), 575–590.
Elias, G. A., Gulick, D., Wilkinson, D. S., & Gould, T. J. (2010). Nicotine and extinction of fear conditioning. *Neuroscience*, *165*(4), 1063–1073.
English, B. A., Hahn, M. K., Gizer, I. R., Mazei-Robison, M., Steele, A., Kurnik, D. M., et al. (2009). Choline transporter gene variation is associated with attention-deficit hyperactivity disorder. *Journal of Neurodevelopmental Disorders*, *1*(4), 252–263.
Ericson, M., Olausson, P., Engel, J. A., & Söderpalm, B. (2000). Nicotine induces disinhibitory behavior in the rat after subchronic peripheral nicotinic acetylcholine receptor blockade. *European Journal of Pharmacology*, *397*(1), 103–111.
Feldner, M. T., Babson, K. A., & Zvolensky, M. J. (2007). Smoking, traumatic event exposure, and post-traumatic stress: A critical review of the empirical literature. *Clinical Psychology Review*, *27*(1), 14–45.
Fergusson, D. M., Goodwin, R. D., & Horwood, L. J. (2003). Major depression and cigarette smoking: Results of a 21-year longitudinal study. *Psychological Medicine*, *33*(08), 1357–1367.
File, S. E., Kenny, P. J., & Ouagazzal, A. M. (1998). Bimodal modulation by nicotine of anxiety in the social interaction test: Role of the dorsal hippocampus. *Behavioral Neuroscience*, *112*(6), 1423.
Fu, S. S., McFall, M., Saxon, A. J., Beckham, J. C., Carmody, T. P., Baker, D. G., et al. (2007). Post-traumatic stress disorder and smoking: A systematic review. *Nicotine & Tobacco Research*, *9*(11), 1071–1084.
Fuemmeler, B. F., Kollins, S. H., & McClernon, F. J. (2007). Attention deficit hyperactivity disorder symptoms predict nicotine dependence and progression to regular smoking from adolescence to young adulthood. *Journal of Pediatric Psychology*, *32*(10), 1203–1213.
Gangitano, D., Salas, R., Teng, Y., Perez, E., & De Biasi, M. (2009). Progesterone modulation of α5 nAChR subunits influences anxiety-related behavior during estrus cycle. *Genes, Brain and Behavior*, *8*(4), 398–406.
Gatto, G. J., Bohme, G. A., Caldwell, W. S., Letchworth, S. R., Traina, V. M., Obinu, M. C., et al. (2004). TC-1734: An orally active neuronal nicotinic acetylcholine receptor modulator with antidepressant, neuroprotective and long-lasting cognitive effects. *CNS Drug Reviews*, *10*(2), 147–166.

Gehricke, J. G., Loughlin, S. E., Whalen, C. K., Potkin, S. G., Fallon, J. H., Jamner, L. D., et al. (2007). Smoking to self-medicate attentional and emotional dysfunctions. *Nicotine & Tobacco Research*, *9*(Suppl. 4), S523–S536.

Glassman, A. H., Covey, L. S., Stetner, F., & Rivelli, S. (2001). Smoking cessation and the course of major depression: A follow-up study. *The Lancet*, *357*(9272), 1929–1932.

Glassman, A. H., Helzer, J. E., Covey, L. S., Cottler, L. B., Stetner, F., Tipp, J. E., et al. (1990). Smoking, smoking cessation, and major depression. *JAMA*, *264*(12), 1546–1549.

Glassman, A. H., Stetner, F., Walsh, T., Raizman, P. S., Fleiss, J. L., Cooper, T. B., et al. (1988). Heavy smokers, smoking cessation, and clonidine. *Journal of the American Medical Association*, *259*, 2863–2866.

Goodman, E., & Capitman, J. (2000). Depressive symptoms and cigarette smoking among teens. *Pediatrics*, *106*(4), 748–755.

Goodwin, R. D., Lewinsohn, P. M., & Seeley, J. R. (2005). Cigarette smoking and panic attacks among young adults in the community: The role of parental smoking and anxiety disorders. *Biological Psychiatry*, *58*(9), 686–693.

Gotti, C., Clementi, F., Fornari, A., Gaimarri, A., Guiducci, S., Manfredi, I., et al. (2009). Structural and functional diversity of native brain neuronal nicotinic receptors. *Biochemical Pharmacology*, *78*, 703–711.

Gould, T. J. (2003). Nicotine produces a within-subject enhancement of contextual fear conditioning in C57BL/6 mice independent of sex. *Integrative Physiological & Behavioral Science*, *38*(2), 124–132.

Gould, T. J., Feiro, O., & Moore, D. (2004). Nicotine enhances trace cued fear conditioning but not delay cued fear conditioning in C57BL/6 mice. *Behavioural Brain Research*, *155*(1), 167–173.

Gould, T. J., & Higgins, J. (2003). Nicotine enhances contextual fear conditioning in C57BL/6J mice at 1 and 7 days post-training. *Neurobiology of Learning and Memory*, *80*(2), 147–157.

Gould, T. J., & Leach, P. T. (2014). Cellular, molecular, and genetic substrates underlying the impact of nicotine on learning. *Neurobiology of Learning and Memory*, *107*, 108–132.

Gould, T. J., Portugal, G. S., Andre, J. M., Tadman, M. P., Marks, M. J., Kenney, J. W., et al. (2012). The duration of nicotine withdrawal-associated deficits in contextual fear conditioning parallels changes in hippocampal high affinity nicotinic acetylcholine receptor upregulation. *Neuropharmacology*, *62*, 2118–2125.

Gould, T. J., & Wehner, J. M. (1999). Nicotine enhancement of contextual fear conditioning. *Behavioural Brain Research*, *102*(1), 31–39.

Gould, T. J., Wilkinson, D. S., Yildirim, E., Blendy, J. A., & Adoff, M. D. (2014). Dissociation of tolerance and nicotine withdrawal-associated deficits in contextual fear. *Brain Research*, *1559*, 1–10.

Grant, B. F., Hasin, D. S., Chou, S. P., Stinson, F. S., & Dawson, D. A. (2004). Nicotine dependence and psychiatric disorders in the United States: Results from the national epidemiologic survey on alcohol and related conditions. *Archives of General Psychiatry*, *61*(11), 1107–1115.

Grant, B. F., Stinson, F. S., Dawson, D. A., Chou, S. P., Dufour, M. C., Compton, W., et al. (2004). Prevalence and co-occurrence of substance use disorders and independent mood and anxiety disorders: Results from the national epidemiologic survey on alcohol and related conditions. *Archives of General Psychiatry*, *61*(8), 807–816.

Greenberg, J. B., Ameringer, K. J., Trujillo, M. A., Sun, P., Sussman, S., Brightman, M., et al. (2012). Associations between posttraumatic stress disorder symptom clusters and cigarette smoking. *Psychology of Addictive Behaviors*, *26*(1), 89.

Greenberg, P. E., Sisitsky, T., Kessler, R. C., Finkelstein, S. N., Berndt, E. R., Davidson, J. R., et al. (1999). The economic burden of anxiety disorders in the 1990s. *The Journal of Clinical Psychiatry*, *60*(7), 427–435.

Grottick, A. J., & Higgins, G. A. (2000). Effect of subtype selective nicotinic compounds on attention as assessed by the five-choice serial reaction time task. *Behavioural Brain Research*, *117*(1), 197–208.
Hahn, B., Shoaib, M., & Stolerman, I. (2002). Nicotine-induced enhancement of attention in the five-choice serial reaction time task: The influence of task demands. *Psychopharmacology*, *162*(2), 129–137.
Hahn, B., & Stolerman, I. P. (2002). Nicotine-induced attentional enhancement in rats: Effects of chronic exposure to nicotine. *Neuropsychopharmacology*, *27*(5), 712–722.
Hall, S. M., Muñoz, R. F., Reus, V. I., & Sees, K. L. (1993). Nicotine, negative affect, and depression. *Journal of Consulting and Clinical Psychology*, *61*(5), 761.
Hammerness, P., Joshi, G., Doyle, R., Georgiopoulos, A., Geller, D., Spencer, T., et al. (2013). Do stimulants reduce the risk for cigarette smoking in youth with attention-deficit hyperactivity disorder? A prospective, long-term, open-label study of extended-release methylphenidate. *The Journal of Pediatrics*, *162*(1), 22–27.
Hampson, M., Driesen, N. R., Skudlarski, P., Gore, J. C., & Constable, R. T. (2006). Brain connectivity related to working memory performance. *The Journal of Neuroscience*, *26*(51), 13338–13343.
Hapke, U., Schumann, A., Rumpf, H. J., John, U., Konerding, U., & Meyer, C. (2005). Association of smoking and nicotine dependence with trauma and posttraumatic stress disorder in a general population sample. *The Journal of Nervous and Mental Disease*, *193*(12), 843–846.
Hogg, R. C., Raggenbass, M., & Bertrand, D. (2003). Nicotinic acetylcholine receptors: from structure to brain function. *Reviews of Physiology, Biochemistry and Pharmacology*, *147*, 1–46.
Howe, W. M., Berry, A. S., Francois, J., Gilmour, G., Carp, J. M., Tricklebank, M., et al. (2013). Prefrontal cholinergic mechanisms instigating shifts from monitoring of cues to cues-guided performance: Converging electrochemical and fMRI evidence from rats and humans. *The Journal of Neuroscience*, *33*(20), 8742–8752.
Howe, W. M., Ji, J., Parikh, V., Williams, S., Mocaër, E., Trocmé-Thibierge, C., et al. (2010). Enhancement of attentional performance by selective stimulation of $\alpha 4\beta 2^*$ nAChRs: Underlying cholinergic mechanisms. *Neuropsychopharmacology*, *35*(6), 1391–1401.
Hoyle, E., Genn, R. F., Fernandes, C., & Stolerman, I. P. (2006). Impaired performance of alpha7 nicotinic receptor knockout mice in the five-choice serial reaction time task. *Psychopharmacology*, *189*(2), 211–223.
Hughes, J. R., Keenan, R. M., & Yellin, A. (1989). Effect of tobacco withdrawal on sustained attention. *Addictive Behaviors*, *14*(5), 577–580.
Indredavik, M. S., Brubakk, A. M., Romundstad, P., & Vik, T. (2007). Prenatal smoking exposure and psychiatric symptoms in adolescence. *Acta Paediatrica*, *96*(3), 377–382.
Irvine, E. E., Cheeta, S., & File, S. E. (1999). Time-course of changes in the social interaction test of anxiety following acute and chronic administration of nicotine. *Behavioural Pharmacology*, *10*(6–7), 691–697.
Irvine, E. E., Cheeta, S., & File, S. E. (2001). Tolerance to nicotine's effects in the elevated plus-maze and increased anxiety during withdrawal. *Pharmacology Biochemistry and Behavior*, *68*(2), 319–325.
Ischaki, E., & Gratziou, C. (2009). Smoking and depression: Is smoking cessation effective? *Therapeutic Advances in Respiratory Disease*, *3*(1), 31–38.
Jacobsen, L. K., Krystal, J. H., Mencl, W. E., Westerveld, M., Frost, S. J., & Pugh, K. R. (2005). Effects of smoking and smoking abstinence on cognition in adolescent tobacco smokers. *Biological Psychiatry*, *57*(1), 56–66.

John, U., Meyer, C., Rumpf, H. J., & Hapke, U. (2004). Smoking, nicotine dependence and psychiatric comorbidity—A population-based study including smoking cessation after three years. *Drug and Alcohol Dependence*, *76*(3), 287–295.

Jones, S., Sudweeks, S., & Yakel, J. L. (1999). Nicotinic receptors in the brain: Correlating physiology with function. *Trends in Neurosciences*, *22*(12), 555–561.

Jovanovic, T., Kazama, A., Bachevalier, J., & Davis, M. (2012). Impaired safety signal learning may be a biomarker of PTSD. *Neuropharmacology*, *62*(2), 695–704.

Jovanovic, T., Norrholm, S. D., Blanding, N. Q., Davis, M., Duncan, E., Bradley, B., et al. (2010). Impaired fear inhibition is a biomarker of PTSD but not depression. *Depression and Anxiety*, *27*(3), 244–251.

Jovanovic, T., Norrholm, S. D., Fennell, J. E., Keyes, M., Fiallos, A. M., Myers, K. M., et al. (2009). Posttraumatic stress disorder may be associated with impaired fear inhibition: Relation to symptom severity. *Psychiatry Research*, *167*(1), 151–160.

Kenney, J. W., Raybuck, J. D., & Gould, T. J. (2012). Nicotinic receptors in the dorsal and ventral hippocampus differentially modulate contextual fear conditioning. *Hippocampus*, *22*(8), 1681–1690.

Kessler, R. C. (1997). The prevalence of psychiatric comorbidity. In S. Wetzler & W. C. Sanderson (Eds.), *Treatment strategies for patients with psychiatric comorbidity*. New York: John Wiley & Sons.

Kessler, R. C., Adler, L., Barkley, R., Biederman, J., Conners, C. K., Demler, O., et al. (2006). The prevalence and correlates of adult ADHD in the United States: Results from the National Comorbidity Survey Replication. *The American Journal of Psychiatry*, *163*(4), 716–723.

Kessler, R. C., Chiu, W. T., Demler, O., & Walters, E. E. (2005). Prevalence, severity, and comorbidity of 12-month DSM-IV disorders in the National Comorbidity Survey Replication. *Archives of General Psychiatry*, *62*(6), 617–627.

Killen, J. D., Robinson, T. N., Ammerman, S., Hayward, C., Rogers, J., Samuels, D., et al. (2004). Major depression among adolescent smokers undergoing treatment for nicotine dependence. *Addictive Behaviors*, *29*(8), 1517–1526.

Klungsøyr, O., Nygård, J. F., Sørensen, T., & Sandanger, I. (2006). Cigarette smoking and incidence of first depressive episode: An 11-year, population-based follow-up study. *American Journal of Epidemiology*, *163*(5), 421–432.

Koenen, K. C., Hitsman, B., Lyons, M. J., Niaura, R., McCaffery, J., Goldberg, J., et al. (2005). A twin registry study of the relationship between posttraumatic stress disorder and nicotine dependence in men. *Archives of General Psychiatry*, *62*(11), 1258–1265.

Kollins, S. H., English, J. S., Roley, M. E., O'Brien, B., Blair, J., Lane, S. D., et al. (2013). Effects of smoking abstinence on smoking-reinforced responding, withdrawal, and cognition in adults with and without attention deficit hyperactivity disorder. *Psychopharmacology*, *227*(1), 19–30.

Kuryatov, A., Luo, J., Cooper, J., & Lindstrom, J. (2005). Nicotine acts as a pharmacological chaperone to up-regulate human alpha4beta2 acetylcholine receptors. *Molecular Pharmacology*, *68*, 1839–1851.

Kutlu, M. G., & Gould, T. J. (2014). Acute nicotine delays extinction of contextual fear in mice. *Behavioural Brain Research*, *263*, 133–137.

Kutlu, M. G., Oliver, C., & Gould, T. J. (2014). The effects of acute nicotine on contextual safety discrimination. *Journal of Psychopharmacology*, *28*(11), 1064–1070. http://dx.doi.org/10.1177/0269881114552743.

Labarca, C., Schwarz, J., Deshpande, P., Schwarz, S., Nowak, M. W., Fonck, C., et al. (2001). Point mutant mice with hypersensitive α4 nicotinic receptors show dopaminergic deficits and increased anxiety. *Proceedings of the National Academy of Sciences of the United States of America*, *98*(5), 2786–2791.

Lambert, N. M., & Hartsough, C. S. (1998). Prospective study of tobacco smoking and substance dependencies among samples of ADHD and non-ADHD participants. *Journal of Learning Disabilities*, *31*(6), 533–544.

Lanius, R. A., Williamson, P. C., Densmore, M., Boksman, K., Gupta, M. A., Neufeld, R. W., et al. (2001). Neural correlates of traumatic memories in posttraumatic stress disorder: A functional MRI investigation. *American Journal of Psychiatry*, *158*(11), 1920–1922.

Lasser, K., Boyd, J. W., Woolhandler, S., Himmelstein, D. U., McCormick, D., & Bor, D. H. (2000). Smoking and mental illness: A population-based prevalence study. *JAMA*, *284*(20), 2606–2610.

Leach, P. T., Cordero, K. A., & Gould, T. J. (2013). The effects of acute nicotine, chronic nicotine, and withdrawal from chronic nicotine on performance of a cued appetitive response. *Behavioral Neuroscience*, *127*(2), 303.

Lee, C. T., Fuemmeler, B. F., McClernon, F. J., Ashley-Koch, A., & Kollins, S. H. (2013). Nicotinic receptor gene variants interact with attention deficient hyperactive disorder symptoms to predict smoking trajectories from early adolescence to adulthood. *Addictive Behaviors*, *38*(11), 2683–2689.

Lerman, C., Audrain, J., Orleans, C. T., Boyd, R., Gold, K., Main, D., et al. (1996). Investigation of mechanisms linking depressed mood to nicotine dependence. *Addictive Behaviors*, *21*, 9–19.

Levin, E. D., Conners, C. K., Sparrow, E., Hinton, S. C., Erhardt, D., Meck, W. H., et al. (1996). Nicotine effects on adults with attention-deficit/hyperactivity disorder. *Psychopharmacology*, *123*(1), 55–63.

Lissek, S., Powers, A. S., McClure, E. B., Phelps, E. A., Woldehawariat, G., Grillon, C., et al. (2005). Classical fear conditioning in the anxiety disorders: A meta-analysis. *Behaviour Research and Therapy*, *43*(11), 1391–1424.

Lister, R. G. (1990). Ethologically-based animal models of anxiety disorders. *Pharmacology & Therapeutics*, *46*(3), 321–340.

Logue, S. F., Paylor, R., & Wehner, J. M. (1997). Hippocampal lesions cause learning deficits in inbred mice in the Morris water maze and conditioned-fear task. *Behavioral Neuroscience*, *111*(1), 104.

Lotfipour, S., Byun, J. S., Leach, P., Fowler, C. D., Murphy, N. P., Kenny, P. J., et al. (2013). Targeted deletion of the mouse α2 nicotinic acetylcholine receptor subunit gene (Chrna2) potentiates nicotine-modulated behaviors. *The Journal of Neuroscience*, *33*(18), 7728–7741.

Markou, A., Kosten, T. R., & Koob, G. F. (1998). Neurobiological similarities in depression and drug dependence: A self-medication hypothesis. *Neuropsychopharmacology*, *18*(3), 135–174.

Marks, M. J., Burch, J. B., & Collins, A. C. (1983). Effects of chronic nicotine infusion on tolerance development and nicotinic receptors. *Journal of Pharmacology and Experimental Therapeutics*, *226*, 817–825.

Marks, M. J., & Collins, A. C. (1982). Characterization of nicotine binding in mouse brain and comparison with the binding of alpha-bungarotoxin and quinuclidinyl benzilate. *Molecular Pharmacology*, *22*(3), 554–564.

Marks, M. J., Grady, S. R., & Collins, A. C. (1993). Downregulation of nicotinic receptor function after chronic nicotine infusion. *Journal of Pharmacology and Experimental Therapeutics*, *266*(3), 1268–1276.

Matta, S. G., Balfour, D. J., Benowitz, N. L., Boyd, R. T., Buccafusco, J. J., Caggiula, A. R., et al. (2007). Guidelines on nicotine dose selection for in vivo research. *Psychopharmacology*, *190*(3), 269–319.

Matthews, K., Christmas, D., Swan, J., & Sorrell, E. (2005). Animal models of depression: Navigating through the clinical fog. *Neuroscience & Biobehavioral Reviews*, *29*(4), 503–513.

McClernon, F. J., Van Voorhees, E. E., English, J., Hallyburton, M., Holdaway, A., & Kollins, S. H. (2011). Smoking withdrawal symptoms are more severe among smokers with ADHD and independent of ADHD symptom change: Results from a 12-day contingency-managed abstinence trial. *Nicotine & Tobacco Research, 13*(9), 784–792.
McEchron, M. D., Bouwmeester, H., Tseng, W., Weiss, C., & Disterhoft, J. F. (1998). Hippocampectomy disrupts auditory trace fear conditioning and contextual fear conditioning in the rat. *Hippocampus, 8*(6), 638–646.
McGaughy, J., Kaiser, T., & Sarter, M. (1996). Behavioral vigilance following infusions of 192 IgG-saporin into the basal forebrain: Selectivity of the behavioral impairment and relation to cortical AChE-positive fiber density. *Behavioral Neuroscience, 110*(2), 247–265.
McGehee, D. S. (1999). Molecular diversity of neuronal nicotinic acetylcholine receptors. *Annals of the New York Academy of Sciences, 868*(1), 565–577.
McGranahan, T. M., Patzlaff, N. E., Grady, S. R., Heinemann, S. F., & Booker, T. K. (2011). α4β2 nicotinic acetylcholine receptors on dopaminergic neurons mediate nicotine reward and anxiety relief. *The Journal of Neuroscience, 31*(30), 10891–10902.
Mendrek, A., Monterosso, J., Simon, S. L., Jarvik, M., Brody, A., Olmstead, R., et al. (2006). Working memory in cigarette smokers: Comparison to non-smokers and effects of abstinence. *Addictive Behaviors, 31*(5), 833–844.
Michael, T., Blechert, J., Vriends, N., Margraf, J., & Wilhelm, F. H. (2007). Fear conditioning in panic disorder: Enhanced resistance to extinction. *Journal of Abnormal Psychology, 116*(3), 612.
Mihailescu, S., & Drucker-Colin, R. (2000). Nicotine, brain nicotinic receptors, and neuropsychiatric disorders. *Archives of Medical Research, 31*, 131–144.
Mihalak, K. B., Carroll, F. I., & Luetje, C. W. (2006). Varenicline is a partial agonist at α4β2 and a full agonist at α7 neuronal nicotinic receptors. *Molecular Pharmacology, 70*(3), 801–805.
Milad, M. R., Pitman, R. K., Ellis, C. B., Gold, A. L., Shin, L. M., Lasko, N. B., et al. (2009). Neurobiological basis of failure to recall extinction memory in posttraumatic stress disorder. *Biological Psychiatry, 66*(12), 1075–1082.
Milberger, S., Biederman, J., Faraone, S. V., Chen, L., & Jones, J. (1996). Is maternal smoking during pregnancy a risk factor for attention deficit hyperactivity disorder in children? *The American Journal of Psychiatry, 153*(9), 1138–1142.
Milberger, S., Biederman, J., Faraone, S. V., & Jones, J. (1998). Further evidence of an association between maternal smoking during pregnancy and attention deficit hyperactivity disorder: Findings from a high-risk sample of siblings. *Journal of Clinical Child Psychology, 27*(3), 352–358.
Mineka, S., & Zinbarg, R. (2006). A contemporary learning theory perspective on the etiology of anxiety disorders: It's not what you thought it was. *American Psychologist, 61*(1), 10.
Mineur, Y. S., & Picciotto, M. R. (2010). Nicotine receptors and depression: Revisiting and revising the cholinergic hypothesis. *Trends in Pharmacological Sciences, 31*(12), 580–586.
Mirza, N. R., & Bright, J. L. (2001). Nicotine-induced enhancements in the five-choice serial reaction time task in rats are strain-dependent. *Psychopharmacology, 154*(1), 8–12.
Moore, J. D., & Bona, J. R. (2001). Advances in the pathophysiology and treatment of psychiatric disorders for internal medicine: Depression and dysthymia. *Medical Clinics of North America, 85*, 631–644.
Morrell, H. E., & Cohen, L. M. (2006). Cigarette smoking, anxiety, and depression. *Journal of Psychopathology and Behavioral Assessment, 28*(4), 281–295.
Myers, K. M., & Davis, M. (2006). Mechanisms of fear extinction. *Molecular Psychiatry, 12*(2), 120–150.
Nelson, M. E., Kuryatov, A., Choi, C. H., Zhou, Y., & Lindstrom, J. (2003). Alternate stoichiometries of alpha4beta2 nicotinic acetylcholine receptors. *Molecular Pharmacology, 63*, 332–341.

Nunes, E. V., Covey, L. S., Brigham, G., Hu, M. C., Levin, F. R., Somoza, E., et al. (2013). Treating nicotine dependence by targeting attention deficit hyperactivity disorder (ADHD): The role of ADHD severity and treatment response in a randomized controlled trial. *The Journal of Clinical Psychiatry*, *74*(10), 983.

Olale, F., Gerzanich, V., Kuryatov, A., Wang, F., & Lindstrom, J. (1997). Chronic nicotine exposure differentially affects the function of human alpha3, alpha4, and alpha7 neuronal nicotinic receptor subtypes. *Journal of Pharmacology and Experimental Therapeutics*, *283*, 675–683.

O'Neill, A. B., & Brioni, J. D. (1994). Benzodiazepine receptor mediation of the anxiolytic-like effect of (−)-nicotine in mice. *Pharmacology Biochemistry and Behavior*, *49*(3), 755–757.

Ortega, L. A., Tracy, B. A., Gould, T. J., & Parikh, V. (2013). Effects of chronic low-and high-dose nicotine on cognitive flexibility in C57BL/6J mice. *Behavioural Brain Research*, *238*, 134–145.

Ouagazzal, A. M., Kenny, P. J., & File, S. E. (1999). Modulation of behaviour on trials 1 and 2 in the elevated plus-maze test of anxiety after systemic and hippocampal administration of nicotine. *Psychopharmacology*, *144*(1), 54–60.

Pandya, A. A., & Yakel, J. L. (2013). Activation of the α7 nicotinic ACh receptor induces anxiogenic effects in rats which is blocked by a 5-HT_{1a} receptor antagonist. *Neuropharmacology*, *70*, 35–42.

Paperwalla, K. N., Levin, T. T., Weiner, J., & Saravay, S. M. (2004). Smoking and depression. *Medical Clinics of North America*, *88*(6), 1483–1494.

Papke, R. L., & Heinemann, S. F. (1994). Partial agonist properties of cytisine on neuronal nicotinic receptors containing the beta 2 subunit. *Molecular Pharmacology*, *45*(1), 142–149.

Parikh, V., Kozak, R., Martinez, V., & Sarter, M. (2007). Prefrontal acetylcholine release cue detection on multiple time scales. *Neuron*, *56*(1), 141–154.

Parikh, V., Peters, M. S., Blakely, R. D., & Sarter, M. (2013). The presynaptic choline transporter imposes limits on sustained cortical acetylcholine release and attention. *The Journal of Neuroscience*, *33*(6), 2326–2337.

Paylor, R., Nguyen, M., Crawley, J. N., Patrick, J., Beaudet, A., & Orr-Urtreger, A. (1998). α7 nicotinic receptor subunits are not necessary for hippocampal-dependent learning or sensorimotor gating: A behavioral characterization of Acra7-deficient mice. *Learning & Memory*, *5*(4), 302–316.

Phillips, R. G., & LeDoux, J. E. (1992). Differential contribution of amygdala and hippocampus to cued and contextual fear conditioning. *Behavioral Neuroscience*, *106*(2), 274.

Picciotto, M. R., Zoli, M., Rimondini, R., Léna, C., Marubio, L. M., Pich, E. M., et al. (1998). Acetylcholine receptors containing the β2 subunit are involved in the reinforcing properties of nicotine. *Nature*, *391*(6663), 173–177.

Picciotto, M. R., Caldarone, B. J., King, S. L., & Zachariou, V. (2000). Nicotinic receptors in the brain: links between molecular biology and behavior. *Neuropsychopharmacology*, *22*(5), 451–465.

Polina, E. R., Rovaris, D. L., de Azeredo, L. A., Mota, N. R., Vitola, E. S., Silva, K. L., et al. (2014). ADHD diagnosis may influence the association between polymorphisms in nicotinic acetylcholine receptor genes and tobacco smoking. *Neuromolecular Medicine*, *16*(2), 389–397.

Pomerleau, O. F., Downey, K. K., Stelson, F. W., & Pomerleau, C. S. (1995). Cigarette smoking in adult patients diagnosed with attention deficit hyperactivity disorder. *Journal of Substance Abuse*, 7(3), 373–378.

Pomerleau, O. F., & Pomerleau, C. S. (1985). Neuroregulators and the reinforcement of smoking: Towards a biobehavioral explanation. *Neuroscience & Biobehavioral Reviews*, *8*(4), 503–513.

Pomerleau, O. F., Pomerleau, C. S., Marks, J. L., Snedecor, S. M., Mehringer, A. M., Brouwer, R. J. N., et al. (2003). Prolonged nicotine patch use in quitters with past abstinence-induced depressed mood. *Journal of Substance Abuse Treatment*, *24*(1), 13–18.

Portugal, G. S., & Gould, T. J. (2009). Nicotine withdrawal disrupts new contextual learning. *Pharmacology Biochemistry and Behavior*, *92*(1), 117–123.

Portugal, G. S., Kenney, J. W., & Gould, T. J. (2008). β2 subunit containing acetylcholine receptors mediate nicotine withdrawal deficits in the acquisition of contextual fear conditioning. *Neurobiology of Learning and Memory*, *89*(2), 106–113.

Portugal, G. S., Wilkinson, D. S., Kenney, J. W., Sullivan, C., & Gould, T. J. (2012). Strain-dependent effects of acute, chronic, and withdrawal from chronic nicotine on fear conditioning. *Behavior Genetics*, *42*(1), 133–150.

Potter, A. S., Bucci, D. J., & Newhouse, P. A. (2012). Manipulation of nicotinic acetylcholine receptors differentially affects behavioral inhibition in human subjects with and without disordered baseline impulsivity. *Psychopharmacology*, *220*(2), 331–340.

Potter, A. S., & Newhouse, P. A. (2008). Acute nicotine improves cognitive deficits in young adults with attention-deficit/hyperactivity disorder. *Pharmacology Biochemistry and Behavior*, *88*(4), 407–417.

Quirk, G. J., & Mueller, D. (2007). Neural mechanisms of extinction learning and retrieval. *Neuropsychopharmacology*, *33*(1), 56–72.

Rabenstein, R. L., Caldarone, B. J., & Picciotto, M. R. (2006). The nicotinic antagonist mecamylamine has antidepressant-like effects in wild-type but not β2-or α7-nicotinic acetylcholine receptor subunit knockout mice. *Psychopharmacology*, *189*(3), 395–401.

Raybuck, J. D., & Gould, T. J. (2009). Nicotine withdrawal-induced deficits in trace fear conditioning in C57BL/6 mice—A role for high-affinity β2 subunit-containing nicotinic acetylcholine receptors. *European Journal of Neuroscience*, *29*(2), 377–387.

Rezvani, A. H., Cauley, M. C., Johnson, E. C., Gatto, G. J., & Levin, E. D. (2012). Effects of AZD3480, a neuronal nicotinic acetylcholine receptor agonist, and donepezil on dizocilpine-induced attentional impairment in rats. *Psychopharmacology*, *223*(3), 251–258.

Rezvani, A. H., Cauley, M., Sexton, H., Xiao, Y., Brown, M. L., Paige, M. A., et al. (2011). Sazetidine-A, a selective α4β2 nicotinic acetylcholine receptor ligand: Effects on dizocilpine and scopolamine-induced attentional impairments in female Sprague–Dawley rats. *Psychopharmacology*, *215*(4), 621–630.

Rezvani, A. H., Cauley, M., Xiao, Y., Kellar, K. J., & Levin, E. D. (2013). Effects of chronic sazetidine-A, a selective α4β2 neuronal nicotinic acetylcholine receptors desensitizing agent on pharmacologically-induced impaired attention in rats. *Psychopharmacology*, *226*(1), 35–43.

Rodgers, R. J. (1997). Animal models of 'anxiety': Where next? *Behavioural Pharmacology*, *8*(6–7), 477–496.

Rollema, H., Guanowsky, V., Mineur, Y. S., Shrikhande, A., Coe, J. W., Seymour, P. A., et al. (2009). Varenicline has antidepressant-like activity in the forced swim test and augments sertraline's effect. *European Journal of Pharmacology*, *605*(1), 114–116.

Ross, S. A., Wong, J. Y., Clifford, J. J., Kinsella, A., Massalas, J. S., Horne, M. K., et al. (2000). Phenotypic characterization of an α4 neuronal nicotinic acetylcholine receptor subunit knock-out mouse. *The Journal of Neuroscience*, *20*(17), 6431–6441.

Rukstalis, M., Jepson, C., Patterson, F., & Lerman, C. (2005). Increases in hyperactive-impulsive symptoms predict relapse among smokers in nicotine replacement therapy. *Journal of Substance Abuse Treatment*, *28*(4), 297–304.

Salas, R., Pieri, F., Fung, B., Dani, J. A., & De Biasi, M. (2003). Altered anxiety-related responses in mutant mice lacking the β4 subunit of the nicotinic receptor. *The Journal of Neuroscience*, *23*(15), 6255–6263.

Salminen, O., Drapeau, J. A., McIntosh, J. M., Collins, A. C., Marks, M. J., & Grady, S. R. (2007). Pharmacology of α-conotoxin MII-sensitive subtypes of nicotinic acetylcholine receptors isolated by breeding of null mutant mice. *Molecular Pharmacology*, *71*(6), 1563–1571.

Salminen, O., Murphy, K. L., McIntosh, J. M., Drago, J., Marks, M. J., Collins, A. C., et al. (2004). Subunit composition and pharmacology of two classes of striatal presynaptic nicotinic acetylcholine receptors mediating dopamine release in mice. *Molecular Pharmacology*, *65*(6), 1526–1535.

Schleicher, H. E., Harris, K. J., Catley, D., & Nazir, N. (2009). The role of depression and negative affect regulation expectancies in tobacco smoking among college students. *Journal of American College Health*, *57*(5), 507–512.

Schneider, T., Ilott, N., Brolese, G., Bizarro, L., Asherson, P. J., & Stolerman, I. P. (2011). Prenatal exposure to nicotine impairs performance of the 5-choice serial reaction time task in adult rats. *Neuropsychopharmacology*, *36*(5), 1114–1125.

Schoenfelder, E. N., Faraone, S. V., & Kollins, S. H. (2014). Stimulant treatment of ADHD and cigarette smoking: A meta-analysis. *Pediatrics*, *133*(6), 1070–1080.

Schwartz, R. D., & Kellar, K. J. (1983). Nicotinic cholinergic receptor binding sites in the brain: Regulation in vivo. *Science*, *220*, 214–216.

Semba, J. I., Mataki, C., Yamada, S., Nankai, M., & Toru, M. (1998). Antidepressantlike effects of chronic nicotine on learned helplessness paradigm in rats. *Biological Psychiatry*, *43*(5), 389–391.

Semenova, S., Stolerman, I. P., & Markou, A. (2007). Chronic nicotine administration improves attention while nicotine withdrawal induces performance deficits in the 5-choice serial reaction time task in rats. *Pharmacology Biochemistry and Behavior*, *87*(3), 360–368.

Shin, L. M., Rauch, S. L., & Pitman, R. K. (2006). Amygdala, medial prefrontal cortex, and hippocampal function in PTSD. *Annals of the New York Academy of Sciences*, *1071*(1), 67–79.

Shoaib, M., & Bizarro, L. (2005). Deficits in a sustained attention task following nicotine withdrawal in rats. *Psychopharmacology*, *178*(2–3), 211–222.

Slotkin, T. A., Seidler, F. J., Qiao, D., Aldridge, J. E., Tate, C. A., Cousins, M. M., et al. (2005). Effects of prenatal nicotine exposure on primate brain development and attempted amelioration with supplemental choline or vitamin C: Neurotransmitter receptors, cell signaling and cell development biomarkers in fetal brain regions of rhesus monkeys. *Neuropsychopharmacology*, *30*(1), 129–144.

Solomon, P. R., Vander Schaaf, E. R., Thompson, R. F., & Weisz, D. J. (1986). Hippocampus response. *Behavioral Neuroscience*, *100*(5), 729.

Sonntag, H., Wittchen, H. U., Höfler, M., Kessler, R. C., & Stein, M. B. (2000). Are social fears and DSM-IV social anxiety disorder associated with smoking and nicotine dependence in adolescents and young adults? *European Psychiatry*, *15*(1), 67–74.

Stone, E. A., & Lin, Y. (2011). Open-space forced swim model of depression for mice. *Current Protocols in Neuroscience*, *54*, 9.36.1–9.36.8.

Strine, T., Mokdad, A., Balluz, L., Gonzalez, O., Crider, R., Berry, J., et al. (2008). Depression and anxiety in the United States: Findings from the 2006 behavioral risk factor surveillance system. *Psychiatric Services*, *59*(12), 1383–1390.

Suemaru, K., Yasuda, K., Cui, R., Li, B., Umeda, K., Amano, M., et al. (2006). Antidepressant-like action of nicotine in forced swimming test and brain serotonin in mice. *Physiology & Behavior*, *88*(4), 545–549.

Swan, G. E., Ward, M. M., & Jack, L. M. (1996). Abstinence effects as predictors of 28-day relapse in smokers. *Addictive Behaviors*, *21*, 481–490.

Tarantino, L. M., Gould, T. J., Druhan, J. P., & Bucan, M. (2000). Behavior and mutagenesis screens: The importance of baseline analysis of inbred strains. *Mammalian Genome*, *11*(7), 555–564.

Tercyak, K. P., Lerman, C., & Audrain, J. (2002). Association of attention-deficit/hyperactivity disorder symptoms with levels of cigarette smoking in a community sample of adolescents. *Journal of the American Academy of Child & Adolescent Psychiatry, 41*(7), 799–805.

Thakur, G. A., Sengupta, S. M., Grizenko, N., Schmitz, N., Pagé, V., & Joober, R. (2013). Maternal smoking during pregnancy and ADHD: A comprehensive clinical and neurocognitive characterization. *Nicotine & Tobacco Research, 15*(1), 149–157.

Thorndike, F. P., Wernicke, R., Pearlman, M. Y., & Haaga, D. A. (2006). Nicotine dependence, PTSD symptoms, and depression proneness among male and female smokers. *Addictive Behaviors, 31*(2), 223–231.

Tian, S., Gao, J., Han, L., Fu, J., Li, C., & Li, Z. (2008). Prior chronic nicotine impairs cued fear extinction but enhances contextual fear conditioning in rats. *Neuroscience, 153*(4), 935–943.

Tizabi, Y., Getachew, B., Rezvani, A. H., Hauser, S. R., & Overstreet, D. H. (2009). Antidepressant-like effects of nicotine and reduced nicotinic receptor binding in the Fawn-Hooded rat, an animal model of co-morbid depression and alcoholism. *Progress in Neuro-Psychopharmacology and Biological Psychiatry, 33*(3), 398–402.

Tizabi, Y., Overstreet, D. H., Rezvani, A. H., Louis, V. A., Clark, E., Jr., Janowsky, D. S., et al. (1999). Antidepressant effects of nicotine in an animal model of depression. *Psychopharmacology, 142*(2), 193–199.

Treur, J. L., Willemsen, G., Bartels, M., Geels, L. M., van Beek, J. H., Huppertz, C., et al. (in press). Smoking during adolescence as a risk factor for attention problems. *Biological Psychiatry.*

Tronson, N. C., Schrick, C., Guzman, Y. F., Huh, K. H., Srivastava, D. P., Penzes, P., et al. (2009). Segregated populations of hippocampal principal CA1 neurons mediating conditioning and extinction of contextual fear. *The Journal of Neuroscience, 29*(11), 3387–3394.

Turner, J. R., Castellano, L. M., & Blendy, J. A. (2010). Nicotinic partial agonists varenicline and sazetidine-A have differential effects on affective behavior. *Journal of Pharmacology and Experimental Therapeutics, 334*(2), 665–672.

Vansickel, A. R., Stoops, W. W., Glaser, P. E., Poole, M. M., & Rush, C. R. (2011). Methylphenidate increases cigarette smoking in participants with ADHD. *Psychopharmacology, 218*(2), 381–390.

Vázquez-Palacios, G., Bonilla-Jaime, H., & Velázquez-Moctezuma, J. (2005). Antidepressant effects of nicotine and fluoxetine in an animal model of depression induced by neonatal treatment with clomipramine. *Progress in Neuro-Psychopharmacology and Biological Psychiatry, 29*(1), 39–46.

Viñals, X., Molas, S., Gallego, X., Fernández-Montes, R. D., Robledo, P., Dierssen, M., et al. (2012). Overexpression of α3/α5/β4 nicotinic receptor subunits modifies impulsive-like behavior. *Drug and Alcohol Dependence, 122*(3), 247–252.

Wehner, J. M., Keller, J. J., Keller, A. B., Picciotto, M. R., Paylor, R., Booker, T. K., et al. (2004). Role of neuronal nicotinic receptors in the effects of nicotine and ethanol on contextual fear conditioning. *Neuroscience, 129*(1), 11–24.

Weiss, M., & Murray, C. (2003). ADHD and driving safety. *Canadian Medical Association Journal, 169*(1), 16–17.

West, R. J., Hajek, P., & Belcher, M. (1989). Severity of withdrawal symptoms as a predictor of outcome of an attempt to quit smoking. *Psychological Medicine, 19*, 981–985.

Whiteaker, P., Davies, A. R., Marks, M. J., Blagbrough, I. S., Potter, B. V., Wolstenholme, A. J., et al. (1999). An autoradiographic study of the distribution of binding sites for the novel α7-selective nicotinic radioligand [3H]-methyllycaconitine in the mouse brain. *European Journal of Neuroscience, 11*(8), 2689–2696.

Wigestrand, M. B., Mineur, Y. S., Heath, C. J., Fonnum, F., Picciotto, M. R., & Walaas, S. I. (2011). Decreased α4β2 nicotinic receptor number in the absence of mRNA changes suggests post-transcriptional regulation in the spontaneously hypertensive rat model of ADHD. *Journal of Neurochemistry, 119*(1), 240–250.

Wilens, T. E., Vitulano, M., Upadhyaya, H., Adamson, J., Sawtelle, R., Utzinger, L., et al. (2008). Cigarette smoking associated with attention deficit hyperactivity disorder. *The Journal of Pediatrics, 153*(3), 414–419.

Wilkinson, D. S., & Gould, T. J. (2013). Withdrawal from chronic nicotine and subsequent sensitivity to nicotine challenge on contextual learning. *Behavioural Brain Research, 250*, 58–61.

Wu, L. T., & Anthony, J. C. (1999). Tobacco smoking and depressed mood in late childhood and early adolescence. *American Journal of Public Health, 89*(12), 1837–1840.

Xiao, Y., Fan, H., Musachio, J. L., Wei, Z. L., Chellappan, S. K., Kozikowski, A. P., et al. (2006). Sazetidine-A, a novel ligand that desensitizes α4β2 nicotinic acetylcholine receptors without activating them. *Molecular Pharmacology, 70*(4), 1454–1460.

Yochum, C., Doherty-Lyon, S., Hoffman, C., Hossain, M. M., Zelikoff, J. T., & Richardson, J. R. (2014). Prenatal cigarette smoke exposure causes hyperactivity and aggressive behavior: Role of altered catecholamines and BDNF. *Experimental Neurology, 254*, 145–152.

Yohn, N. L., Turner, J. R., & Blendy, J. A. (2014). Activation of α4β2*/α6β2* nicotinic receptors alleviates anxiety during nicotine withdrawal without upregulating nicotinic receptors. *Journal of Pharmacology and Experimental Therapeutics, 349*(2), 348–354.

Young, J. W., Finlayson, K., Spratt, C., Marston, H. M., Crawford, N., Kelly, J. S., et al. (2004). Nicotine improves sustained attention in mice: Evidence for involvement of the α7 nicotinic acetylcholine receptor. *Neuropsychopharmacology, 29*(5), 891–900.

Young, J. W., Meves, J. M., & Geyer, M. A. (2013). Nicotinic agonist-induced improvement of vigilance in mice in the 5-choice continuous performance test. *Behavioural Brain Research, 240*, 119–133.

Zhu, J., Lee, K. P., Spencer, T. J., Biederman, J., & Bhide, P. G. (2014). Transgenerational transmission of hyperactivity in a mouse model of ADHD. *The Journal of Neuroscience, 34*(8), 2768–2773.

Zhu, J., Zhang, X., Xu, Y., Spencer, T. J., Biederman, J., & Bhide, P. G. (2012). Prenatal nicotine exposure mouse model showing hyperactivity, reduced cingulate cortex volume, reduced dopamine turnover, and responsiveness to oral methylphenidate treatment. *The Journal of Neuroscience, 32*(27), 9410–9418.

Ziedonis, D., Hitsman, B., Beckham, J. C., Zvolensky, M., Adler, L. E., Audrain-McGovern, J., et al. (2008). Tobacco use and cessation in psychiatric disorders: National Institute of Mental Health report. *Nicotine & Tobacco Research, 10*(12), 1691–1715.

Zwart, R., & Vijverberg, H. P. (1998). Four pharmacologically distinct subtypes of alpha4beta2 nicotinic acetylcholine receptor expressed in Xenopus laevis oocytes. *Molecular Pharmacology, 54*, 1124–1131.

CHAPTER EIGHT

The Role of Depression in the Uptake and Maintenance of Cigarette Smoking

Janet Audrain-McGovern*[,1], Adam M. Leventhal[†], David R. Strong[‡]
*Department of Psychiatry, Perelman School of Medicine, University of Pennsylvania, Philadelphia, Pennsylvania, USA
[†]Departments of Preventive Medicine and Psychology, Keck School of Medicine, University of Southern California, Los Angeles, California, USA
[‡]Department of Family Medicine and Public Health, University of California, San Diego, California, USA
[1]Corresponding author: e-mail address: audrain@mail.med.upenn.edu

Contents

Abstract

Cigarette smoking and depression both account for significant morbidity, mortality, and economic burden. The prevalence of both smoking and depression increase across mid-to-late adolescence and show high rates of comorbidity. While little is known about why smoking is disproportionately higher among depressed adolescents than adolescents without depression, emerging research has begun to offer some initial insights. The high rates of comorbidity between depression and smoking emphasize the importance of identifying intervention targets to inform smoking prevention efforts for this high-risk group. Interventions during adolescence may lessen the prevalence of depression-prone adult smokers. Depression is over-represented among adult smokers and

International Review of Neurobiology, Volume 124
ISSN 0074-7742
http://dx.doi.org/10.1016/bs.irn.2015.07.004

contributes to lower smoking cessation rates. Negative mood management and pharmacotherapy have been the central focus of smoking cessation interventions for depression-prone populations to date. Converging lines of research highlight novel smoking cessation targets such as the maintenance of positive mood and reward regulation. Smoking cessation research in depression-prone smokers is critical to identify efficacious treatments that will ultimately decrease the excess smoking burden for this population.

1. THE SMOKING AND DEPRESSION PROBLEM

Cigarette smoking is the leading preventable cause of morbidity and mortality in the UnitedStates, accounting for almost 500,000 deaths each year (CDC, 2008). About 20% of adults in the general population smoke cigarettes, exacting an annual economic toll of ~193 billion dollars (CDC, 2008). About 20% of adolescents regularly smoke cigarettes (CDC, 2014), with the percentage of regular smokers doubling from mid-to-late adolescence (CDC, 2014). Of those adults who have ever smoked regularly, the majority began smoking during adolescence (USDHHS, 2012).

Depression is one of the most common mental health disorders among adolescents and adults and accounts for more morbidity, mortality, and economic burden than any other psychiatric diagnosis (Lewinsohn, Hops, Roberts, Seeley, & Andrews, 1993; Murray & Lopez, 1996, 1997; Reinherz et al., 1993). About 20% of adolescents have at least one major depressive episode by 18 years of age (USDHHS, 2012). The initial onset of major depression usually occurs during mid-adolescence (Lewinsohn, Rohde, & Seeley, 1998; Lewinsohn, Rohde, Seeley, Klein, & Gotlib, 2000; Williamson, Birmaher, Axelson, Ryan, & Dahl, 2004). Another 20–30% of adolescents experience subthreshold depressive symptoms (depressive episode that does not meet diagnostic criteria for a major depression) and are considered to be at significant risk for the development of major depression, or its recurrence (Audrain-McGovern, Rodriguez, Rodgers, et al., 2012; Fergusson, Horwood, Ridder, & Beautrais, 2005; Georgiades, Lewinsohn, Monroe, & Seeley, 2006; Lewinsohn, Shankman, Gau, & Klein, 2004; Saluja et al., 2004). Research has consistently shown that the prevalence of subthreshold and major depression increase significantly during mid-to-late adolescence (Clarke et al., 2001; Costello et al., 2002; Weller & Weller, 2000).

Cigarette smoking is also over-represented in adult smokers prone to depression. About 40% of individuals with Major Depressive Disorder (MDD) smoke cigarettes (Grant, Hasin, Chou, Stinson, & Dawson, 2004), and up to 60% of smokers seeking to enroll in smoking cessation programs have had at least one lifetime episode of MDD (Brown et al., 2001; Cinciripini et al., 2005; Ginsberg, Hall, Reus, & Muñoz, 1995; Glassman et al., 1988; Hall, Munoz, & Reus, 1994; Hall et al., 1998), compared to 15% of the general population (SAMHSA, 2004). Further, ~50% of smokers who enroll in smoking cessation programs have elevated depression symptoms, but do not meet diagnostic criteria for MDD (Lerman et al., 1996; Niaura et al., 2001).

Both smoking and depression increase across mid-to-late adolescence and show high rates of comorbidity. Section 2 summarizes the evidence base regarding how depression influences smoking behavior, how smoking may influence the development of depression, and the possibility that depression and smoking are associated because of common variables. The chapter also describes how individual and environmental mechanisms might shed light on a more complex relationship between adolescent depression and smoking. Parsing depression into its distinct dimensions and accounting for individual differences has the potential to further our understanding of the comorbidity. Finally, we discuss how insights from recent research might be used to develop more effective adolescent smoking prevention and intervention approaches.

Because cigarette smoking is over-represented in adults with past and current MDD, it is critical to evaluate smoking cessation approaches in this population. Section 3 summarizes the difficulties with smoking cessation among this population and the smoking cessation treatment research among adults with depression. By integrating preclinical and clinical research, we highlight potentially novel targets to improve the smoking cessation outcomes among adults with depression.

2. ADOLESCENT SMOKING UPTAKE AND DEPRESSION

Research indicates that depression contributes to adolescent smoking uptake. Major depression is associated with a 19% increase in the average daily smoking rate and a 75% increase in the odds of being nicotine dependent from mid-adolescence to young adulthood (ages 16–21 years old) (Fergusson, Goodwin, & Horwood, 2003). Adolescents with a history of major depression are two times more likely to experiment with cigarettes

(Rohde, Lewinsohn, Brown, Gau, & Kahler, 2003), 50% more likely to progress to daily smoking, and significantly less likely to quit by age 25 years old compared to those without a history of major depression (Rohde et al., 2003). Longitudinal research also supports an association between smoking and subthreshold depression. Adolescents with higher depressive symptoms are more likely to initiate experimentation and to progress to regular smoking compared to adolescents with lower depressive symptoms (Escobedo, Reddy, & Giovino, 1998; Killen & Fortmann, 1997; Patton et al., 1998; Wang et al., 1999).

Studies also support the opposite direction of influence, that is, smoking appears to influence the development of subthreshold depression among adolescents. Adolescent heavy smokers are almost four times more likely to develop high depressive symptoms compared to adolescent nonsmokers (Goodman & Capitman, 2000). Similarly, smoking initiation and regular smoking have been shown to predict the development of depressive symptoms (Choi, Pierce, Gilpin, Farkas, & Berry, 1997; Goodman & Capitman, 2000; Steuber & Danner, 2006; Wu & Anthony, 1999).

Further, research has documented bidirectional influences for smoking and depression among adolescents. Smoking predicts the development of subthreshold and major depression and subthreshold and major depression predict the development of regular smoking (Brown, Lewinsohn, Seeley, & Wagner, 1996; Wang, Fitzhugh, Turner, Fu, & Westerfield, 1996; Windle & Windle, 2001). Daily smoking and nicotine dependence predict a history of major depression and a history of major depression predicts the onset of daily smoking and progression to nicotine dependence in young adults (Breslau, Kilbey, & Andreski, 1993; Breslau, Peterson, Schultz, Chilcoat, & Andreski, 1998). A more recent study found a slightly different relationship. Depression symptoms predicted smoking progression across mid-to-late adolescence while smoking uptake mitigated depression symptoms (Audrain-McGovern, Rodriguez, & Kassel, 2009). Although some studies have not found support for a reciprocal relationship (Fergusson et al., 2003; Wang et al., 1999; Wu & Anthony, 1999), a meta-analysis of 15 longitudinal studies of adolescents utilizing various measures of smoking and depression syndrome and symptoms found evidence of a bidirectional relation, with the effect of depression status on smoking likelihood (pooled Odds Ratio = 1.73) being larger than the effect of smoking on depression (pooled Odds Ratio = 1.41; Chaiton, Cohen, O'Loughlin, & Rehm, 2009).

While the studies noted above have examined whether the nature of the relationship between adolescent depression and smoking is causal (either unidirectional or bidirectional), the relationship may not be causal, but rather due to a common set of genetic or environmental factors that contribute to both depression and smoking. However, the available support for common genetic influences is limited (Dierker, Avenevoli, Stolar, & Merikangas, 2002; Kendler et al., 1993; Leventhal, Ray, Rhee, & Unger, 2012; Lyons et al., 2008) and controlling for potential confounds is insufficient to explain the association between depression and smoking (Breslau et al., 1993, 1998; Chaiton et al., 2009; Fergusson et al., 2003; Martini, Wagner, & Anthony, 2002; Munafo, Hitsman, Rende, Metcalfe, & Niaura, 2008).

Methodological variability may help to account for some of the disparate findings in this body of research (Dierker et al., 2002; Johnson, Rhee, Chase, & Breslau, 2004; Kassel, Stroud, & Paronis, 2003; Munafo et al., 2008; Windle & Windle, 2001). Some of the methodological issues include different definitions of smoking or stage of smoking (initiation, regular smoking, nicotine dependence), inconsistent measures of depression and depression symptoms, large measurement intervals without repeated assessments of smoking and depression, lack of control for potential confounding variables, and only assessing one directional causal path. More recent research suggests that evaluating indirect or mediated effects between depression and smoking, heterogeneity in the impact of smoking on depression, and parsing depression into its composite dimensions may provide a richer understanding of the comorbidity.

2.1 How Does Adolescent Depression Influence Smoking Uptake?

While the comorbidity between adolescent depression and smoking has been well documented, very little is known about why smoking is disproportionately higher among depressed adolescents compared to adolescents without depression. The mechanisms that account for the comorbidity have received very little attention. While studies have begun to account for why depression influences adolescent smoking uptake, we are aware of only one study that investigated how smoking may influence the development of adolescent depression. Below, we review the available studies which provide a preliminary framework for understanding how depression may foster smoking uptake by considering the challenges associated with depression and the role of cigarette smoking in lessening these challenges. Elucidating mechanisms responsible for the comorbidity may shed light on novel smoking prevention approaches for this at-risk population.

For example, depression may make an adolescent more vulnerable to social influences to smoke, such as peers who smoke, which in turn, promotes smoking uptake. Adolescent depression is often accompanied by interpersonal difficulties including making and retaining friends (Connell & Dishion, 2006; Fergusson et al., 2003). The impact of peer smoking on adolescent smoking acquisition is well documented (Kobus, 2003). Adolescents with higher levels of depression may be more sensitive to peer behavior, more likely to select nonconventional peers, or both (Hoffman, Monge, Chou, & Valente, 2007; Kobus, 2003; Wang, Eddy, & Fitzhugh, 2000; Wills & Cleary, 1999). A greater number of smoking peers provides better access to cigarettes, promotes a normative perception of smoking, and may be a source of peer approval (Chaiton & Zhang, 2007; Ritt-Olson et al., 2005). These issues may be especially salient for youth with higher levels of depression symptoms. A prospective cohort study examined whether peer smoking accounted for the effects of depression on adolescent smoking uptake (Audrain-McGovern, Rodriguez, & Kassel, 2009). The study found that depression symptoms measured at mid-adolescence (age 14) predicted smoking progression across mid-to-late adolescence (ages 14–18 years old). Peer smoking mediated these developmental influences such that higher depression symptoms predicted an increase in the number of smoking peers, which in turn predicted smoking progression (Audrain-McGovern, Rodriguez, & Kassel, 2009). While most smoking prevention programs address social influences to smoke, considering the unique peer-related issues faced by depressed adolescents may be necessary to protect this vulnerable group.

Converging research offers further insight into reward-related mechanisms that may increase our understanding of the link between smoking and depression among adolescents. Depression, even at the level of elevated symptoms, is often accompanied by withdrawal, less involvement in activities, and diminished reward from usual activities (Jacobson et al., 1996; Lewinsohn et al., 1998; MacPherson et al., 2010). Behavioral theory suggests that this decline in overall reinforcement creates a vulnerability for substance use, especially when substances are readily available (Higgins, Heil, & Lussier, 2004; Jacobson et al., 1996; Rogers et al., 2008). As such, behavioral treatment for depression addresses the frequency of reinforcing activities (Jacobson et al., 1996), given that depressed mood prompts a withdrawal from pleasant activities that then exacerbates depression (Lewinsohn, Youngren, & Grosscup, 1979). Research has documented the clinical relevance of alternative reinforcers (e.g., hobbies, physical activity, special interests) for substance use in general and for smoking specifically (Higgins, Heil, & Lussier, 2004). For example, a lower level of alternative reinforcers has been shown to predict adolescent smoking progression, young adult smoking status, and abstinence from other substances of abuse (Audrain-McGovern, Rodriguez, Epstein, et al., 2009; Audrain-McGovern, Rodriguez, & Kassel, 2009; Audrain-McGovern et al., 2004; Rogers et al., 2008; Van Etten, Higgins, Budney, & Badger, 1998). Based on the literature, it is plausible that depression symptoms contribute to a decline in alternative reinforcers, which in turn, increase the likelihood of smoking uptake in youth.

Within a prospective cohort study, we examined the impact that declines in alternative reinforcers can have on smoking escalation in depressed youth. We observed that higher depression symptoms in late adolescence predicted declines in alternative reinforcers across time, which in turn, predicted increases in smoking uptake and smoking rate in young adulthood (Audrain-McGovern, Rodriguez, Rodgers, & Cuevas, 2011). The results of the study documented that depression is indeed a risk factor for subsequent smoking because of its negative influence on alternative reinforcers. As alternative reinforcers decline in those with heightened depression symptoms, the reinforcing value smoking may increase as smoking may be an easy way to increase the level of pleasurable experiences (Green & Fisher, 2000; Perkins, Hickcox, & Grobe, 2000). In addition to being an easily available reinforcer, nicotine may mitigate reward deficits by enhancing the reward from those alternatives that are available in the environment (Audrain-McGovern, Wileyto, Ashare, Cuevas, & Strasser, 2014; Caggiula et al., 2009; Chaudhri et al., 2006; Spring et al., 2008). Animal models suggest that nicotine potentiates reward from drug and nondrug reinforcers by increasing the sensitivity of brain reward systems (Kenny & Markou, 2006). Promoting greater alternative reinforcers or mitigating declines may prevent smoking uptake and increases in smoking rate.

In an effort to better understand reward-related events that may occur early in the smoking acquisition process for youth with elevated depression symptoms, we examined expectations of rewards from smoking in a subsequent study (Audrain-McGovern, Rodriguez, Rodgers, et al., 2012). As elevated depression symptoms precipitate declines in involvement in other reinforcers or the pleasure derived from those reinforcers, adolescents may develop expectations that other activities, such as smoking confer reward. We hypothesized that that adolescents with elevated depression symptoms develop greater smoking reward expectations that then influence smoking progression. The study revealed that expectations of smoking reward facilitated smoking uptake among depressed adolescents. Higher depression symptoms across mid-to-late adolescence predicted a 17% increase in smoking reward expectations, which in turn predicted a 23% increase in the odds of smoking progression (Audrain-McGovern, Rodriguez, Rodgers, et al., 2012). Similarly, a study in college students found that the cross-sectional association between depression history and smoking initiation was mediated by reward expectations (McChargue, Spring, Cook, & Neumann, 2004b).

Research has shown that expectations that smoking would remove displeasure, such as reducing negative affect, predict smoking escalation in adolescents (Heinz, Kassel, Berbaum, & Mermelstein, 2010). Our study of smoking reward expectations added to this research by showing that expectations that smoking will provide pleasure motivate smoking among depressed youth. Expectations that smoking confers rewards (e.g., enjoyable, something to do when bored, stay thin) increase the likelihood that depressed adolescents will choose to smoke and, as a result, learn that smoking is rewarding. As such, their smoking experience may then serve to validate their expectations. Smoking reward expectations may identify depressed adolescents at risk of smoking. Addressing alternative ways to meet the reward expectations rather than smoking may be an important component to include in adolescent smoking prevention and cessation interventions.

2.2 Which Adolescents Are More Vulnerable?

Most multifactorial models of depression suggest that depression can be parsed into two primary affect dimensions of low positive affect (i.e., diminished levels of happiness, enjoyment, and positive moods) and high negative affect (i.e., sadness, distress) (Clark & Watson, 1991; Shafer, 2006b). Although conceptualizations of the relationship between adolescent smoking and depression have focused on smoking to relieve negative affect, there has been growing interest in the relationship between smoking and positive affect (Forbes, 2009). Adolescent smokers experience increases in positive affect and decreases in negative affect following smoking (Hedeker, Mermelstein, Berbaum, & Campbell, 2009) and greater overall affective changes following smoking predict smoking escalation (Memelstein, Hedeker, Flay, & Shiffman, 2007). Thus, low positive affect and high negative affect reflect empirically distinct processes that may both play a role in adolescent smoking.

Recent studies have shown that negative affect predicted smoking escalation among boys, but only among boys whose smoking motives were to manage affect (Weinstein & Mermelstein, 2013b). Rapid smoking escalation was associated with improved negative affect among boys (Weinstein & Mermelstein, 2013a). However, this effect was not observed at the 15-month follow-up for boys who had a higher smoking rate and had symptoms of nicotine dependence (Weinstein & Mermelstein, 2013a). Our own findings indicate that adolescent males with higher levels of negative affect were 17% more likely to progress in their smoking across mid-to-late adolescence (Audrain-McGovern, Rodriguez, & Leventhal, 2015). Positive affect had no relation with smoking progression for males.

In contrast to males, this study found that lower positive affect coupled with higher negative affect increased the odds of smoking uptake for adolescent females by 25% (Audrain-McGovern et al., 2015). Recent studies have shown that variability in negative affect predicted smoking escalation among girls (Weinstein & Mermelstein, 2013b), and that rapid smoking escalation was associated with less negative affect variability (Weinstein & Mermelstein, 2013a). In an effort to regulate positive affect, adolescent females with may seek out potent reinforcers like smoking in order to enhance positive affect (Audrain-McGovern, Rodriguez, Leventhal, et al., 2012; Franken, Zijlstra, & Muris, 2006). While cigarettes alone may produce a mild increment in positive emotions, the ability of nicotine to enhance pleasure derived from other reinforcers (Caggiula et al., 2009; Chaudhri et al., 2006; Kenny & Markou, 2006; Spring et al., 2008) may have an important impact on positive affect among girls. These findings emphasize the importance of higher negative affect in smoking uptake for all adolescents, and highlight a unique role for positive affect in smoking uptake for female adolescents.

Hedonic capacity is another variable that may explain individual differences in adolescent smoking uptake among affectively vulnerable adolescents. Hedonic capacity is a stable dispositional ability to experience positive emotion in response to events that are typically considered rewarding (Bogdan & Pizzagalli, 2009; Meehl, 1987, 2001). The capacity to derive pleasure from natural reinforcers is a hallmark characteristic of clinical depression (anhedonia) (Shafer, 2006a; Watson, 2005). As a trait that exists on a continuum, hedonic capacity shows considerable variability among the general population (Fawcett, Clark, Scheftner, & Hedeker, 1983; Harvey, Pruessner, Czechowska, & Lepage, 2007; Meehl, 1975,

1987). At one end of the continuum are individuals who find a broad array of life experiences as rewarding and experience a high degree of pleasure in response to rewards. At the other end of the continuum are individuals who have a blunted response to typical rewarding experiences. Blunted responsiveness to natural reinforcers may lead to declines in those activities and a greater willingness to try pharmacological reinforcers for pleasure, such as cigarettes, to stimulate an under-responsive reward system. Indeed, hedonic capacity has been shown to moderate nicotine's ability to increase positive affect in adults such that smoking a nicotinized cigarette heightened positive mood in smokers low in hedonic capacity compared to a denicotinized cigarette, but had no effect on mood in smokers higher in hedonic capacity (Cook, Spring, & McChargue, 2007). Those lower in hedonic capacity tend to place higher priority on smoking in comparison to other reinforcers (Leventhal et al., 2014; Leventhal, Waters, Kahler, Ray, & Sussman, 2009).

In a prospective cohort study, we measured hedonic capacity at age 15 years old and examined its impact on adolescent smoking 18 months later (Audrain-McGovern, Rodriguez, Leventhal, et al., 2012). We hypothesized that reduced hedonic capacity might predispose adolescents to smoke and to eventually become regular smokers. Our results revealed that adolescents low in hedonic capacity were over two and a half times more likely to have smoked a cigarette in the past month at age 15 years old and to show a 90% increase in the rate of smoking escalation across the following 18 months compared to adolescents with high hedonic capacity. In a separate study, we found that 14-year-old never smokers reported greater expectations that smoking produces pleasure and was associated with lifetime smoking (Stone & Leventhal, 2014). Hedonic capacity appears to be a marker for a preexisting vulnerability for the early progression of smoking. Given that nicotine stimulates dopamine neurotransmission and other brain systems associated with reward, it may help adolescents with low hedonic capacity to overcome pleasure deficits by stimulating an under-responsive reward system (Barrett, Boileau, Okker, Pihl, & Dagher, 2004; Brody et al., 2004, 2009; Epping-Jordan, Watkins, Koob, & Markou, 1998). Nicotine may allow adolescents with low hedonic capacity to increase their hedonic tone and derive more pleasure from natural reinforcers in their environment. As such, this trait may be an important variable to identify youth at risk for smoking and an important population to target for smoking prevention and smoking cessation efforts.

2.3 Clinical Implications

Adolescent smoking prevention programs to date have been largely ineffective at preventing adolescent smoking (Skara & Sussman, 2003; Sussman, Lichtman, Ritt, & Pallonen, 1999). While population-based policies have been relatively more successful, almost 20% of adolescents still smoke (CDC, 2012). The adolescent smoking rate, coupled with the limited effectiveness of smoking prevention approaches to date, underscores the need to identify novel factors that drive smoking and their interactions with other variables to inform more effective smoking prevention interventions (Skara & Sussman, 2003; Sussman et al., 1999).

The high rates of comorbidity between depression and smoking emphasize the importance of targeting smoking prevention efforts to this high-risk group. There is an evidence base to support a shift in the long-standing focus of smoking prevention models from social intervention targets to affect-related targets. A shift in prevention targets would be consistent with the over-representation of smoking among affectively vulnerable populations (Audrain-McGovern, Rodriguez, & Kassel, 2009; Audrain-McGovern, Rodriguez, Rodgers, et al., 2012; Windle & Windle, 2001). The observation that affective influences on smoking are already present at age 14 years old highlights mid-adolescence as a critical period for intervention and emphasizes the importance of affect regulation skills to reduce the risk of smoking during this developmental period. Earlier interventions may mitigate the comorbidity that appears to track well into adulthood thereby decreasing the disproportionate smoking attributable morbidity and mortality in a population prone to depression.

Interventions that have depression prevention and management components could have an important impact on smoking uptake as well as subsequent depression. Social influence-based models of smoking prevention or intervention address peer influences to smoke (e.g., cigarette offer refusal skills). However, research suggests that in order to lessen peer influences on smoking behavior among depressed adolescents, it may also be important to address peer issues from the standpoint of adolescent depression (Audrain-McGovern, Rodriguez, & Kassel, 2009). For example, addressing coping and negative mood management skills, limited social networks, need for peer approval, and accessing nonsmoking peer groups may be especially beneficial for adolescents with elevated depression symptoms.

Moreover, addressing alternative ways to meet the reward expectations of smoking may prove to be an important part of adolescent smoking prevention and cessation interventions for depressed youth (Audrain-McGovern, Rodriguez, Rodgers, et al., 2012). The expectations of smoking include a source of pleasure, something to do, a way to handle problems, a method to stay thin and look mature, and a way to relax. Although relatively less attention has been paid to these features in adolescent smoking prevention programs, counteracting these beliefs may be critical to preventing smoking uptake among youth who have elevated depression symptoms. Likewise, promoting greater alternative reinforcers to smoking or mitigating declines in alternative reinforcers may prevent smoking uptake and increases in smoking rate (Audrain-McGovern et al., 2011), as has been shown in the prevention of alcohol problems in young adults (Murphy, Dennhardt, et al., 2012; Murphy, Skidmore, et al., 2012; Reynolds, Macpherson, Tull, Baruch, & Lejuez, 2011).

For adolescents low in hedonic capacity, finding suitable alternative reinforcers may require greater effort as they will likely have a diminished response to typical alternative rewards (Audrain-McGovern, Rodriguez, Leventhal, et al., 2012). If smoking's role is to permit an adolescent to derive greater reward from natural reinforcers, it may be more effective to help these adolescents acquire skills to enhance pleasure from typical reinforcers rather than identifying reinforcers potent enough to elicit feelings of pleasure. Using Positive Psychology approaches to the treatment of depression, behavioral skills to extract greater enjoyment derived from daily reinforcers—such as mindful savoring of pleasant experiences to extend their duration and intensity—may help ameliorate the pleasure deficit associated with lower hedonic capacity (Lee Duckworth, Steen, & Seligman, 2005; Seligman, Steen, Park, & Peterson, 2005).

Increasing our understanding of the link between affect and adolescent smoking for males and females is needed as the field moves to develop the next generation of smoking prevention interventions. Our research suggests that the mitigation of negative mood for all adolescents, and the maximization of positive mood for female adolescents could prevent the reliance on cigarettes to perform these functions (Audrain-McGovern et al., 2014, 2015). Different strategies may be required to offset the smoking risk carried by low positive affect versus high negative affect (Seligman et al., 2005). Further research of gender differences in how affect influences smoking uptake is meaningful, especially in light of the fact that men and women smoke for different reasons (Perkins, 2009) and because they respond differently to specific smoking cessation pharmacotherapies (Perkins, 2001).

2.4 Conclusions

The prevalence of smoking, depression, and their comorbidity increases across mid-to-late adolescence, highlighting an important period to evaluate the onset, and the inter-relationship between smoking and depression over time. This period of adolescence may determine whether a smoking habit is established and may offer a unique opportunity to intervene upon smoking behavior as well as depression, at an optimal time, to prevent lifelong health habits associated with morbidity and mortality

Greater understanding of the comorbidity may provide distinct etiological or intervention implications (i.e., depression is a risk factor for smoking or smoking is a risk factor for depression, or they both influence each other). If smoking is a risk factor for depression, prevention programs would need to focus on smoking to effectively reduce risk for both depression and smoking. If depression is a risk factor for smoking, prevention programs would need to focus on depression to effectively reduce the risk for both depression and smoking. Such research will help determine which one (depression or smoking) needs to be the focus of prevention efforts to effectively impact both depression and smoking or if targeting both is important.

While we are beginning to understand the mechanisms that account for the comorbidity between depression and smoking, more research is necessary. Identifying mechanisms will provide targets for smoking and/or depression prevention efforts. Evaluating moderators of the comorbidity between smoking and depression will aid in the identification of vulnerable adolescents to target for intervention. Research examining the conjoint developmental heterogeneity between smoking and depression could yield a refined understanding of the relationship between smoking and depression for specific adolescents (who needs an intervention and what type), who is at risk for adopting and maintaining smoking within the context of depression and who is at risk for developing depression within the context of smoking, when the relationship is established (optimize intervention timing), and what variables characterize at-risk groups (potential intervention content).

3. SMOKING CESSATION AND DEPRESSION

Because persistent cigarette smoking is over-represented in individuals with past and current MDD, it is critical to evaluate smoking cessation approaches in this population. Unfortunately, smokers with current MDD, and a past history of MDD are often excluded from smoking cessation treatment studies, especially those involving pharmacotherapy (Hitsman, Borrelli, McChargue, Spring, & Niaura, 2003). However, research suggests that a significant percentage of these smokers are interested in quitting and MDD severity, duration of MDD history, and recurrent MDD episodes of depression are unrelated to readiness to quit and treatment enrollment (Haug et al., 2005; Prochaska et al., 2004). In fact, depressed smokers will enter into smoking cessation interventions (if permitted) while they are in mental health treatment without adversely affecting their mental health recovery (Hall et al., 2006; Prochaska et al., 2008). Unfortunately, readiness to quit and enrollment in smoking cessation programs among smokers with past or current MDD have not consistently translated to improved smoking cessation outcomes for treatment approaches to date (Hall & Prochaska, 2009).

A history of MDD lowers the odds of short-term and long-term smoking abstinence by about 20% (Hitsman et al., 2013). In addition, pretreatment negative mood and increases in negative mood in the early stages of treatment are predictive of inability to quit and of smoking relapse (Berlin & Covey, 2006; Hall et al., 1994, 1996; Strong et al., 2009). These mood disturbances tend to be more common among those smokers with a history of MDD (Ginsberg et al., 1995; Hall et al., 1994, 1996; Killen et al., 2004; Levine, Marcus, & Perkins, 2003; Tsoh & Hall, 2004). This is not surprising given that 30–50% of those who remit from an episode of MDD will continue to experience depressive symptomatology (Bothwell & Scott, 1997; Fawcett, 1994; Nierenberg et al., 1999). As such, several studies have focused on treating depression symptoms and negative mood as well as preventing their emergence in smokers with a past history of MDD (single and recurrent) in order to optimize smoking cessation outcomes.

3.1 Smoking Cessation Treatment Outcomes in Smokers with Depression

The hypothesis that smokers with a history of MDD may be less likely to quit smoking successfully because of smoking cessation related exacerbations of negative mood has led to a series of investigations of mood management smoking cessation interventions for this population, with and without pharmacotherapy. These mood management smoking cessation interventions focused on preventing increases in negative mood following cessation or mitigating the impact of negative mood on smoking relapse. These studies are reviewed below.

3.2 Behavioral Therapy for Smoking Cessation: Mood Management

Hall and colleagues were among the first to investigate whether mood management smoking cessation treatment (MM) was more beneficial for smokers with a history of MDD compared to smokers without a history of MDD. The initial study compared 10 sessions of MM to 5 sessions of standard smoking cessation treatment (ST) in 46 smokers with a positive history of MDD compared to 103 smokers with no history of MDD (Hall et al., 1994). All smokers received nicotine gum therapy. Smokers with a history of MDD who received the MM treatment achieved higher rates of continuous abstinence than those in the ST group (34% vs. 18%), although MM did not impact negative mood after quitting. In a follow-up study with 201 smokers (44 with a past history of MDD), where treatment contact time was equated (i.e., 10 sessions for MM and control), the MM treatment was not significantly better than a control group among those smokers with a history of MDD (29% vs. 26% at 6-month follow-up) (Hall et al., 1996). A third study by Hall and colleagues compared the antidepressant nortriptyline to placebo in a group of smokers with ($n=65$) and without ($n=134$) a history of MDD who either received 10 sessions of MM or 5 sessions of a health education control condition (Hall et al., 1998). Nortriptyline was not more effective for those with a past history of MDD, but the MM condition was more effective at promoting smoking cessation than the control condition at end of treatment (41% vs. 19%) but not at the 26-week follow-up (18% vs. 13%). The mixed results of these studies suggest that smokers with a history of a single episode of MDD have better smoking cessation outcomes if they receive more behavioral treatment, but the benefits of mood management were more advantageous among smokers with recurrent MDD (Haas, Muñoz, Humfleet, Reus, & Hall, 2004).

Using a sample consisting solely of smokers with a history of MDD ($n=179$), Brown and colleagues (2001) evaluated whether MM enhanced the effects of ST on smoking cessation rates. Both groups received eight, 2-h sessions over 6 weeks. There were no significant differences in smoking abstinence at the end of treatment or at any of the posttreatment time points among the treatment groups (e.g., 25% of ST participants and 24% of MM participants were quit at the 6-month follow-up). MM was more effective at promoting smoking cessation among those smokers ($n=98$) with a history of recurrent MDD in secondary analyses (Brown et al., 2001), but not effective for those with a history of single episode of MD. MM did not influence negative mood, and in fact was associated with increased depressive symptoms, negative mood, risk of subsequent MDD episode, and withdrawal symptoms after smoking cessation (Kahler et al., 2002).

In a subsequent study, Brown and colleagues examined whether the combination of bupropion, an antidepressant, and MM was more effective at promoting cessation among smokers with a history of MDD ($N=92$) and smokers with elevated depression symptoms ($N=\sim 60$) compared to MM and placebo, ST and placebo or ST and bupropion (Brown et al., 2007). Neither MM nor bupropion, alone or in combination, was more effective at promoting smoking cessation for smokers with a history of MDD (single or recurrent) or smokers with elevated depression symptoms compared to smokers with no past history of MDD and smokers with low depression symptoms ($N=\sim 372$). Risk of lapse or relapse did not differ among the ST and MM conditions. In a recent meta-analysis (Gierisch, Bastian, Calhoun, McDuffie, & Williams, 2012; van der Meer, Willemsen, Smit, & Cuijpers, 2013), adding mood management to standard behavioral treatments led to significantly improved cessation outcomes among smokers with past depression (13 trials, $N=1496$, RR 1.41, 95% CI 1.13–1.77). Thus, there is promise for mood management approaches to improve cessation outcomes and a critical need to develop and evaluate specialized behavioral smoking cessation interventions that address the unique needs of smokers with a history of MDD, current MDD, or elevated depression symptoms. Several clinical trials of behavioral interventions that address various affective targets have shown more promising findings (Cinciripini et al., 2010; MacPherson et al., 2010; van der Meer et al., 2013; van der Meer, Willemsen, Smit, Cuijpers, & Schippers, 2010).

3.3 Pharmacotherapy for Smoking Cessation

Smokers with a history of MDD or depression symptoms have been treated with various pharmacological interventions, with a focus on antidepressant therapy. The explicit assumption is that antidepressant medication would offset any increases in negative mood or depression symptoms that may arise following smoking cessation, and therefore promote successful quitting. The implicit assumption is that antidepressants have the same functional role as nicotine, and therefore, may replace nicotine's mood regulation effects (Hall, 2007). Several antidepressants have been evaluated (e.g., sertraline, nortriptyline, bupropion). Overall, these medications have a modest benefit for smoking cessation in smokers with depression (see van der Meer et al., 2013 for a meta-analysis), and none have demonstrated greater efficacy for smokers with a history of MDD compared to smokers without a history of MDD (Catley et al., 2005; Covey, Glassman, Stetner, Rivelli, & Stage, 2002; Hall et al., 1994, 1998, 2002; Hughes, Stead, & Lancaster, 2007; Prochaska et al., 2004). That is, antidepressants do not appear to offset the increased risk of relapse in depressed versus non-depressed smokers. In addition, the effect of antidepressant therapy on negative mood does not necessarily translate to improved smoking cessation outcomes (Brown et al., 2007; Cinciripini et al., 2005; Evins et al., 2008; Hall et al., 1998); however, see Strong et al. (2009).

Surprisingly, there are few larger-scale smoking cessation investigations that included transdermal nicotine (TN) in smokers with current or past history of MDD (Evins et al., 2008; Hall et al., 2006; Smith et al., 2003). Hall and colleagues recruited 322 smokers with current MDD and randomized these smokers to either receive a brief contact control intervention (i.e., self-help guide, smoking cessation treatment referral list) or a stepped care intervention involving computerized motivational smoking feedback, 6 sessions of psychological counseling which included mood monitoring and management, and TN (Hall et al., 2006). Participants in the staged care condition were only slightly more likely to be smoking abstinent at 3-months (16% vs. 12%) and 6 months (21% vs. 18%) posttreatment compared to participants in the brief contact control intervention.

Nicotine gum use combined with brief therapy has been shown to help smokers with elevated depression symptoms achieve 12-month smoking cessation rates that are comparable to smokers with fewer depressive symptoms (15% vs. 20%) and more than doubled the 3-month quit rates (30% vs. 13%) and the 12-months quit rates (15% vs. 6%) compared to smokers with elevated depression symptoms who received placebo gum (Kinnunen, Doherty, Militello, & Garvey, 1996; Kinnunen, Korhonen, & Garvey, 2008). Nicotine gum use was also associated with a reduction in depression symptoms postcessation among smokers with elevated depression symptoms. It is unknown what proportion of smokers with high depression symptoms had a history of MD, as MDD history was not assessed.

Lastly, a recent study evaluated whether bupropion enhanced smoking cessation rates and reduced abstinence-induced depressive symptoms when added to TN therapy and standard smoking cessation counseling in 199 smokers with current or past MDD (Evins et al., 2008). About 30% of participants were abstinent at the end of 13 weeks of treatment. Bupropion did not improve the efficacy of TN and smoking cessation counseling, nor did bupropion mitigate the abstinence-induced increase in depression symptoms. Importantly, however, TN use was associated with abstinence, such that each patch used increased the likelihood of abstinence by 6% (Evins et al., 2008).

3.4 Summary

Evidence suggests that most negative mood management interventions may not significantly improve smoking cessation rates or mitigate negative mood among smokers with a history of MDD or smokers with higher levels of negative mood (Brown et al., 1996; Haas et al., 2004; Hall et al., 1994; Kahler et al., 2002; Strong et al., 2009). In fact, some negative mood management interventions can exacerbate negative mood, depression symptoms, withdrawal symptoms and increase risk of a major depression episode (Kahler et al., 2002). Antidepressants tend not to offset negative mood or promote smoking cessation among smokers with current or historic MDD (Brown et al., 2007; Cinciripini et al., 2005; Evins et al., 2008; Hall et al., 1998).

It has been assumed that smoking cessation will result in increases in depressive symptoms in all smokers with a history of MDD, but there appears to be significant heterogeneity in depressive symptoms during quitting. About 40% of smokers with a history of MDD have increased depressive symptoms during the 2 weeks following quitting, while other smokers have stable or decreasing depression symptoms (Burgess et al., 2002). The receipt of negative mood management smoking cessation interventions appears to be unrelated to the course of depressive symptoms during smoking cessation (Burgess et al., 2002) and increased depressive symptoms are associated with lower smoking abstinence rates (Burgess et al., 2002). Further, depression-prone smokers experience lower positive affect and higher negative affect on quit day (Strong et al., 2009). As we discuss below, smoking cessation interventions for this vulnerable population may be significantly more effective if positive mood, rather than negative mood is the focus of treatment (Miller et al., 1999; Miller & Miller, 2009; Nutt et al., 2007). In addition, the benefits of nicotine replacement therapy or varenicline have been understudied in this population, but may hold promise in promoting smoking cessation among smokers with past and current MDD, as well as smokers with high levels of depression symptoms (Cook, Spring, McChargue, & Hedeker, 2004; Fiore et al., 2008).

4. TARGETING POSITIVE MOOD MAY INCREASE SMOKING CESSATION TREATMENT EFFECTIVENESS

The relationship between changes in positive mood and the ability to quit smoking and avoid relapse has received far less attention than negative mood. Positive mood is distinct from negative mood and not simply the opposite end of a single mood continuum (Cook et al., 2004; Davidson et al., 2002; Watson, Clark, & Carey, 1988). We are aware of only one mood management smoking cessation treatment study that assessed positive mood as well as negative mood (Strong et al., 2009) in smokers with a past history of MDD. Participants received negative mood management or standard smoking cessation counseling in combination with bupropion or placebo (Brown et al., 2007). Positive mood declined significantly prior to the quit date, irrespective of receiving bupropion or negative mood management smoking cessation treatment. Decreases in positive mood prior to quit day and lower quit day levels of positive mood were significant predictors of smoking lapse, shorter time to relapse, and overall relapse risk (Strong et al., 2009). Increases in negative mood also predicted smoking lapse (Strong et al., 2009). Low positive mood and high negative mood after quitting were found to predict relapse in a study of smokers making an unaided quit attempt (Strong et al., 2011). Although not yet investigated, the dysregulation of positive mood may exacerbate the relapse risk associated with increased negative mood. Depression treatment research has shown that targeting positive mood increases positive mood and decreases negative mood in outpatients with MDD, although the converse does not appear to be true (Seligman, Rashid, & Parks, 2006; Seligman et al., 2005).

Declines in positive mood may be especially relevant for smokers with current or a past history of MDD. These individuals tend to be less attentive to sources of reinforcement (Forbes et al., 2009; Pizzagalli et al., 2009; Pizzagalli, Iosifescu, Hallett, Ratner, & Fava, 2008), have fewer alternative reinforcers (Audrain-McGovern et al., 2014; Jacobson et al., 1996; MacPhillamy & Lewinsohn, 1974), and derive less reward from natural reinforcers in their environment that typically increase positive mood (Dunn, Dalgleish, Lawrence, Cusack, & Ogilvie, 2004; Forbes, 2009; Forbes, Shaw, & Dahl, 2007; Pizzagalli et al., 2008; Shankman, Klein, Tenke, & Bruder, 2007; Sloan, Strauss, Quirk, & Sajatovic, 1997; Wichers et al., 2009). As such, positive mood is usually deficient in depression-prone persons (Clark, Watson, & Mineka, 1994; Forbes et al., 2006, 2009; Nutt et al., 2007; Wichers et al., 2009).

Research suggests three mechanisms by which nicotine ameliorates these deficits that characterize depression to increase positive mood (Audrain-McGovern et al., 2014; Perkins et al., 2006; Stein et al., 1998; Warburton & Mancuso, 1998); each of these can be considered a target for a smoking cessation intervention. One, nicotine increases the salience of rewarding stimuli in the environment (Mancuso, Warburton, Melen, Sherwood, & Tirelli, 1999; Powell, Tait, & Lessiter, 2002) and general attention (Evans & Drobes, 2009; Heishman, Kleykamp, & Singleton, 2010). Nicotine withdrawal is associated with declines in general attention (Patterson et al., 2010; Rukstalis, Jepson, Patterson, & Lerman, 2005). Two, smoking is a reinforcer itself and can compensate for limited alternative reinforcers (Audrain-McGovern, Rodriguez, Epstein, et al., 2009; Audrain-McGovern et al., 2011, 2014; Perkins et al., 2000). Depression prone smokers may find smoking especially reinforcing as evidenced by greater smoking-induced dopamine release compared to smokers not prone to depression (Brody et al., 2009). Three, nicotine enhances reward from available drug and nondrug reinforcers by increasing reward sensitivity or the ability to experience pleasure (Audrain-McGovern et al., 2015; Barr, Pizzagalli, Culhane, Goff, & Evins, 2008; Dawkins, Acaster, & Powell, 2007; Kenny & Markou, 2006; Perkins, Karelitz, Giedgowd, & Conklin, 2013; Spring et al., 2008). Nicotine withdrawal decreases reward sensitivity (Epping-Jordan et al., 1998; Johnson, Hollander, & Kenny, 2008). Thus, DP+ smokers may be less able to sustain smoking abstinence because they lose the reward-related effects that were supported by nicotine and experience diminished positive mood. In fact, expectations of positive effects of nicotine have been shown to mediate the relationship between history of depression and smoking behaviors (McChargue, Spring, Cook, & Neumann, 2004a). Declines in positive mood after quitting smoking have been observed in smokers (Cook et al., 2004, 2007; McChargue et al., 2004a) and linked to smoking relapse (al'Absi, Hatsukami, Davis, & Wittmers, 2004; Strong et al., 2009). Furthermore, anhedonia has been shown to predict preference for smoking versus alternative rewards (Cook et al., 2007; Leventhal, Ameringer, Osborn, Zvolensky, & Langdon, 2013; Leventhal, Munafo, et al., 2012; Leventhal et al., 2009, 2014). In addition, anhedonic smokers exhibit greater abstinence-provoked decreases in positive mood (Cook et al., 2007; Leventhal et al., 2013), diminution in reward processing (Leventhal, Ray, et al., 2012), and escalation in urge to smoke for pleasure (Leventhal et al., 2009).

Thus, smoking cessation intervention for depression-prone smokers that target behavioral skills to increase attention to opportunities for reward, identify and engage in reinforcing activities, and maximize enjoyment of those activities to maintain or enhance positive mood may increase the likelihood of quitting success. Such an intervention would integrate the functional role of nicotine (i.e., increases positive mood, greater salience to environmental rewards, increases enjoyment from available rewards) and the phenomenology of depression (i.e., less positive mood, less attention to opportunities for reward, fewer alternative reinforcers, difficulty deriving enjoyment from rewards). Preliminary research has shown that scheduling pleasant events can facilitate smoking cessation in smokers with elevated depression symptoms (Brown et al., 2001; Hall et al., 1996; MacPherson et al., 2010); however, long-term abstinence may require attention to the quality of the alternative reinforcers selected and the level of enjoyment derived from these alternative reinforcers. Initial research suggests that positive psychology interventions that enhance skills for obtaining positive affect and enhancing the duration and intensity of pleasurable and meaningful experiences may benefit smoking cessation outcomes (Kahler et al., 2014, 2015). Skills to magnify or savor enjoyment derived from alternative reinforcers may help replicate the reward enhancement that accompanies nicotine.

Smoking cessation treatments that incorporate nicotine replacement therapy or varenicline while concurrently addressing the positive mood changes through behavioral counseling may prove especially effective for promoting smoking cessation among smokers with current and a past history of MDD. Behavioral skills to increase attention to opportunities for reward, identify, and engage in alternative reinforcers, and maximize enjoyment derived from those alternative reinforcers may prevent a decline in positive mood and increase likelihood of successful short-term and long-term smoking cessation for smokers with current as well as past history of MD.

5. INDIVIDUAL DIFFERENCES IN CESSATION OUTCOME AND TREATMENT EFFECTIVENESS

In terms of depression as a prognosticative factor of cessation outcome irrespective of treatment type, results from a recent meta-analysis suggest that past depression carries modest risk of elevated cessation failure

(Hitsman et al., 2013). Evidence that current or recent depression or depressive symptoms predict poor cessation outcomes is more consistent (al'Absi, Hatsukami, & Davis, 2005; al'Absi et al., 2004; Anda et al., 1990; Brown et al., 2001; Catley, Ahluwalia, Resnicow, & Nazir, 2003; Catley et al., 2005; Cinciripini et al., 2003; Japuntich et al., 2007; Leventhal, Japuntich, et al., 2012; Leventhal, Ramsey, Brown, LaChance, & Kahler, 2008; Piper et al., 2010; Schnoll, Leone, & Hitsman, 2013). Furthermore, evidence suggests that variation at subclinical levels of depression, even the distinction between zero versus one symptom (Niaura et al., 2001), predict poorer cessation outcomes (Leventhal et al., 2008) in smokers without current major depression (al'Absi et al., 2004, 2005; Brown et al., 2001; Cinciripini et al., 2003; Ginsberg et al., 1995; Haas et al., 2004; Killen, Fortmann, Davis, Strausberg, & Varady, 1999; Swan et al., 2003). Importantly, there is heterogeneity among smokers with depressive characteristics that explains variance in smoking cessation outcomes, as relapse risk is higher in individuals with recurrent (vs. single episode) forms of depression (Brown et al., 2001, 2007; Haas et al., 2004; Leventhal et al., 2014), assessed using brief depression screeners versus clinical interviews (Hitsman et al., 2013), and with anhedonia versus other types of depressive symptoms (Cook, Spring, McChargue, & Doran, 2010; Leventhal et al., 2008, 2014; Niaura et al., 2001). With regard to the specific symptom types that predict outcome, anhedonia and low positive affect have shown incremental prediction over and above other factors, including depressed mood, depression proneness, somatic features, personality traits, anxiety disorders, and substance use disorders (Cook et al., 2010; Japuntich et al., 2007; Leventhal, Munafo, et al., 2012; Leventhal et al., 2014). Thus, absence of reward and positive affect may be an important facet of depression that contributes to difficulty quitting smoking.

Individual differences in response to smoking cessation treatments in smokers with current or a past history of MDD have gone largely unexplored. One potential influence on smoking cessation treatment effectiveness in smokers with a history of MDD is hedonic capacity. As noted earlier, hedonic capacity is a stable dispositional ability to experience positive mood in response to things that are typically rewarding (Clark, Fawcett, Salazar-Grueso, & Fawcett, 1984; Cook et al., 2004; Meehl, 1987). Low hedonic capacity, measured by validated questionnaire predicts heightened urge to smoke after smoking abstinence (Cook et al., 2004) and moderates nicotine's ability to increase positive mood (Cook et al., 2007). Research indicates variability in hedonic capacity among depression-prone smokers

(Cook et al., 2007; Leventhal, Chasson, Tapia, Miller, & Pettit, 2006), with lower levels predicting inability to maintain abstinence for greater than 24 h (Leventhal et al., 2009).

One could also speculate that aspects of MDD itself may differentially influence response to smoking cessation treatment. There is evidence for only one variable, MDD recurrence appears to modify treatment effects. Mood management smoking cessations interventions appear to be more effective at promoting smoking cessation among smokers with a history of recurrent MDD than those with single episode (Brown et al., 2001; Haas et al., 2004). Severity of MDD or a recent MDD episode at smoking treatment onset does not predict smoking outcomes (Hall et al., 2006).

6. CONCLUSIONS

The comorbidity between smoking and MDD is well established (Grant et al., 2004; Hitsman et al., 2013). Smokers with current or past MDD are often excluded from smoking cessation treatment studies, especially those involving pharmacotherapy. Only a few studies have included samples comprised solely of smokers with current or a past history of MDD. As a result, we lack studies to guide smoking cessation treatment decision making for this under-served population. In order to determine the most effective elements of pharmacological and psychological smoking cessation interventions for smokers with current or past MDD, research needs to focus on this population (Wilhelm, Wedgwood, Niven, & Kay-Lambkin, 2006).

The practice guidelines for smoking cessation (Fiore et al., 2008), recommend that smokers with comorbid psychiatric conditions be provided with the same smoking cessation treatments as smokers in the general population, but acknowledges that psychiatric comorbidity could affect the acceptability, use and effectiveness of smoking cessation treatment. These guidelines also emphasize that current evidence is insufficient to determine whether smokers with psychiatric conditions benefit more from tobacco use treatments focused on unique aspects of their condition or symptoms than from standard smoking cessation treatments (Fiore & Jaen, 2008; Hall, 2007). Negative mood management smoking cessation interventions and antidepressant pharmacotherapy have shown modest effects on smoking cessation rates, at best and tend not to mitigate negative mood among smokers with a history of MDD or smokers with higher levels of negative mood

(Brown et al., 2007; Covey et al., 2002; Evins et al., 2008; Hall et al., 1994; Kahler et al., 2002; Strong et al., 2009).

Thus, although negative mood after quitting smoking has been a hypothesized pathway through which smokers with MDD are unable to sustain smoking abstinence, converging lines of research suggest that smokers with MDD are unable to become or sustain smoking abstinence because they experience a loss in positive mood and reward regulation that was supported by nicotine (Audrain-McGovern et al., 2014). Smoking cessation interventions focused on positive mood and reward enhancement would answer the call to shift the emphasis of addiction treatment from the suppression of negative mood and substance use behavior to the promotion of positive mood and alternative nondrug reinforcers (Miller & Miller, 2009). It is thought that the latter holds greater promise for making substance use a less likely choice among several rewarding alternatives (Miller & Miller, 2009). Such interventions should evaluate mechanisms of treatment effects and explore who may benefit most from such an intervention. Evidence-based treatments will help inform treatment planning by evaluating the effectiveness of standard smoking cessation counseling compared to novel smoking cessation approaches that integrate the role of nicotine and smoking in normalizing symptoms that are associated with depression.

REFERENCES

al'Absi, M., Hatsukami, D., & Davis, G. L. (2005). Attenuated adrenocorticotropic responses to psychological stress are associated with early smoking relapse. *Psychopharmacology, 181*, 107–117.

al'Absi, M., Hatsukami, D., Davis, G. L., & Wittmers, L. E. (2004). Prospective examination of effects of smoking abstinence on cortisol and withdrawal symptoms as predictors of early smoking relapse. *Drug and Alcohol Dependence, 73*, 267–278.

Anda, R. F., Williamson, D. F., Escobedo, L. G., Mast, E. E., Giovino, G. A., & Remington, P. L. (1990). Depression and the dynamics of smoking. A national perspective. *The Journal of the American Medical Association, 264*, 1541–1545.

Audrain-McGovern, J., Rodriguez, D., Epstein, L. H., Rodgers, K., Cuevas, J., & Wileyto, E. P. (2009). Young adult smoking: What factors differentiate ex-smokers, smoking cessation treatment seekers and nontreatment seekers? *Addictive Behaviors, 34*, 1036–1041.

Audrain-McGovern, J., Rodriguez, D., & Kassel, J. D. (2009). Adolescent smoking and depression: Evidence for self-medication and peer smoking mediation. *Addiction, 104*, 1743–1756.

Audrain-McGovern, J., Rodriguez, D., & Leventhal, A. M. (2015). Gender differences in the relationship between affect and adolescent smoking uptake. *Addiction, 110*, 519–529.

Audrain-McGovern, J., Rodriguez, D., Leventhal, A. M., Cuevas, J., Rodgers, K., & Sass, J. (2012). Where is the pleasure in that? Low hedonic capacity predicts smoking onset and escalation. *Nicotine & Tobacco Research, 14*, 1187–1196.

Audrain-McGovern, J., Rodriguez, D., Rodgers, K., & Cuevas, J. (2011). Declining alternative reinforcers link depression to young adult smoking. *Addiction, 106*, 178–187.

Audrain-McGovern, J., Rodriguez, D., Rodgers, K., Cuevas, J., Sass, J., & Riley, T. (2012). Reward expectations lead to smoking uptake among depressed adolescents. *Drug and Alcohol Dependence, 120*, 181–189.

Audrain-McGovern, J., Rodriguez, D., Tercyak, K. P., Cuevas, J., Rodgers, K., & Patterson, F. (2004). Identifying and characterizing adolescent smoking trajectories. *Cancer Epidemiology, Biomarkers & Prevention, 13*, 2023–2034.

Audrain-McGovern, J., Wileyto, E. P., Ashare, R., Cuevas, J., & Strasser, A. A. (2014). Reward and affective regulation in depression-prone smokers. *Biological Psychiatry, 76*, 689–697.

Barr, R. S., Pizzagalli, D. A., Culhane, M. A., Goff, D. C., & Evins, A. E. (2008). A single dose of nicotine enhances reward responsiveness in nonsmokers: Implications for development of dependence. *Biological Psychiatry, 63*, 1061–1065.

Barrett, S. P., Boileau, I., Okker, J., Pihl, R. O., & Dagher, A. (2004). The hedonic response to cigarette smoking is proportional to dopamine release in the human striatum as measured by positron emission tomography and [11C]raclopride. *Synapse, 54*, 65–71.

Berlin, I., & Covey, L. S. (2006). Pre-cessation depressive mood predicts failure to quit smoking: The role of coping and personality traits. *Addiction, 101*, 1814–1821.

Bogdan, R., & Pizzagalli, D. A. (2009). The heritability of hedonic capacity and perceived stress: A twin study evaluation of candidate depressive phenotypes. *Psychological Medicine, 39*, 211–218.

Bothwell, R., & Scott, J. (1997). The influence of cognitive variables on recovery in depressed inpatients. *Journal of Affective Disorders, 43*, 207–212.

Breslau, N., Kilbey, M. M., & Andreski, P. (1993). Nicotine dependence and major depression. New evidence from a prospective investigation. *Archives of General Psychiatry, 50*, 31–35.

Breslau, N., Peterson, E. L., Schultz, L. R., Chilcoat, H. D., & Andreski, P. (1998). Major depression and stages of smoking. A longitudinal investigation. *Archives of General Psychiatry, 55*, 161–166.

Brody, A. L., Olmstead, R. E., Abrams, A. L., Costello, M. R., Khan, A., Kozman, D., et al. (2009). Effect of a history of major depressive disorder on smoking-induced dopamine release. *Biological Psychiatry, 66*, 898–901.

Brody, A. L., Olmstead, R. E., London, E. D., Farahi, J., Meyer, J. H., Grossman, P., et al. (2004). Smoking-induced ventral striatum dopamine release. *The American Journal of Psychiatry, 161*, 1211–1218.

Brown, R. A., Kahler, C. W., Niaura, R., Abrams, D. B., Sales, S. D., Ramsey, S. E., et al. (2001). Cognitive-behavioral treatment for depression in smoking cessation. *Journal of Consulting and Clinical Psychology, 69*, 471–480.

Brown, R. A., Lewinsohn, P. M., Seeley, J. R., & Wagner, E. F. (1996). Cigarette smoking, major depression, and other psychiatric disorders among adolescents. *Journal of the American Academy of Child & Adolescent Psychiatry, 35*, 1602–1610.

Brown, R. A., Niaura, R., Lloyd-Richardson, E. E., Strong, D. R., Kahler, C. W., Abrantes, A. M., et al. (2007). Bupropion and cognitive-behavioral treatment for depression in smoking cessation. *Nicotine & Tobacco Research, 9*, 721–730.

Burgess, E. S., Brown, R. A., Kahler, C. W., Niaura, R., Abrams, D. B., Goldstein, M. G., et al. (2002). Patterns of change in depressive symptoms during smoking cessation: Who's at risk for relapse? *Journal of Consulting and Clinical Psychology, 70*, 356–361.

Caggiula, A. R., Donny, E. C., Palmatier, M. I., Liu, X., Chaudhri, N., & Sved, A. F. (2009). The role of nicotine in smoking: A dual-reinforcement model. *Nebraska Symposium on Motivation, 55*, 91–109.

Catley, D., Ahluwalia, J. S., Resnicow, K., & Nazir, N. (2003). Depressive symptoms and smoking cessation among inner-city African Americans using the nicotine patch. *Nicotine & Tobacco Research, 5*, 61–68.

Catley, D., Harris, K. J., Okuyemi, K. S., Mayo, M. S., Pankey, E., & Ahluwalia, J. S. (2005). The influence of depressive symptoms on smoking cessation among African Americans in a randomized trial of bupropion. *Nicotine & Tobacco Research, 7*, 859–870.

CDC. (2008). Smoking attributable mortality, years of potential life lost, and productivity losses. United States, 2000–2004. *Morbidity and Mortality Weekly Report, 57*, 1226–1228.

CDC. (2012). Youth risk behavior surveillance—United States, 2011. *Morbidity and Mortality Weekly Report. Surveillance Summaries, 61*, 1–162.

CDC. (2014). *National Youth Tobacco Survey (NYTS)*. http://www.cdc.gov/TOBACCO/data_statistics/surveys/NYTS/index.htm. accessed 25.02.15.

Chaiton, M. O., Cohen, J. E., O'Loughlin, J., & Rehm, J. (2009). A systematic review of longitudinal studies on the association between depression and smoking in adolescents. *BMC Public Health, 9*, 356.

Chaiton, M. O., & Zhang, B. (2007). Environment modifies the association between depression symptoms and smoking among adolescents. *Psychology of Addictive Behaviors, 21*, 420–424.

Chaudhri, N., Caggiula, A. R., Donny, E. C., Palmatier, M. I., Liu, X., & Sved, A. F. (2006). Complex interactions between nicotine and nonpharmacological stimuli reveal multiple roles for nicotine in reinforcement. *Psychopharmacology, 184*, 353–366.

Choi, W. S., Pierce, J. P., Gilpin, E. A., Farkas, A. J., & Berry, C. C. (1997). Which adolescent experimenters progress to established smoking in the United States. *American Journal of Preventive Medicine, 13*, 385–391.

Cinciripini, P. M., Blalock, J. A., Minnix, J. A., Robinson, J. D., Brown, V. L., Lam, C., et al. (2010). Effects of an intensive depression-focused intervention for smoking cessation in pregnancy. *Journal of Consulting and Clinical Psychology, 78*, 44–54.

Cinciripini, P. M., Tsoh, J. Y., Wetter, D. W., Lam, C., de Moor, C., Cinciripini, L., et al. (2005). Combined effects of venlafaxine, nicotine replacement, and brief counseling on smoking cessation. *Experimental and Clinical Psychopharmacology, 13*, 282–292.

Cinciripini, P. M., Wetter, D. W., Fouladi, R. T., Blalock, J. A., Carter, B. L., Cinciripini, L. G., et al. (2003). The effects of depressed mood on smoking cessation: Mediation by postcessation self-efficacy. *Journal of Consulting and Clinical Psychology, 71*, 292–301.

Clark, D. C., Fawcett, J., Salazar-Grueso, E., & Fawcett, E. (1984). Seven-month clinical outcome of anhedonic and normally hedonic depressed inpatients. *The American Journal of Psychiatry, 141*, 1216–1220.

Clark, L. A., & Watson, D. (1991). Tripartite model of anxiety and depression: Psychometric evidence and taxonomic implications. *Journal of Abnormal Psychology, 100*, 316–336.

Clark, L. A., Watson, D., & Mineka, S. (1994). Temperament, personality, and the mood and anxiety disorders. *Journal of Abnormal Psychology, 103*, 103–116.

Clarke, G. N., Hornbrook, M., Lynch, F., Polen, M., Gale, J., Beardslee, W., et al. (2001). A randomized trial of a group cognitive intervention for preventing depression in adolescent offspring of depressed parents. *Archives of General Psychiatry, 58*, 1127–1134.

Connell, A. M., & Dishion, T. J. (2006). The contribution of peers to monthly variation in adolescent depressed mood: A short-term longitudinal study with time-varying predictors. *Development and Psychopathology, 18*, 139–154.

Cook, J. W., Spring, B., & McChargue, D. (2007). Influence of nicotine on positive affect in anhedonic smokers. *Psychopharmacology, 192*, 87–95.

Cook, J., Spring, B., McChargue, D., & Doran, N. (2010). Effects of anhedonia on days to relapse among smokers with a history of depression: A brief report. *Nicotine & Tobacco Research, 12*, 978–982.

Cook, J. W., Spring, B., McChargue, D., & Hedeker, D. (2004). Hedonic capacity, cigarette craving, and diminished positive mood. *Nicotine & Tobacco Research, 6*, 39–47.

Costello, E. J., Pine, D. S., Hammen, C., March, J. S., Plotsky, P. M., Weissman, M. M., et al. (2002). Development and natural history of mood disorders. *Biological Psychiatry, 52*, 529–542.

Covey, L. S., Glassman, A. H., Stetner, F., Rivelli, S., & Stage, K. (2002). A randomized trial of sertraline as a cessation aid for smokers with a history of major depression. *The American Journal of Psychiatry, 159*, 1731–1737.

Davidson, R. J., Lewis, D. A., Alloy, L. B., Amaral, D. G., Bush, G., Cohen, J. D., et al. (2002). Neural and behavioral substrates of mood and mood regulation. *Biological Psychiatry, 52*, 478–502.

Dawkins, L., Acaster, S., & Powell, J. H. (2007). The effects of smoking and abstinence on experience of happiness and sadness in response to positively valenced, negatively valenced, and neutral film clips. *Addictive Behaviors, 32*, 425–431.

Dierker, L. C., Avenevoli, S., Stolar, M., & Merikangas, K. R. (2002). Smoking and depression: An examination of mechanisms of comorbidity. *The American Journal of Psychiatry, 159*, 947–953.

Dunn, B. D., Dalgleish, T., Lawrence, A. D., Cusack, R., & Ogilvie, A. D. (2004). Categorical and dimensional reports of experienced affect to emotion-inducing pictures in depression. *Journal of Abnormal Psychology, 113*, 654–660.

Epping-Jordan, M. P., Watkins, S. S., Koob, G. F., & Markou, A. (1998). Dramatic decreases in brain reward function during nicotine withdrawal. *Nature, 393*, 76–79.

Escobedo, L. G., Reddy, M., & Giovino, G. A. (1998). The relationship between depressive symptoms and cigarette smoking in US adolescents. *Addiction, 93*, 433–440.

Evans, D. E., & Drobes, D. J. (2009). Nicotine self-medication of cognitive-attentional processing. *Addiction Biology, 14*, 32–42.

Evins, A. E., Culhane, M. A., Alpert, J. E., Pava, J., Liese, B. S., Farabaugh, A., et al. (2008). A controlled trial of bupropion added to nicotine patch and behavioral therapy for smoking cessation in adults with unipolar depressive disorders. *Journal of Clinical Psychopharmacology, 28*, 660–666.

Fawcett, J. (1994). Overview of mood disorders: Diagnosis, classification, and management. *Clinical Chemistry, 40*, 273–278.

Fawcett, J., Clark, D. C., Scheftner, W. A., & Hedeker, D. (1983). Differences between anhedonic and normally hedonic depressive states. *The American Journal of Psychiatry, 140*, 1027–1030.

Fergusson, D. M., Goodwin, R. D., & Horwood, L. J. (2003). Major depression and cigarette smoking: Results of a 21-year longitudinal study. *Psychological Medicine, 33*, 1357–1367.

Fergusson, D. M., Horwood, L. J., Ridder, E. M., & Beautrais, A. L. (2005). Subthreshold depression in adolescence and mental health outcomes in adulthood. *Archives of General Psychiatry, 62*, 66–72.

Fiore, M. C., & Jaen, C. R. (2008). A clinical blueprint to accelerate the elimination of tobacco use. *The Journal of the American Medical Association, 299*, 2083–2085.

Fiore, M. C., Jaen, C. R., Baker, T. B., Bailey, W. C., Benowitz, N., & Curry, S. J. (2008). Treating tobacco use and dependence: 2008 update U.S. Public Health Service Clinical Practice Guideline executive summary. *Respiratory Care, 53*, 1217–1222.

Forbes, E. E. (2009). Where's the fun in that? Broadening the focus on reward function in depression. *Biological Psychiatry, 66*, 199–200.

Forbes, E. E., Christopher May, J., Siegle, G. J., Ladouceur, C. D., Ryan, N. D., Carter, C. S., et al. (2006). Reward-related decision-making in pediatric major depressive disorder: An fMRI study. *The Journal of Child Psychology and Psychiatry, 47*, 1031–1040.

Forbes, E. E., Hariri, A. R., Martin, S. L., Silk, J. S., Moyles, D. L., Fisher, P. M., et al. (2009). Altered striatal activation predicting real-world positive affect in adolescent major depressive disorder. *The American Journal of Psychiatry, 166*, 64–73.
Forbes, E. E., Shaw, D. S., & Dahl, R. E. (2007). Alterations in reward-related decision making in boys with recent and future depression. *Biological Psychiatry, 61*, 633–639.
Franken, I. H., Zijlstra, C., & Muris, P. (2006). Are nonpharmacological induced rewards related to anhedonia? A study among skydivers. *Progress in Neuropsychopharmacology & Biological Psychiatry, 30*, 297–300.
Georgiades, K., Lewinsohn, P. M., Monroe, S. M., & Seeley, J. R. (2006). Major depressive disorder in adolescence: The role of subthreshold symptoms. *Journal of the American Academy of Child & Adolescent Psychiatry, 45*, 936–944.
Gierisch, J. M., Bastian, L. A., Calhoun, P. S., McDuffie, J. R., & Williams, J. W., Jr. (2012). Smoking cessation interventions for patients with depression: A systematic review and meta-analysis. *Journal of General Internal Medicine, 27*, 351–360.
Ginsberg, D., Hall, S. M., Reus, V. I., & Muñoz, R. F. (1995). Mood and depression diagnosis in smoking cessation. *Experimental and Clinical Psychopharmacology, 3*, 389–395.
Glassman, A. H., Stetner, F., Walsh, B. T., Raizman, P. S., Fleiss, J. L., Cooper, T. B., et al. (1988). Heavy smokers, smoking cessation, and clonidine. Results of a double-blind, randomized trial. *The Journal of the American Medical Association, 259*, 2863–2866.
Goodman, E., & Capitman, J. (2000). Depressive symptoms and cigarette smoking among teens. *Pediatrics, 106*, 748–755.
Grant, B. F., Hasin, D. S., Chou, S. P., Stinson, F. S., & Dawson, D. A. (2004). Nicotine dependence and psychiatric disorders in the United States: Results from the national epidemiologic survey on alcohol and related conditions. *Archives of General Psychiatry, 61*, 1107–1115.
Green, L., & Fisher, E. B., Jr. (2000). Economic substitutability: Some implications for health behavior. In W. K. Bickel & R. E. Vuchinich (Eds.), *Reframing health behavior change with behavior economics* (pp. 115–144). Mahway: Lawrence Erlbaum Associates.
Haas, A. L., Muñoz, R. F., Humfleet, G. L., Reus, V. I., & Hall, S. M. (2004). Influences of mood, depression history, and treatment modality on outcomes in smoking cessation. *Journal of Consulting and Clinical Psychology, 72*, 563–570.
Hall, S. M. (2007). Nicotine interventions with comorbid populations. *American Journal of Preventive Medicine, 33*, S406–S413.
Hall, S. M., Humfleet, G. L., Reus, V. I., Munoz, R. F., Hartz, D. T., & Maude-Griffin, R. (2002). Psychological intervention and antidepressant treatment in smoking cessation. *Archives of General Psychiatry, 59*, 930–936.
Hall, S. M., Munoz, R. F., & Reus, V. I. (1994). Cognitive-behavioral intervention increases abstinence rates for depressive-history smokers. *Journal of Consulting and Clinical Psychology, 62*, 141–146.
Hall, S. M., Munoz, R. F., Reus, V. I., Sees, K. L., Duncan, C., Humfleet, G. L., et al. (1996). Mood management and nicotine gum in smoking treatment: A therapeutic contact and placebo-controlled study. *Journal of Consulting and Clinical Psychology, 64*, 1003–1009.
Hall, S. M., & Prochaska, J. J. (2009). Treatment of smokers with co-occurring disorders: Emphasis on integration in mental health and addiction treatment settings. *Annual Review of Clinical Psychology, 5*, 409–431.
Hall, S. M., Reus, V. I., Munoz, R. F., Sees, K. L., Humfleet, G., Hartz, D. T., et al. (1998). Nortriptyline and cognitive-behavioral therapy in the treatment of cigarette smoking. *Archives of General Psychiatry, 55*, 683–690.
Hall, S. M., Tsoh, J. Y., Prochaska, J. J., Eisendrath, S., Rossi, J. S., Redding, C. A., et al. (2006). Treatment for cigarette smoking among depressed mental health outpatients: A randomized clinical trial. *American Journal of Public Health, 96*, 1808–1814.

Harvey, P. O., Pruessner, J., Czechowska, Y., & Lepage, M. (2007). Individual differences in trait anhedonia: A structural and functional magnetic resonance imaging study in non-clinical subjects. *Molecular Psychiatry*, *12*(703), 767–775.

Haug, N. A., Hall, S. M., Prochaska, J. J., Rosen, A. B., Tsoh, J. Y., Humfleet, G., et al. (2005). Acceptance of nicotine dependence treatment among currently depressed smokers. *Nicotine & Tobacco Research*, *7*, 217–224.

Hedeker, D., Mermelstein, R. J., Berbaum, M. L., & Campbell, R. T. (2009). Modeling mood variation associated with smoking: An application of a heterogeneous mixed-effects model for analysis of ecological momentary assessment (EMA) data. *Addiction*, *104*, 297–307.

Heinz, A. J., Kassel, J. D., Berbaum, M., & Mermelstein, R. (2010). Adolescents' expectancies for smoking to regulate affect predict smoking behavior and nicotine dependence over time. *Drug and Alcohol Dependence*, *111*, 128–135.

Heishman, S. J., Kleykamp, B. A., & Singleton, E. G. (2010). Meta-analysis of the acute effects of nicotine and smoking on human performance. *Psychopharmacology*, *210*, 453–469.

Higgins, S. T., Heil, S. H., & Lussier, J. P. (2004). Clinical implications of reinforcement as a determinant of substance use disorders. *Annual Review of Psychology*, *55*, 431–461.

Hitsman, B., Borrelli, B., McChargue, D. E., Spring, B., & Niaura, R. (2003). History of depression and smoking cessation outcome: A meta-analysis. *Journal of Consulting and Clinical Psychology*, *71*, 657–663.

Hitsman, B., Papandonatos, G. D., McChargue, D. E., DeMott, A., Herrera, M. J., Spring, B., et al. (2013). Past major depression and smoking cessation outcome: A systematic review and meta-analysis update. *Addiction*, *108*, 294–306.

Hoffman, B. R., Monge, P. R., Chou, C. P., & Valente, T. W. (2007). Perceived peer influence and peer selection on adolescent smoking. *Addictive Behaviors*, *32*, 1546–1554.

Hughes, J. R., Stead, L. F., & Lancaster, T. (2007). Antidepressants for smoking cessation. *The Cochrane Database of Systematic Reviews*, *1*, CD000031.

Jacobson, N. S., Dobson, K. S., Truax, P. A., Addis, M. E., Koerner, K., Gollan, J. K., et al. (1996). A component analysis of cognitive-behavioral treatment for depression. *Journal of Consulting and Clinical Psychology*, *64*, 295–304.

Japuntich, S. J., Smith, S. S., Jorenby, D. E., Piper, M. E., Fiore, M. C., & Baker, T. B. (2007). Depression predicts smoking early but not late in a quit attempt. *Nicotine & Tobbacco Research*, *9*, 677–686.

Johnson, P. M., Hollander, J. A., & Kenny, P. J. (2008). Decreased brain reward function during nicotine withdrawal in C57BL6 mice: Evidence from intracranial self-stimulation (ICSS) studies. *Pharmacology, Biochemistry and Behavior*, *90*, 409–415.

Johnson, E. O., Rhee, S. H., Chase, G. A., & Breslau, N. (2004). Comorbidity of depression with levels of smoking: An exploration of the shared familial risk hypothesis. *Nicotine & Tobacco Research*, *6*, 1029–1038.

Kahler, C. W., Brown, R. A., Ramsey, S. E., Niaura, R., Abrams, D. B., Goldstein, M. G., et al. (2002). Negative mood, depressive symptoms, and major depression after smoking cessation treatment in smokers with a history of major depressive disorder. *Journal of Abnormal Psychology*, *111*, 670–675.

Kahler, C. W., Spillane, N. S., Day, A. M., Cioe, P. A., Parks, A., Leventhal, A. M., et al. (2015). Positive psychotherapy for smoking cessation: A pilot randomized controlled trial. *Nicotine & Tobacco Research*. Epub ahead of print.

Kahler, C. W., Spillane, N. S., Day, A., Clerkin, E., Parks, A., Leventhal, A. M., et al. (2014). Positive psychotherapy for smoking cessation: Treatment development, feasibility and preliminary results. *The Journal of Positive Psychology*, *9*, 19–29.

Kassel, J. D., Stroud, L. R., & Paronis, C. A. (2003). Smoking, stress, and negative affect: Correlation, causation, and context across stages of smoking. *Psychological Bulletin*, *129*, 270–304.

Kendler, K. S., Neale, M. C., MacLean, C. J., Heath, A. C., Eaves, L. J., & Kessler, R. C. (1993). Smoking and major depression. A causal analysis. *Archives of General Psychiatry, 50*, 36–43.
Kenny, P. J., & Markou, A. (2006). Nicotine self-administration acutely activates brain reward systems and induces a long-lasting increase in reward sensitivity. *Neuropsychopharmacology, 31*, 1203–1211.
Killen, J. D., & Fortmann, S. P. (1997). Craving is associated with smoking relapse: Findings from three prospective studies. *Experimental and Clinical Psychopharmacology, 5*, 137–142.
Killen, J. D., Fortmann, S. P., Davis, L., Strausberg, L., & Varady, A. (1999). Do heavy smokers benefit from higher dose nicotine patch therapy? *Experimental and Clinical Psychopharmacology, 7*, 226–233.
Killen, J. D., Robinson, T. N., Ammerman, S., Hayward, C., Rogers, J., Samuels, D., et al. (2004). Major depression among adolescent smokers undergoing treatment for nicotine dependence. *Addictive Behaviors, 29*, 1517–1526.
Kinnunen, T., Doherty, K., Militello, F. S., & Garvey, A. J. (1996). Depression and smoking cessation: Characteristics of depressed smokers and effects of nicotine replacement. *Journal of Consulting and Clinical Psychology, 64*, 791–798.
Kinnunen, T., Korhonen, T., & Garvey, A. J. (2008). Role of nicotine gum and pretreatment depressive symptoms in smoking cessation: Twelve-month results of a randomized placebo controlled trial. *International Journal of Psychiatry in Medicine, 38*, 373–389.
Kobus, K. (2003). Peers and adolescent smoking. *Addiction, 98*(Suppl. 1), 37–55.
Lee Duckworth, A., Steen, T. A., & Seligman, M. E. (2005). Positive psychology in clinical practice. *Annual Review of Clinical Psychology, 1*, 629–651.
Lerman, C., Audrain, J., Orleans, C. T., Boyd, R., Gold, K., Main, D., et al. (1996). Investigation of mechanisms linking depressed mood to nicotine dependence. *Addictive Behaviors, 21*, 9–19.
Leventhal, A. M., Ameringer, K. J., Osborn, E., Zvolensky, M. J., & Langdon, K. J. (2013). Anxiety and depressive symptoms and affective patterns of tobacco withdrawal. *Drug and Alcohol Dependence, 133*, 324–329.
Leventhal, A. M., Chasson, G. S., Tapia, E., Miller, E. K., & Pettit, J. W. (2006). Measuring hedonic capacity in depression: A psychometric analysis of three anhedonia scales. *The Journal of Clinical Psychiatry, 62*, 1545–1558.
Leventhal, A. M., Japuntich, S. J., Piper, M. E., Jorenby, D. E., Schlam, T. R., & Baker, T. B. (2012). Isolating the role of psychological dysfunction in smoking cessation: Relations of personality and psychopathology to attaining cessation milestones. *Psychology of Addictive Behaviors, 26*, 838–849.
Leventhal, A. M., Munafo, M., Tidey, J. W., Sussman, S., Monterosso, J. R., Sun, P., et al. (2012). Anhedonia predicts altered processing of happy faces in abstinent cigarette smokers. *Psychopharmacology, 222*, 343–351.
Leventhal, A. M., Ramsey, S. E., Brown, R. A., LaChance, H. R., & Kahler, C. W. (2008). Dimensions of depressive symptoms and smoking cessation. *Nicotine & Tobacco Research, 10*, 507–517.
Leventhal, A. M., Ray, L. A., Rhee, S. H., & Unger, J. B. (2012). Genetic and environmental influences on the association between depressive symptom dimensions and smoking initiation among Chinese adolescent twins. *Nicotine & Tobacco Research, 14*, 559–568.
Leventhal, A. M., Trujillo, M., Ameringer, K. J., Tidey, J. W., Sussman, S., & Kahler, C. W. (2014). Anhedonia and the relative reward value of drug and nondrug reinforcers in cigarette smokers. *Journal of Abnormal Psychology, 123*, 375–386.
Leventhal, A. M., Waters, A. J., Kahler, C. W., Ray, L. A., & Sussman, S. (2009). Relations between anhedonia and smoking motivation. *Nicotine & Tobacco Research, 11*, 1047–1054.

Levine, M. D., Marcus, M. D., & Perkins, K. A. (2003). A history of depression and smoking cessation outcomes among women concerned about post-cessation weight gain. *Nicotine & Tobacco Research, 5*, 69–76.

Lewinsohn, P. M., Hops, H., Roberts, R. E., Seeley, J. R., & Andrews, J. A. (1993). Adolescent psychopathology: I. Prevalence and incidence of depression and other DSM-III-R disorders in high school students. *Journal of Abnormal Psychology, 102*, 133–144.

Lewinsohn, P. M., Rohde, P., & Seeley, J. R. (1998). Major depressive disorder in older adolescents: Prevalence, risk factors, and clinical implications. *Clinical Psychology Review, 18*, 765–794.

Lewinsohn, P. M., Rohde, P., Seeley, J. R., Klein, D. N., & Gotlib, I. H. (2000). Natural course of adolescent major depressive disorder in a community sample: Predictors of recurrence in young adults. *The American Journal of Psychiatry, 157*, 1584–1591.

Lewinsohn, P. M., Shankman, S. A., Gau, J. M., & Klein, D. N. (2004). The prevalence and co-morbidity of subthreshold psychiatric conditions. *Psychological Medicine, 34*, 613–622.

Lewinsohn, P. M., Youngren, M. A., & Grosscup, S. J. (1979). Reinforcement and depression. In R. A. Dupue (Ed.), *The psychobiology of depressive disorders: Implications for the effects of stress* (pp. 291–316). New York: Academic Press.

Lyons, M., Hitsman, B., Xian, H., Panizzon, M. S., Jerskey, B. A., Santangelo, S., et al. (2008). A twin study of smoking, nicotine dependence, and major depression in men. *Nicotine & Tobacco Research, 10*, 97–108.

MacPherson, L., Tull, M. T., Matusiewicz, A. K., Rodman, S., Strong, D. R., Kahler, C. W., et al. (2010). Randomized controlled trial of behavioral activation smoking cessation treatment for smokers with elevated depressive symptoms. *Journal of Consulting and Clinical Psychology, 78*, 55–61.

MacPhillamy, D. J., & Lewinsohn, P. M. (1974). Depression as a function of levels of desired and obtained pleasure. *Journal of Abnormal Psychology, 83*, 651–657.

Mancuso, G., Warburton, D. M., Melen, M., Sherwood, N., & Tirelli, E. (1999). Selective effects of nicotine on attentional processes. *Psychopharmacology, 146*, 199–204.

Martini, S., Wagner, F. A., & Anthony, J. C. (2002). The association of tobacco smoking and depression in adolescence: Evidence from the United States. *Substance Use & Misuse, 37*, 1853–1867.

McChargue, D. E., Spring, B., Cook, J. W., & Neumann, C. A. (2004a). Reinforcement expectations explain the relationship between depressive history and smoking status in college students. *Addictive Behaviors, 29*, 991–994.

McChargue, D. E., Spring, B., Cook, J. W., & Neumann, C. A. (2004b). Reinforcement expectations explain the relationship between depressive history and smoking status in college students. *Addictive Behavior, 29*, 991–994.

Meehl, P. E. (1975). Hedonic capacity: Some conjectures. *Bulletin of the Menninger Clinic, 39*, 295–307.

Meehl, P. E. (1987). "Hedonic capacity" ten years later: Some clarifications. In D. C. Clark & J. Fawcett (Eds.), *Anhedonia and affect deficit states* (pp. 47–50). New York, NY: PMA Publishing.

Meehl, P. E. (2001). Primary and secondary hypohedonia. *Journal of Abnormal Psychology, 110*, 188–193.

Memelstein, R. J., Hedeker, D., Flay, B., & Shiffman, S. (2007). Real-time data capture and adolescent cigarette smoking: Moods and smoking. In A. Stone, S. Shiffman, A. Atienza, & L. Nebeling (Eds.), *The science of real-time data capture: Self-report in health research* (pp. 117–135). New York: Oxford University Press.

Miller, N. L., Markowitz, J. C., Kocsis, J. H., Leon, A. C., Brisco, S. T., & Garno, J. L. (1999). Cost effectiveness of screening for clinical trials by research assistants versus senior investigators. *Journal of Psychiatric Research, 33*, 81–85.

Miller, P. G., & Miller, W. R. (2009). What should we be aiming for in the treatment of addiction? *Addiction, 104*, 685–686.

Munafo, M. R., Hitsman, B., Rende, R., Metcalfe, C., & Niaura, R. (2008). Effects of progression to cigarette smoking on depressed mood in adolescents: Evidence from the National Longitudinal Study of Adolescent Health. *Addiction, 103*, 162–171.

Murphy, J. G., Dennhardt, A. A., Skidmore, J. R., Borsari, B., Barnett, N. P., Colby, S. M., et al. (2012). A randomized controlled trial of a behavioral economic supplement to brief motivational interventions for college drinking. *Journal of Consulting and Clinical Psychology, 80*, 876–886.

Murphy, J. G., Skidmore, J. R., Dennhardt, A. A., Martens, M. P., Borsari, B., Barnett, N. P., et al. (2012). A behavioral economic supplement to brief motivational interventions for college drinking. *Addiction Research & Theory, 20*, 456–465.

Murray, C. J., & Lopez, A. D. (1996). Evidence-based health policy—Lessons from the Global Burden of Disease Study. *Science, 274*, 740–743.

Murray, C. J., & Lopez, A. D. (1997). Mortality by cause for eight regions of the world: Global Burden of Disease Study. *Lancet, 349*, 1269–1276.

Niaura, R., Britt, D. M., Shadel, W. G., Goldstein, M., Abrams, D., & Brown, R. (2001). Symptoms of depression and survival experience among three samples of smokers trying to quit. *Psychology of Addictive Behaviors, 15*, 13–17.

Nierenberg, A. A., Keefe, B. R., Leslie, V. C., Alpert, J. E., Pava, J. A., Worthington, J. J., 3rd, et al. (1999). Residual symptoms in depressed patients who respond acutely to fluoxetine. *The Journal of Clinical Psychiatry, 60*, 221–225.

Nutt, D., Demyttenaere, K., Janka, Z., Aarre, T., Bourin, M., Canonico, P. L., et al. (2007). The other face of depression, reduced positive affect: The role of catecholamines in causation and cure. *Journal of Psychopharmacology (Oxford, England), 21*, 461–471.

Patterson, F., Jepson, C., Loughead, J., Perkins, K., Strasser, A. A., Siegel, S., et al. (2010). Working memory deficits predict short-term smoking resumption following brief abstinence. *Drug and Alcohol Dependence, 106*, 61–64.

Patton, G. C., Carlin, J. B., Coffey, C., Wolfe, R., Hibbert, M., & Bowes, G. (1998). Depression, anxiety, and smoking initiation: A prospective study over 3 years. *American Journal of Public Health, 88*, 1518–1522.

Perkins, K. A. (2001). Smoking cessation in women. Special considerations. *CNS Drugs, 15*, 391–411.

Perkins, K. A. (2009). Sex differences in nicotine reinforcement and reward: Influences on the persistence of tobacco smoking. *Nebraska Symposium on Motivation, 55*, 143–169.

Perkins, K. A., Doyle, T., Ciccocioppo, M., Conklin, C., Sayette, M., & Caggiula, A. (2006). Sex differences in the influence of nicotine dose instructions on the reinforcing and self-reported rewarding effects of smoking. *Psychopharmacology, 184*, 600–607.

Perkins, K. A., Hickcox, M. E., & Grobe, J. E. (2000). Behavioral economics of tobacco smoking. In W. K. Bickel & R. E. Vuchinich (Eds.), *Reframing health behavior change with behavior economics* (pp. 293–311). Mahway: Lawrence Erlbaum Associates, Inc.

Perkins, K. A., Karelitz, J. L., Giedgowd, G. E., & Conklin, C. A. (2013). Negative mood effects on craving to smoke in women versus men. *Addictive Behaviors, 38*, 1527–1531.

Piper, M. E., Smith, S. S., Schlam, T. R., Fleming, M. F., Bittrich, A. A., Brown, J. L., et al. (2010). Psychiatric disorders in smokers seeking treatment for tobacco dependence: Relations with tobacco dependence and cessation. *Journal of Consulting and Clinical Psychology, 78*, 13–23.

Pizzagalli, D. A., Holmes, A. J., Dillon, D. G., Goetz, E. L., Birk, J. L., Bogdan, R., et al. (2009). Reduced caudate and nucleus accumbens response to rewards in unmedicated individuals with major depressive disorder. *The American Journal of Psychiatry, 166*, 702–710.

Pizzagalli, D. A., Iosifescu, D., Hallett, L. A., Ratner, K. G., & Fava, M. (2008). Reduced hedonic capacity in major depressive disorder: Evidence from a probabilistic reward task. *Journal of Psychiatric Research, 43*, 76–87.

Powell, J., Tait, S., & Lessiter, J. (2002). Cigarette smoking and attention to signals of reward and threat in the Stroop paradigm. *Addiction, 97*, 1163–1170.

Prochaska, J. J., Hall, S. M., Tsoh, J. Y., Eisendrath, S., Rossi, J. S., Redding, C. A., et al. (2008). Treating tobacco dependence in clinically depressed smokers: Effect of smoking cessation on mental health functioning. *American Journal of Public Health, 98*, 446–448.

Prochaska, J. J., Rossi, J. S., Redding, C. A., Rosen, A. B., Tsoh, J. Y., Humfleet, G. L., et al. (2004). Depressed smokers and stage of change: Implications for treatment interventions. *Drug and Alcohol Dependence, 76*, 143–151.

Reinherz, H. Z., Giaconia, R. M., Pakiz, B., Silverman, A. B., Frost, A. K., & Lefkowitz, E. S. (1993). Psychosocial risks for major depression in late adolescence: A longitudinal community study. *Journal of the American Academy of Child & Adolescent Psychiatry, 32*, 1155–1163.

Reynolds, E. K., Macpherson, L., Tull, M. T., Baruch, D. E., & Lejuez, C. W. (2011). Integration of the brief behavioral activation treatment for depression (BATD) into a college orientation program: Depression and alcohol outcomes. *Journal of Counseling Psychology, 58*, 555–564.

Ritt-Olson, A., Unger, J., Valente, T., Nezami, E., Chou, C. P., Trinidad, D., et al. (2005). Exploring peers as a mediator of the association between depression and smoking in young adolescents. *Substance Use & Misuse, 40*, 77–98.

Rogers, R. E., Higgins, S. T., Silverman, K., Thomas, C. S., Badger, G. J., Bigelow, G., et al. (2008). Abstinence-contingent reinforcement and engagement in non-drug-related activities among illicit drug abusers. *Psychology of Addictive Behaviors, 22*, 544–550.

Rohde, P., Lewinsohn, P. M., Brown, R. A., Gau, J. M., & Kahler, C. W. (2003). Psychiatric disorders, familial factors and cigarette smoking: I. Associations with smoking initiation. *Nicotine & Tobacco Research, 5*, 85–98.

Rukstalis, M., Jepson, C., Patterson, F., & Lerman, C. (2005). Increases in hyperactive-impulsive symptoms predict relapse among smokers in nicotine replacement therapy. *Journal of Substance Abuse Treatment, 28*, 297–304.

Saluja, G., Iachan, R., Scheidt, P. C., Overpeck, M. D., Sun, W., & Giedd, J. N. (2004). Prevalence of and risk factors for depressive symptoms among young adolescents. *Archives of Pediatrics & Adolescent Medicine, 158*, 760–765.

SAMHSA (2004). *Office of Applied Studies, National Survey on Drug Use and Health. Subsample*. http://www.samhsa.gov/data/nsduh/2k4nsduh/2k4results/2k4results.pdf. accessed 25.02.15.

Schnoll, R. A., Leone, F. T., & Hitsman, B. (2013). Symptoms of depression and smoking behaviors following treatment with transdermal nicotine patch. *Journal of Addictive Diseases, 32*, 46–52.

Seligman, M. E., Rashid, T., & Parks, A. C. (2006). Positive psychotherapy. *The American Psychologist, 61*, 774–788.

Seligman, M. E., Steen, T. A., Park, N., & Peterson, C. (2005). Positive psychology progress: Empirical validation of interventions. *American Psychologist, 60*, 410–421.

Shafer, A. B. (2006a). Meta-analysis of the factor structures of four depression questionnaires: Beck, CES-D, Hamilton, and Zung. *Journal of Clinical Psychology, 62*, 123–146.

Shafer, A. B. (2006b). Meta-analysis of the factor structures of four depression questionnaires: Beck, CES-D, Hamilton, and Zung. *The Journal of Clinical Psychiatry, 62*, 123–146.

Shankman, S. A., Klein, D. N., Tenke, C. E., & Bruder, G. E. (2007). Reward sensitivity in depression: A biobehavioral study. *Journal of Abnormal Psychology, 116*, 95–104.

Skara, S., & Sussman, S. (2003). A review of 25 long-term adolescent tobacco and other drug use prevention program evaluations. *Preventive Medicine, 37*, 451–474.

Sloan, D. M., Strauss, M. E., Quirk, S. W., & Sajatovic, M. (1997). Subjective and expressive emotional responses in depression. *Journal of Affective Disorders, 46*, 135–141.

Smith, S. S., Jorenby, D. E., Leischow, S. J., Nides, M. A., Rennard, S. I., Johnston, J. A., et al. (2003). Targeting smokers at increased risk for relapse: Treating women and those with a history of depression. *Nicotine & Tobacco Research, 5*, 99–109.

Spring, B., Cook, J. W., Appelhans, B., Maloney, A., Richmond, M., Vaughn, J., et al. (2008). Nicotine effects on affective response in depression-prone smokers. *Psychopharmacology, 196*, 461–471.

Stein, E. A., Pankiewicz, J., Harsch, H. H., Cho, J. K., Fuller, S. A., Hoffmann, R. G., et al. (1998). Nicotine-induced limbic cortical activation in the human brain: A functional MRI study. *The American Journal of Psychiatry, 155*, 1009–1015.

Steuber, T. L., & Danner, F. (2006). Adolescent smoking and depression: Which comes first? *Addictive Behaviors, 31*, 133–136.

Stone, M. D., & Leventhal, A. M. (2014). Associations between anhedonia and risk factors for smoking initiation among adolescent never smokers. In *Poster presented at the 20th annual meeting of the Society for Research on Nicotine and Tobacco, February 2014, Seattle, WA.*

Strong, D. R., Kahler, C. W., Leventhal, A. M., Abrantes, A. M., Lloyd-Richardson, E., Niaura, R., et al. (2009). Impact of bupropion and cognitive-behavioral treatment for depression on positive affect, negative affect, and urges to smoke during cessation treatment. *Nicotine & Tobacco Research, 11*, 1142–1153.

Strong, D. R., Leventhal, A. M., Evatt, D. P., Haber, S., Greenberg, B. D., Abrams, D., et al. (2011). Positive reactions to tobacco predict relapse after cessation. *Journal of Abnormal Psychology, 120*, 999–1005.

Sussman, S., Lichtman, K., Ritt, A., & Pallonen, U. E. (1999). Effects of thirty-four adolescent tobacco use cessation and prevention trials on regular users of tobacco products. *Substance Use & Misuse, 34*, 1469–1503.

Swan, G. E., Jack, L. M., Curry, S., Chorost, M., Javitz, H., McAfee, T., et al. (2003). Bupropion SR and counseling for smoking cessation in actual practice: Predictors of outcome. *Nicotine & Tobacco Research, 5*, 911–921.

Tsoh, J. Y., & Hall, S. M. (2004). Depression and smoking: From the transtheoretical model of change perspective. *Addictive Behaviors, 29*, 801–805.

USDHHS. (2012). *Preventing tobacco use among young people: A report of the Surgeon General.* Atlanta: U.S. Department of Health and Human Services, Center for Disease Control and Prevention, Office on Smoking and Health. accessed 25.02.15.

van der Meer, R. M., Willemsen, M. C., Smit, F., & Cuijpers, P. (2013). Smoking cessation interventions for smokers with current or past depression. *The Cochrane Database of Systematic Reviews, 8*. CD006102.

van der Meer, R. M., Willemsen, M. C., Smit, F., Cuijpers, P., & Schippers, G. M. (2010). Effectiveness of a mood management component as an adjunct to a telephone counselling smoking cessation intervention for smokers with a past major depression: A pragmatic randomized controlled trial. *Addiction, 105*, 1991–1999.

Van Etten, M. L., Higgins, S. T., Budney, A. J., & Badger, G. J. (1998). Comparison of the frequency and enjoyability of pleasant events in cocaine abusers vs. non-abusers using a standardized behavioral inventory. *Addiction, 93*, 1669–1680.

Wang, M. Q., Eddy, J. M., & Fitzhugh, E. C. (2000). Smoking acquisition: Peer influence and self-selection. *Psychological Reports, 86*, 1241–1246.

Wang, M. Q., Fitzhugh, E. C., Green, B. L., Turner, L. W., Eddy, J. M., & Westerfield, R. C. (1999). Prospective social-psychological factors of adolescent smoking progression. *Journal of Adolescent Health, 24*, 2–9.

Wang, M. Q., Fitzhugh, E. C., Turner, L., Fu, Q., & Westerfield, R. C. (1996). Association of depressive symptoms and school adolescents' smoking: A cross-lagged analysis. *Psychological Reports*, *79*, 127–130.

Warburton, D. M., & Mancuso, G. (1998). Evaluation of the information processing and mood effects of a transdermal nicotine patch. *Psychopharmacology*, *135*, 305–310.

Watson, D. (2005). Rethinking the mood and anxiety disorders: A quantitative hierarchical model for DSM-V. *Journal of Abnormal Psychology*, *114*, 522–536.

Watson, D., Clark, L. A., & Carey, G. (1988). Positive and negative affectivity and their relation to anxiety and depressive disorders. *Journal of Abnormal Psychology*, *97*, 346–353.

Weinstein, S. M., & Mermelstein, R. J. (2013a). Dynamic associations of negative mood and smoking across the development of smoking in adolescence. *Journal of Clinical Child & Adolescent Psychology*, *42*, 629–642.

Weinstein, S. M., & Mermelstein, R. J. (2013b). Influences of mood variability, negative moods, and depression on adolescent cigarette smoking. *Psychology of Addictive Behaviors*, *27*, 1068–1078.

Weller, E. B., & Weller, R. A. (2000). Depression in adolescents growing pains or true morbidity? *Journal of Affective Disorders*, *61*(Suppl. 1), 9–13.

Wichers, M. C., Barge-Schaapveld, D. Q., Nicolson, N. A., Peeters, F., de Vries, M., Mengelers, R., et al. (2009). Reduced stress-sensitivity or increased reward experience: The psychological mechanism of response to antidepressant medication. *Neuropsychopharmacology*, *34*, 923–931.

Wilhelm, K., Wedgwood, L., Niven, H., & Kay-Lambkin, F. (2006). Smoking cessation and depression: Current knowledge and future directions. *Drug and Alcohol Review*, *25*, 97–107.

Williamson, D. E., Birmaher, B., Axelson, D. A., Ryan, N. D., & Dahl, R. E. (2004). First episode of depression in children at low and high familial risk for depression. *Journal of the American Academy of Child and Adolescent Psychiatry*, *43*, 291–297.

Wills, T. A., & Cleary, S. D. (1999). Peer and adolescent substance use among 6th-9th graders: Latent growth analyses of influence versus selection mechanisms. *Health Psychology*, *18*, 453–463.

Windle, M., & Windle, R. C. (2001). Depressive symptoms and cigarette smoking among middle adolescents: Prospective associations and intrapersonal and interpersonal influences. *Journal of Consulting and Clinical Psychology*, *69*, 215–226.

Wu, L. T., & Anthony, J. C. (1999). Tobacco smoking and depressed mood in late childhood and early adolescence. *American Journal of Public Health*, *89*, 1837–1840.

PART 4

Parkinson's disease

CHAPTER NINE

Nicotine and Nicotinic Receptor Drugs: Potential for Parkinson's Disease and Drug-Induced Movement Disorders

Maryka Quik[1], Tanuja Bordia, Danhui Zhang, Xiomara A. Perez
Center for Health Sciences, SRI International, Menlo Park, California, USA
[1]Corresponding author: e-mail address: maryka.quik@sri.com

Contents

Abstract

Parkinson's disease is a progressive neurodegenerative disorder associated with tremor, rigidity, and bradykinesia, as well as nonmotor symptoms including autonomic impairments, olfactory dysfunction, sleep disturbances, depression, and dementia. Although the major neurological deficit is a loss of nigrostriatal dopaminergic neurons, multiple neurotransmitters systems are compromised in Parkinson's disease. Consistent with this observation, dopamine replacement therapy dramatically improves Parkinson's disease motor symptoms. Additionally, drugs targeting the serotonergic, glutamatergic, adenosine, and other neurotransmitter systems may be beneficial. Recent evidence also

International Review of Neurobiology, Volume 124
ISSN 0074-7742
http://dx.doi.org/10.1016/bs.irn.2015.07.005

indicates that nicotinic cholinergic drugs may be useful for the management of Parkinson's disease. This possibility initially arose from the results of epidemiological studies, which showed that smoking was associated with a decreased incidence of Parkinson's disease, an effect mediated in part by the nicotine in smoke. Further evidence for this idea stemmed from preclinical studies which showed that nicotine administration reduced nigrostriatal damage in parkinsonian rodents and monkeys. In addition to a potential neuroprotective role, emerging work indicates that nicotinic receptor drugs improve the abnormal involuntary movements or dyskinesias that arise as a side effect of L-dopa treatment, the gold standard therapy for Parkinson's disease. Both nicotine and nicotinic receptor drugs reduced L-dopa-induced dyskinesias by over 50% in parkinsonian rodent and monkey models. Notably, nicotine also attenuated the abnormal involuntary movements or tardive dyskinesias that arise with antipsychotic treatment. These observations, coupled with reports that nicotinic receptor drugs have procognitive and antidepressant effects, suggest that central nervous system (CNS) nicotinic receptors may represent useful targets for the treatment of movement disorders.

1. PARKINSON'S DISEASE OVERVIEW

Parkinson's disease is a progressive neurodegenerative disorder characterized by tremor, rigidity, and bradykinesia, as well as autonomic nervous system impairments, olfactory dysfunction, sleep disturbances, cognitive declines, changes in affect, depression, and others (Braak et al., 2002; Connolly & Lang, 2014; Huot, Johnston, Koprich, Fox, & Brotchie, 2013; Pahwa & Lyons, 2014; Stayte & Vissel, 2014). A prominent CNS pathological change that is thought to underlie the motor deficits is a loss of nigrostriatal dopaminergic neurons that project from the substantia nigra to the striatum. Consistent with this contention, drugs affecting the dopaminergic system dramatically improve Parkinson's disease motor symptoms, especially in the early stages. Indeed, a primary approach to treat Parkinson's disease has involved symptomatic treatment to restore or replace dopamine, with L-dopa being the gold standard. However, chronic L-dopa use is associated with side effects such as wearing off, on–off phenomena, and dyskinesias that limit its effectiveness. This led to the development of alternate treatment strategies to enhance dopaminergic function including dopamine agonists, slow release carbidopa/levodopa, and monoamine oxidase type B inhibitors alone or in combination. Additionally, nondopaminergic strategies have proved of benefit, most likely to compensate for deficits in the serotonergic, glutamatergic, adenosine, and other neurotransmitter systems that exist with Parkinson's disease (Connolly & Lang, 2014; Huot et al., 2013; Pahwa & Lyons, 2014; Stayte & Vissel, 2014).

More recent work also indicates a role for the nicotinic cholinergic system in Parkinson's disease. Emerging evidence suggests that drugs targeting nicotinic acetylcholine receptors (nAChRs) may be useful for neuroprotection against nigrostriatal damage and for the treatment of L-dopa-induced dyskinesias (LIDs), as discussed in the remainder of this review.

2. NEUROPROTECTION AND PARKINSON'S DISEASE

Although numerous treatments are available for the symptomatic treatment of Parkinson's disease motor symptoms as mentioned above, a more desirable disease management strategy would be to delay or halt disease progression and ideally restore function. In fact, extensive preclinical and clinical studies have been done to evaluate the neuroprotective potential of drugs targeting different aberrant cellular processes that arise in Parkinson's disease such as mitochondrial dysfunction, oxidative stress, calcium dysregulation, inflammation, protein mishandling, and prion-like processes (Schapira, Olanow, Greenamyre, & Bezard, 2014). However, compounds such as creatine and coenzyme Q, which modulate mitochondrial function, the antioxidant glutathione, trophic factors, immune-modulators, as well as diverse agents such as the calcium channel blocker isradipine, did not effectively delay Parkinson's disease motor symptoms (AlDakheel, Kalia, & Lang, 2014; Quik, Perez, & Bordia, 2012). The epidemiological observation of an inverse association between Parkinson's disease and elevated uric acid levels also led to a clinical trial to test inosine, which elevates urate levels (Davis et al., 1996; de Lau, Koudstaal, Hofman, & Breteler, 2005; Schwarzschild et al., 2008; Weisskopf, O'Reilly, Chen, Schwarzschild, & Ascherio, 2007). In addition, a similar inverse link between Parkinson's disease and the drinking of coffee (which contains the adenosine A2a antagonist caffeine) led to the testing of the adenosine A2a antagonist preladenant. However, these trials have also not yielded a successful neuroprotective strategy (Morelli, Carta, Kachroo, & Schwarzschild, 2010; Prediger, 2010; Quik, Perez, et al., 2012; Szabo, Kincses, & Vecsei, 2011).

2.1 Decreased Incidence of Parkinson's Disease with Smoking

Extensive epidemiological studies, originally conducted to identify risk factors for cardiovascular disease and cancer, have demonstrated a consistent negative correlation between Parkinson's disease and tobacco use. Numerous reports subsequently suggested that the reduced incidence of Parkinson's disease with tobacco use was due to a true biological effect of smoking (Elbaz & Moisan, 2008; Noyce et al., 2012; Searles Nielsen et al., 2012; Tanner, 2010; Wirdefeldt, Adami, Cole, Trichopoulos, & Mandel, 2011). First, the inverse correlation between Parkinson's disease and smoking was not due to a selective mortality (Elbaz & Moisan, 2008; Noyce et al., 2012; Searles Nielsen et al., 2012; Tanner, 2010; Wirdefeldt et al., 2011). Second, the effect was dose and time dependent; that is, the decline in the incidence of Parkinson's disease was greater with the more cigarettes consumed and more years of smoking. Third, the reduced risk was lost with smoking cessation. Fourth, a decreased incidence of Parkinson's disease was observed with other forms of tobacco (Ritz et al., 2007). Last, twin studies showed that Parkinson's disease develops less readily in the twin that smoked (Tanner et al., 2002).

2.2 Role for the Nicotinic Cholinergic System in Neuroprotection

The question that next arose was the identity of the component(s) in tobacco that contributed to this apparent neuroprotective effect. Tobacco contains thousands of compounds any of which may be responsible (Rodgman & Perfetti, 2013). There is a focus on nicotine because of its well-known ability to modulate dopaminergic function in the striatum, a brain region of particular relevance to Parkinson's disease (Giorguieff-Chesselet, Kemel, Wandscheer, & Glowinski, 1979; Grady, Marks, Wonnacott, & Collins, 1992; Rapier, Lunt, & Wonnacott, 1990). Studies using neuronal cells in culture showed that nicotine and nAChR agonists reduced the effect of toxic insults, including neurotoxins that selectively damage mesencephalic dopaminergic neurons (Quik, O'Neill, & Perez, 2007; Quik, Perez, et al., 2012). More importantly, extensive studies showed that nicotine and nAChR agonist administration protected against nigrostriatal damage in parkinsonian monkey, rat, and mouse models (Dajas, Costa, Abin-Carriquiry, McGregor, & Urbanavicius, 2001; Janson et al., 1988; Maggio et al., 1998, 1997; Mudo, Belluardo, Mauro, & Fuxe, 2007; Quik, O'Neill, et al., 2007; Quik et al., 2006; Shimohama, 2009; Stuckenholz et al., 2013; Suzuki et al., 2013). Studies with genetically modified mice and selective nAChR drugs indicate that protection is mediated by various nAChR populations, including the β2* (with the asterisk indicating the presence of other nAChR subunits in the receptor complex) and α7 subtypes (Bordia et al., 2015;

Jeyarasasingam, Tompkins, & Quik, 2002; Park et al., 2007; Ryan, Ross, Drago, & Loiacono, 2001; Stuckenholz et al., 2013).

These preclinical studies formed the basis for a Michael J. Fox Foundation funded clinical trial, currently in progress, to investigate the potential of the nicotine patch to protect against disease progression in early onset Parkinson's disease (ClinicalTrials.gov Identifier NCT01560754).

3. NICOTINE AND nAChR DRUGS DO NOT IMPROVE PARKINSONIAN MOTOR SYMPTOMS

In contrast to the potential of nicotine and nAChR agonists to protect against nigrostriatal damage, they have not consistently been shown to improve motor deficits. There is only variable or no reduction in motor dysfunction in parkinsonian animal models, including mice, rats, and monkeys (Bordia, Campos, Huang, & Quik, 2008; Domino, Ni, & Zhang, 1999; Huang, Grady, & Quik, 2011; Meshul, Kamel, Moore, Kay, & Krentz, 2002; Quik, Cox, et al., 2007; Schneider, Van Velson, Menzaghi, & Lloyd, 1998). Similarly, inconsistencies have been reported in clinical trials which tested the acute or immediate effect of nicotine or a nAChR agonist on Parkinson's disease motor function (Quik, Zhang, Perez, & Bordia, 2014b). Approximately half of the trials demonstrated an improvement in movement in Parkinson's disease patients administered nicotine, the other half demonstrated no change, and one study even reported a worsening. Such conflicting outcomes may be due to variability in the duration of nicotine treatment, differences in the nicotine dosing regimen, the relatively small study sizes, the clinical tests used, and the stage of Parkinson's disease (Clemens, Baron, Coffey, & Reeves, 1995; Ebersbach et al., 1999; Hanagasi, Lees, Johnson, Singleton, & Emre, 2007; Ishikawa & Miyatake, 1993; Kelton, Kahn, Conrath, & Newhouse, 2000; Marshall & Schnieden, 1966; Mitsuoka et al., 2002; Shoulson, 2006; Vieregge, Sieberer, Jacobs, Hagenah, & Vieregge, 2001; Villafane et al., 2007). Of note, however, is that the positive outcomes were generally associated with the open-label studies, while no improvement was observed in Parkinson's disease motor symptoms in the double-blinded trials.

Thus, the clinical data and the preclinical studies with parkinsonian animal models are in agreement in suggesting that nicotine and nAChR agonist treatments do not acutely improve motor symptoms.

4. NICOTINE AND nAChR DRUGS REDUCE LIDs

4.1 LIDs Overview

As mentioned earlier, the gold standard therapy for Parkinson's disease motor symptoms involves treatment with the dopamine precursor L-dopa (Connolly & Lang, 2014; Huot et al., 2013; Pahwa & Lyons, 2014). However, there are drawbacks to its use including a loss of efficacy with disease progression and side effects such as LIDs. These abnormal involuntary movements can be quite incapacitating and are a major challenge in Parkinson's disease management (Connolly & Lang, 2014; Heumann et al., 2014; Huot et al., 2013; Iravani, McCreary, & Jenner, 2012). They may develop after only a few months of treatment and are present in the majority of patients by 10 years of L-dopa treatment (Ahlskog & Muenter, 2001). LIDs are associated with a significant increase in treatment cost to the patient, with a recent study showing that total expenses per patient increased from $18,645 to $26,439 annually after LIDs developed (Suh, Pahwa, & Mallya, 2012). This was due to greater medical resource use, higher L-dopa dosage, and/or the need for alternative Parkinson's disease medications.

Currently, there are only limited options for the management of LIDs. Deep brain stimulation has proven very successful; however, it is a serious surgical procedure with its associated complications and is restricted to only a subset of Parkinson's disease patients (Aviles-Olmos et al., 2014; Perestelo-Perez et al., 2014; Rizzone et al., 2014). Available pharmacological approaches include reduction in L-dopa dose, although this may result in a worsening of Parkinsonism. Amantadine, the only approved drug for LIDs, is of only limited usefulness because of a loss of efficacy with time and development of its own side effects including hallucinations, confusion, and others (Meissner et al., 2011; Obeso et al., 2010; Schapira, 2009; Schapira & Jenner, 2011; Wichmann, DeLong, Guridi, & Obeso, 2011). Although there is a critical need for novel strategies to reduce LIDs, these have proved elusive most likely because of the complexity of neuronal circuitry and the multiple mechanisms that contribute to LIDs. These include the well-established role of the dopaminergic system, as well as the serotonergic, glutamatergic, adenosine, and other neuronal systems (Blandini & Armentero, 2012; Cenci, 2007; Gasparini, Di Paolo, & Gomez-Mancilla, 2013; Huot et al., 2013; Iravani et al., 2012; Rylander, 2012). In agreement with these data, drugs directed to these different systems all improve LIDs to varying extents in parkinsonian animal models, although their effectiveness in clinical trials is not evident or uncertain (Al Dakheel, Beaulieu-Boire, & Fox, 2014; Blandini & Armentero, 2012; Brotchie & Jenner, 2011; Duty, 2012; Fox, Chuang, & Brotchie, 2009; Huot et al., 2013; Kobylecki et al., 2014; Quik, Mallela, Ly, & Zhang, 2013; Rascol et al., 2014; Schaeffer, Pilotto, & Berg, 2014; Sgambato-Faure & Cenci, 2012). Novel approaches to reduce LIDs are therefore essential.

4.2 Nicotine Decreases LIDs

One promising new approach to treat LIDs may involve targeting of the nicotinic cholinergic system. Recent studies show that drugs acting at nAChRs improve LIDs in different parkinsonian animal models (Fig. 1). Initial proof-of-concept studies showed that nicotine ameliorated LIDs in a small group of MPTP-lesioned nonhuman primates (Quik, Cox, et al., 2007). This model was used because the deficits observed after MPTP treatment closely resemble those observed in Parkinson's disease (Duty & Jenner, 2011; Iderberg, Francardo, & Pioli, 2012; Jenner, 2009). Long-term administration of nicotine to parkinsonian monkeys alleviated both peak and total LIDs by ~60% for 1 year, at which time the study was ended (Quik, Cox, et al., 2007; Quik, Mallela, Chin, et al., 2013; Quik, Mallela, Ly, et al., 2013). A decline in LIDs was observed whether nicotine was administered prior

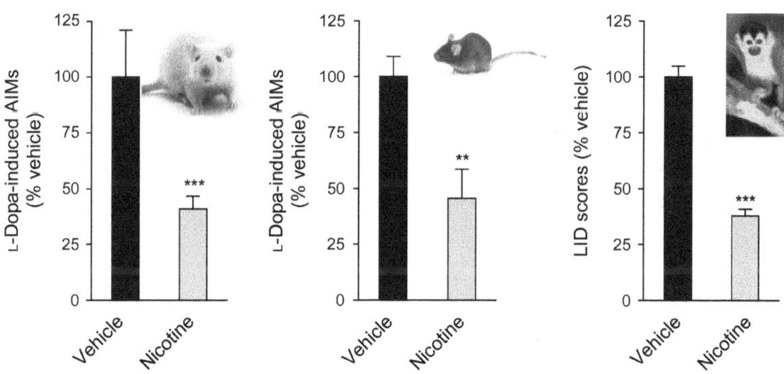

Figure 1 Nicotine treatment reduces LIDs across species. Nicotine consistently reduces LIDs in parkinsonian rats, mice, and monkeys with no worsening of parkinsonism on or off L-dopa. Notably, the maximal decline in LIDs ranges between 60% and 70% across species. Significance of difference from vehicle-treated animals, $^{**}p < 0.01$, $^{***}p < 0.001$. Values are the mean ± SEM of five to six animals. *Data taken in modified form from Bordia et al. (2008), Huang, Grady, et al. (2011), Quik, Cox, et al. (2007), and Quik, Mallela, Ly, et al. (2013).*

to the development of LIDs or in the presence of existing LIDs (Quik, Cox, et al., 2007; Quik, Mallela, Ly, et al., 2013). Importantly, the ability of nicotine to reduce LIDs did not wane with continued treatment (Quik, Cox, et al., 2007; Quik, Mallela, Ly, et al., 2013). Washout experiments demonstrated that the antidyskinetic effect persisted up to 6 weeks after nicotine discontinuation, suggesting that long-term molecular changes underlie nicotine's antidyskinetic effect. These combined studies suggested that nicotine administration may be a viable strategy for the treatment of LIDs in Parkinson's disease patients.

To investigate the robustness of the nicotine-mediated decline in LIDs across species, its effects were investigated in other parkinsonian animal models (Fig. 1) (Bordia et al., 2008; Bordia, Campos, McIntosh, & Quik, 2010; Huang, Campos, Ly, Carroll, & Quik, 2011; Huang, Grady, et al., 2011; Quik, Park, et al., 2012). Nicotine treatment reduced LIDs in parkinsonian rats and mice up to ~60%, similar to the extent of the decline in monkeys. Different modes of nicotine administration resulted in a similar reduction in LIDs, including administration in the drinking water, slow release minipumps, and systemic injection (Bordia et al., 2008, 2010). Thus, multiple modes of nicotine administration could potentially be used in the clinic.

These preclinical studies prompted a small Phase I/II clinical trial to test the effect of nicotine in Parkinson's disease patients with moderate disease. Four months of oral nicotine treatment decreased several measures of LIDs (http://www.neuraltus.com/pages/news_rel12_03_10.html). These findings suggest that a larger scale trial to investigate the antidyskinetic potential of nicotine is warranted.

4.3 nAChRs That Mediate the Antidyskinetic Effect of Nicotine

The question that arises is how nicotine exerts its antidyskinetic effect as such knowledge may allow for the development of drugs that selectively reduce LIDs with a minimum of adverse effects. Nicotine generally acts at CNS nAChRs, ligand-gated ion channels consisting of five transmembrane subunits (Albuquerque, Pereira, Alkondon, & Rogers, 2009; Millar & Gotti, 2009). Numerous nAChR populations exist in the brain, some of which are composed exclusively of α subunits (α7), while others are formed from the combination of α (α2–α6) and β (β2–β4) subunits (Albuquerque et al., 2009; Millar & Gotti, 2009). The primary nAChR populations in the brain contain the α7 or the β2 receptor subunit, with the latter nAChR population primarily consisting of α4β2* and α6β2* receptors (Fig. 2). The asterisk indicates the possible presence of other nAChR subunits in the receptor complex. The α7 nAChRs are very widely distributed throughout the brain. They are not densely expressed in the basal ganglia; however, they are present in numerous other cortical and subcortical regions linked to basal ganglia circuitry (Albuquerque et al., 2009; Millar & Gotti, 2009). By contrast, α4β2* nAChRs are densely present in the basal ganglia, as well as other brain regions, while α6β2* nAChRs are mainly located in the striatum and only a few other regions (Quik & Wonnacott, 2011). The finding that β2* and α7 nAChRs are in the striatum and other brain areas involved in motor control suggests that these subtypes may contribute to the antidyskinetic effects of nicotine (Quik & Wonnacott, 2011).

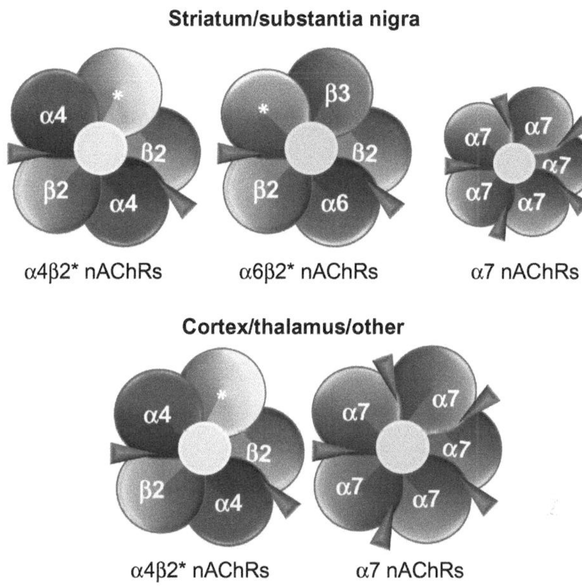

Figure 2 Subunit composition of nAChRs regulating basal ganglia function. Top: The primary nAChRs in the nigrostriatal pathway are the α6β2* and α4β2* subtypes, with a smaller population of α7 nAChRs. α6β2* nAChRs are localized to dopaminergic terminals, α4β2* nAChRs are expressed on dopaminergic terminals, GABAergic interneurons, and medium spiny neurons, while α7 nAChRs are on glutamatergic terminals. Bottom: α4β2* and α7 nAChRs are also widely expressed in other brain areas linked to the basal ganglia such as the cortex and thalamus where they regulate GABA and glutamate transmission, respectively, to influence dopaminergic nigrostriatal function. Two agonist-binding sites (triangles) are depicted at the interface between α and β2 subunits in heteromeric receptors while the homomeric α7 receptors have five binding sites.

4.3.1 General nAChR Drugs Reduce LIDs

To investigate the possibility that the effect of nicotine was nAChR mediated, studies were done with varenicline, a drug that acts at several nAChR populations including α4β2*, α6β2*, and α7 receptors (Fig. 3) (Bordia, Hrachova, Chin, McIntosh, & Quik, 2012; Mihalak, Carroll, & Luetje, 2006; Ween et al., 2010). Studies in 6-hydroxydopamine-lesioned rats showed that varenicline reduced L-dopa-induced abnormal involuntary movements (AIMs) by 40–50% in this parkinsonian animal model (Huang, Campos, et al., 2011). Subsequent work in parkinsonian non-human primates also demonstrated a ~50% reduction in LIDs, without a worsening of the antiparkinsonian effect of L-dopa (Zhang et al., 2013).

The general nAChR antagonist mecamylamine decreased LIDs in parkinsonian rats to a similar extent as nicotine (Bordia et al., 2010). This observation substantiated the data with varenicline that the decline in LIDs is nAChR mediated. Additionally, the observation that both a nAChR antagonist and agonists reduce LIDs suggested that nAChR drugs decrease LIDs via a receptor desensitization blockade. This idea is consistent with other studies indicating that nicotine exerts its antidepressant, cognition, and other effects via nAChR desensitization (Buccafusco, Beach, & Terry, 2009; Mineur & Picciotto, 2010; Rezvani et al., 2012).

Overall, these data confirm that the antidyskinetic effect of nicotine is nAChR mediated and led the way for further pharmacological studies with subtype-selective nAChR drugs.

4.3.2 β2* nAChR Agonists Attenuate LIDs

Subsequent work with agonists that selectivity act at β2* nAChRs, such as ABT-089, ABT-894, A-85380, TC-8831, and others, showed that they all attenuated LIDs by 30–60% in parkinsonian rodents and/or monkeys (Fig. 3) (Huang, Campos, et al., 2011; Johnston et al., 2013; Quik, Campos, Bordia, et al., 2013; Zhang et al., 2013). Notably, none of the drugs worsened Parkinsonism on or off L-dopa (Huang, Campos, et al., 2011; Johnston et al., 2013; Quik, Campos, Bordia, et al., 2013; Zhang et al., 2013).

Studies with nAChR knockout mice support the results of the pharmacological studies and also provide a greater understanding of the specific nAChR populations that regulate the occurrence of LIDs (Table 1) (Quik, Campos, & Grady, 2013). β2 nAChR-null mutant mice, which lack both the α4β2* and α6β2* receptor subtypes, expressed about 50% fewer LIDs than wild-type mice (Huang, Grady, et al., 2011). In addition, nicotine

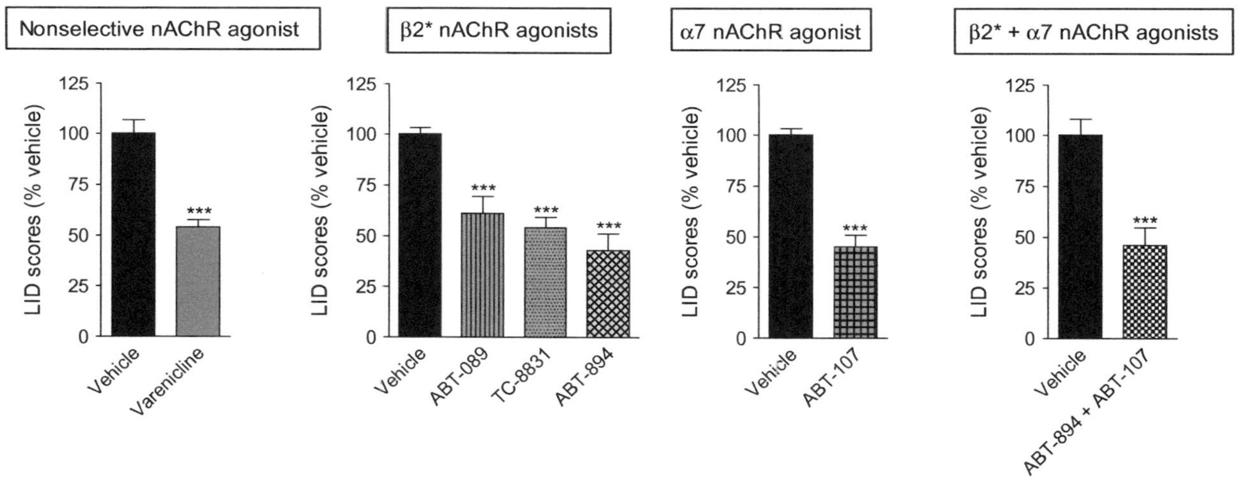

Figure 3 nAChR agonists reduce L-dopa-induced dyskinesias in parkinsonian monkeys. MPTP-lesioned monkeys were gavaged with L-dopa plus carbidopa twice daily 5 days per week, with the agonists given orally 30 min or immediately prior to L-dopa gavage, as described (Zhang, Bordia, et al., 2014; Zhang et al., 2013; Zhang, McGregor, Decker, & Quik, 2014). The data shown are for 0.03 mg/kg varenicline, 0.1 mg/kg ABT-089, 0.1 mg/kg TC-8831, 0.01 mg/kg ABT-894, 0.1 mg/kg ABT-107, 1 mg/kg ABT-126, and 0.01 mg/kg ABT-894 in combination with 0.1 mg/kg ABT-107. Data (expressed as % vehicle) are total dyskinesia scores. Significance of difference from vehicle-treated animals, $^{***}p<0.001$. Values are the mean $\pm$ SEM of 5–12 animals. *Data taken in modified form from Zhang, Bordia, et al. (2014), Zhang et al. (2013), and Zhang, McGregor, et al. (2014).*

Table 1 α4β2*, α6β2*, and α7 nAChRs Modulate Expression of LIDs

nAChR Subunit-Null Mutant Mouse	nAChR Subtype Deleted	Baseline LIDs in Knockout	Nicotine Still Decreases LIDs in Knockout
β2 (−/−)	α4β2* and α6β2*	Reduced	No
α6 (−/−)	α6β2*	Reduced	No
α4 (−/−)	α4β2*	Unaffected	No
α7 (−/−)	α7	Enhanced	Yes

Data from nAChR knockout mice show that nAChR subtypes may modify the level of LIDs (baseline) and/or the ability of nicotine to reduce LIDs (Huang, Grady, et al., 2011; Quik, Park, et al., 2012).

treatment did not reduce LIDs in these mice as it had in wild-type mice. Thus, β2* nAChRs appear essential for the occurrence of LIDs and the antidyskinetic action of nicotine. Similar results were observed with α6 nAChR subunit-null mutant mice that lack α6β2* nAChRs (Quik, Park, et al., 2012). α4* nAChR also regulated the occurrence of LIDs. Although baseline LIDs were not reduced, nicotine did not elicit an antidyskinetic effect in α4 nAChR subunit-null mutant mice (Quik et al., 2013).

These observations indicate that α6β2* and α4β2* nAChRs are both involved in the occurrence of LIDs although in a somewhat differential manner.

4.3.3 α7 nAChR Agonist Decreases LIDs

As mentioned, although α7 nAChRs are not densely expressed in the basal ganglia, they are widely distributed in brain areas closely linked to basal ganglia function (Breese et al., 1997; Han et al., 2003; Pauly, Marks, Gross, & Collins, 1991). Consistent with this observation, the α7 nAChR agonist ABT-107 reduced LIDs by ~50% in nonhuman primates (Fig. 3) (Zhang, McGregor, et al., 2014). Of note, administration of a β2* and α7 nAChR agonist in combination yielded a decline in LIDs similar to that with either drug alone (Zhang, McGregor, et al., 2014). Thus, β2* and α7 nAChR drugs appear to mediate their antidyskinetic effect through a final common mechanism.

An involvement of α7 nAChRs in LIDs is also supported by experiments using α7 nAChR-null mutant mice. α7 nAChR subunit deletion increased the expression of LIDs (Table 1) (Quik et al., 2013), suggesting that α7 receptors partially suppress the occurrence of LIDs. Nicotine was still able to exert its antidyskinetic effect in α7 nAChR-null mutant mice, in contrast to its lack of effect in β2, α4, and α6 nAChR subunit knockout mice (Quik et al., 2013). Altogether, these results indicate that nAChR subtypes differentially regulate the expression of LIDs. This is not that unexpected in view of the very different pharmacological and physiological properties of the α7 and β2* nAChR populations (Albuquerque et al., 2009; Millar & Gotti, 2009).

5. NICOTINE AND nAChR DRUGS REDUCE ANTIPSYCHOTIC-INDUCED TARDIVE DYSKINESIAS

5.1 Tardive Dyskinesia Overview

The finding that nicotine and nAChR drugs reduced LIDs raised the question whether such compounds would also attenuate other drug-induced abnormal involuntary movements, such as tardive dyskinesias. These are involuntary, repetitive, and purposeless movements of the face and limbs that arise with antipsychotic treatment for schizophrenia, psychotic disorders, depression, obsessive compulsive disorders, anxiety, dementia, posttraumatic stress disorder, and autism (Aia, Revuelta, Cloud, & Factor, 2011; Canitano & Scandurra, 2011; Gershanik & Gomez-Arevalo, 2011; Malone & Waheed, 2009; Pickar, Vinik, & Bartko, 2008; Seeman, 2010; Tarsy, Lungu, & Baldessarini, 2011; Zupancic, 2011). Tardive dyskinesia can be very disruptive to patients and may eventually be debilitating (Aia et al., 2011; Correll, Leucht, & Kane, 2004; Gershanik & Gomez-Arevalo, 2011; Jeste & Caligiuri, 1993; Tarsy et al., 2011). Tardive dyskinesia is thought to develop because of the antagonist properties of antipsychotics at striatal dopamine receptors. Approaches to treat tardive dyskinesia include dose reduction and antipsychotic cessation. However, drug removal unmasks supersensitive dopamine receptors with a consequent increase in tardive dyskinesia (Aia et al., 2011; Correll et al., 2004; Gershanik & Gomez-Arevalo, 2011; Tarsy et al., 2011). New strategies are thus important.

5.2 Nicotine Reduces Tardive Dyskinesia by Acting at nAChRs

Studies show that nicotine treatment reduces haloperidol-induced abnormal movements in rats (Bordia, McIntosh, & Quik, 2012). This rodent model shares many characteristics typical of tardive dyskinesia, including a delayed onset (months) and similarities in appearance (Waddington, 1990). Rats administered the antipsychotic haloperidol exhibited vacuous chewing movements (VCMs), an abnormal movement that parallels tardive dyskinesia (Gao, Cooper, Suckow, & Tamminga, 2006; Turrone, Remington, & Nobrega, 2002). Nicotine decreased VCMs in rats when given either before antipsychotic treatment was started or after abnormal movements had developed (Bordia, McIntosh, et al., 2012). Subsequent work also showed that nicotine reduced antipsychotic-induced VCMs in mice, indicating its effectiveness in different models. The nicotine-mediated decline in VCMs was observed in mice given haloperidol either via injection, an intermittent mode of administration, or via subcutaneous pellets, which may resemble depot administration (Quik, Zhang, Perez, & Bordia, 2014a) (Fig. 4). The improvement appeared to be due to a direct effect of nicotine, since its removal for several weeks led to a return of VCMs similar to that in vehicle-treated mice (Quik et al., 2014a) (Fig. 4). Studies with the general nAChR agonist varenicline showed that it reduced VCMs in a dose-dependent fashion (Quik et al., 2014a). These data indicate that the effect was nAChR mediated similar to that for LIDs.

Nicotine and nAChR drugs thus have potential for reducing other forms of abnormal involuntary movements such as antipsychotic-induced tardive dyskinesia.

6. BENEFICIAL ROLE OF NICOTINE AND nAChR DRUGS IN COGNITION, DEPRESSION, AND OTHER NEUROPSYCHIATRIC DISORDERS

nAChRs are widely distributed throughout the brain where they influence numerous physiological processes beyond the modulation of motor control. An extensive preclinical literature shows that these are quite varied and can include effects on cognition, depression, mood, and anxiety, which may be of relevance to other CNS problems that arise with Parkinson's disease (Bacher, Wu, Shytle, & George, 2009; Freedman, 2014; Graef, Schonknecht, Sabri, & Hegerl, 2011; Hurst, Rollema, & Bertrand, 2013; Kutlu & Gould, 2015; Levin, 2013; Mineur et al., 2011; Mineur & Picciotto, 2010; Philip, Carpenter, Tyrka, & Price, 2010; Picciotto, Lewis, van Schalkwyk, & Mineur, 2015; Poorthuis, Goriounova, Couey, & Mansvelder, 2009; Sarter, Parikh, & Howe, 2009). Nicotine and nAChR drugs enhance diverse cognitive functions in experimental animal models, including learning, memory, and attention (Geerts, 2012; Lendvai, Kassai, Szajli, & Nemethy, 2013; Levin, Hall, & Rezvani, 2015; Sarter et al., 2009). These effects extend to the clinic where varenicline has been shown

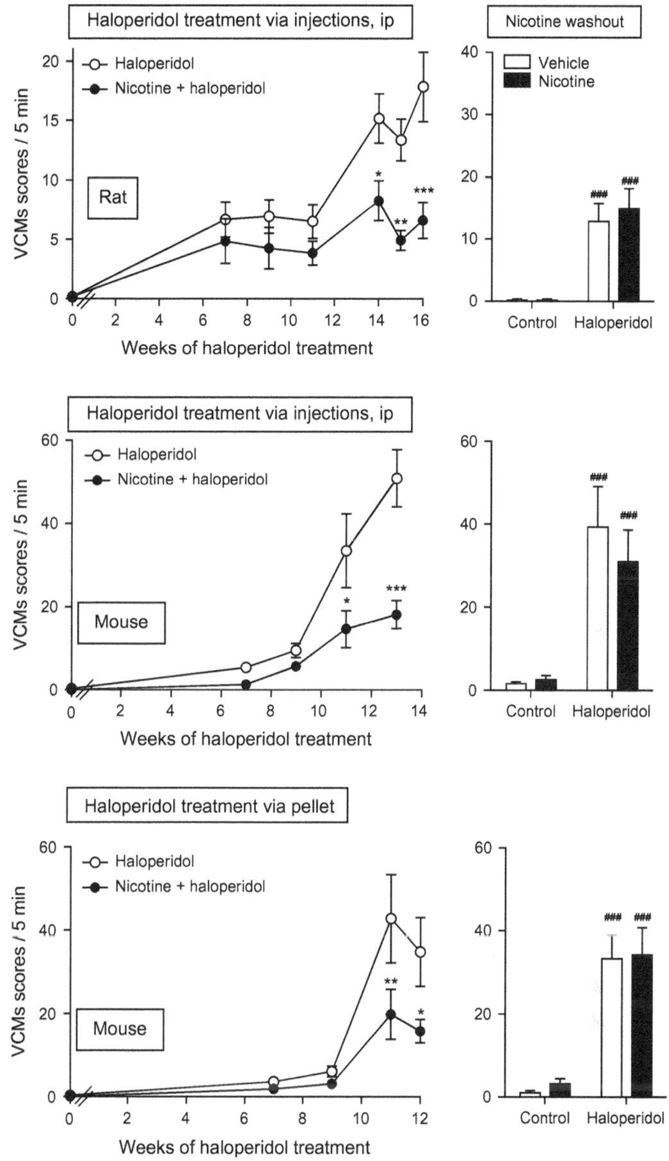

Figure 4 Nicotine treatment reduces abnormal involuntary movements in a rodent model of antipsychotic-induced tardive dyskinesia. Nicotine treatment reduces vacuous chewing movements (VCMs) in both rats and mice by week 11 (left panel) regardless of the mode of haloperidol treatment (injection or pellet). Nicotine discontinuation (2–5 weeks) resulted in a return of VCMs to vehicle-treated levels (right panels), demonstrating a direct effect of the drug. Values are the mean ± SEM of 6–12 animals per group. Significance of difference from the vehicle-treated control, $^{###}p < 0.001$: from the vehicle–haloperidol-treated group, $^{*}p < 0.05$, $^{**}p < 0.01$, $^{***}p < 0.001$. *Data taken in modified form from Bordia, McIntosh, et al. (2012) and Quik et al. (2014b).*

to improve some cognitive impairments in schizophrenics and lapses in attention in treatment-seeking smokers (Rhodes, Hawk, Ashare, Schlienz, & Mahoney, 2012; Shim et al., 2012). In addition, recent clinical reports show that α7 nAChR agonists have procognitive effects in patients with schizophrenia and Alzheimer's disease (Freedman, 2014, Graef et al., 2011, Hurst et al., 2013). Likewise, the β2*-selective nAChR agonist TC-1734 had a positive effect on cognition in older subjects with age-associated memory impairment in a randomized placebo-controlled double-blinded study (Dunbar et al., 2007; Dunbar, Kuchibhatla, & Lee, 2011). Of more direct relevance, nicotine has been shown to improve impaired semantic processing in Parkinson's disease patients (Holmes, Copland, Silburn, & Chenery, 2011).

Extensive preclinical studies, as well as some clinical work, also support a role for nAChRs in depression, mood, and anxiety (Hurst et al., 2013; Mineur & Picciotto, 2010; Philip et al., 2010; Picciotto et al., 2015; Zurkovsky, Taylor, & Newhouse, 2013). The nAChR antagonist mecamylamine improved depression in a double-blind, placebo-controlled study for Tourette's disorder (Silver et al., 2001), although the mecamylamine enantiomer TC-5214 failed to meet the primary endpoint in a more recent trial (Hurst et al., 2013). A small open-label trial with varenicline was also associated with significant improvement in mood in outpatient smokers with persistent depressive symptoms (Philip, Carpenter, Tyrka, Whiteley, & Price, 2009).

Thus, in addition to their potential usefulness for improving motor dysfunctions, such as tardive dyskinesia and LIDs, and for neuroprotection against nigrostriatal damage, nAChR drugs may be helpful for the associated cognitive declines, depression, and changes in affect that arise with neurological disorders.

7. CONCLUDING REMARKS

An accumulating literature suggests that nicotine and nAChR drugs may be useful for the management of Parkinson's disease. One very important role may lie in neuroprotection against the nigrostriatal damage that arises in Parkinson's disease. In addition, nAChR drugs may be useful for combating one of the side effects of L-dopa treatment, that is, the dyskinesias that arise with its continued use. Of note, nicotine and nAChR drugs also reduce the abnormal involuntary movements that result as a side effect of antipsychotic use. The positive effects of nicotine on cognition, depression, anxiety, and mood may be an added benefit. Overall, current findings would suggest that clinical trials with nicotine and/or nAChR drugs may yield novel therapeutic approaches for neurological disorders such as Parkinson's disease.

ACKNOWLEDGMENT

We gratefully acknowledge support from NIH Grant NS59910.

REFERENCES

Ahlskog, J. E., & Muenter, M. D. (2001). Frequency of levodopa-related dyskinesias and motor fluctuations as estimated from the cumulative literature. *Movement Disorders, 16*(3), 448–458.

Aia, P. G., Revuelta, G. J., Cloud, L. J., & Factor, S. A. (2011). Tardive dyskinesia. *Current Treatment Options in Neurology, 13*, 231–241.

Al Dakheel, A., Beaulieu-Boire, I., & Fox, S. H. (2014). Emerging drugs for levodopa-induced dyskinesia. *Expert Opinion on Emerging Drugs, 19*(3), 415–429.

Albuquerque, E. X., Pereira, E. F., Alkondon, M., & Rogers, S. W. (2009). Mammalian nicotinic acetylcholine receptors: From structure to function. *Physiological Reviews, 89*(1), 73–120.

AlDakheel, A., Kalia, L. V., & Lang, A. E. (2014). Pathogenesis-targeted, disease-modifying therapies in Parkinson disease. *Neurotherapeutics, 11*(1), 6–23.

Aviles-Olmos, I., Kefalopoulou, Z., Tripoliti, E., Candelario, J., Akram, H., Martinez-Torres, I., et al. (2014). Long-term outcome of subthalamic nucleus deep brain stimulation for Parkinson's disease using an MRI-guided and MRI-verified approach. *Journal of Neurology, Neurosurgery, and Psychiatry, 85*, 1419–1425.

Bacher, I., Wu, B., Shytle, D. R., & George, T. P. (2009). Mecamylamine—A nicotinic acetylcholine receptor antagonist with potential for the treatment of neuropsychiatric disorders. *Expert Opinion on Pharmacotherapy, 10*(16), 2709–2721.

Blandini, F., & Armentero, M. T. (2012). New pharmacological avenues for the treatment of L-DOPA-induced dyskinesias in Parkinson's disease: Targeting glutamate and adenosine receptors. *Expert Opinion on Investigational Drugs, 21*(2), 153–168.

Bordia, T., Campos, C., Huang, L., & Quik, M. (2008). Continuous and intermittent nicotine treatment reduces L-3,4-dihydroxyphenylalanine (L-DOPA)-induced dyskinesias in a rat model of Parkinson's disease. *The Journal of Pharmacology and Experimental Therapeutics, 327*(1), 239–247.

Bordia, T., Campos, C., McIntosh, J. M., & Quik, M. (2010). Nicotinic receptor-mediated reduction in L-dopa-induced dyskinesias may occur via desensitization. *The Journal of Pharmacology and Experimental Therapeutics, 333*, 929–938.

Bordia, T., Hrachova, M., Chin, M., McIntosh, J. M., & Quik, M. (2012). Varenicline is a potent partial agonist at alpha6beta2* nicotinic acetylcholine receptors in rat and monkey striatum. *The Journal of Pharmacology and Experimental Therapeutics, 342*(2), 327–334.

Bordia, T., McGregor, M., Papke, R. L., Decker, M. W., Michael McIntosh, J., & Quik, M. (2015). The alpha7 nicotinic receptor agonist ABT-107 protects against nigrostriatal damage in rats with unilateral 6-hydroxydopamine lesions. *Experimental Neurology, 263*, 277–284.

Bordia, T., McIntosh, J. M., & Quik, M. (2012). Nicotine reduces antipsychotic-induced orofacial dyskinesia in rats. *The Journal of Pharmacology and Experimental Therapeutics, 340*, 612–619.

Braak, H., Del Tredici, K., Bratzke, H., Hamm-Clement, J., Sandmann-Keil, D., & Rub, U. (2002). Staging of the intracerebral inclusion body pathology associated with idiopathic

Parkinson's disease (preclinical and clinical stages). *Journal of Neurology*, *249*(Suppl. 3). III/1–5.
Breese, C. R., Adams, C., Logel, J., Drebing, C., Rollins, Y., Barnhart, M., et al. (1997). Comparison of the regional expression of nicotinic acetylcholine receptor alpha7 mRNA and [125I]-alpha-bungarotoxin binding in human postmortem brain. *The Journal of Comparative Neurology*, *387*(3), 385–398.
Brotchie, J., & Jenner, P. (2011). New approaches to therapy. *International Review of Neurobiology*, *98*, 123–150.
Buccafusco, J. J., Beach, J. W., & Terry, A. V. (2009). Desensitization of nicotinic acetylcholine receptors as a strategy for drug development. *The Journal of Pharmacology and Experimental Therapeutics*, *328*(2), 364–370.
Canitano, R., & Scandurra, V. (2011). Psychopharmacology in autism: An update. *Progress in Neuro-Psychopharmacology & Biological Psychiatry*, *35*(1), 18–28.
Cenci, M. A. (2007). Dopamine dysregulation of movement control in L-DOPA-induced dyskinesia. *Trends in Neurosciences*, *30*(5), 236–243.
Clemens, P., Baron, J. A., Coffey, D., & Reeves, A. (1995). The short-term effect of nicotine chewing gum in patients with Parkinson's disease. *Psychopharmacology*, *117*(2), 253–256.
Connolly, B. S., & Lang, A. E. (2014). Pharmacological treatment of Parkinson disease: A review. *JAMA*, *311*(16), 1670–1683.
Correll, C. U., Leucht, S., & Kane, J. M. (2004). Lower risk for tardive dyskinesia associated with second-generation antipsychotics: A systematic review of 1-year studies. *The American Journal of Psychiatry*, *161*(3), 414–425.
Dajas, F., Costa, G., Abin-Carriquiry, J. A., McGregor, R., & Urbanavicius, J. (2001). Involvement of nicotinic acetylcholine receptors in the protection of dopamine terminals in experimental parkinsonism. *Functional Neurology*, *16*(Suppl. 4), 113–123.
Davis, J. W., Grandinetti, A., Waslien, C. I., Ross, G. W., White, L. R., & Morens, D. M. (1996). Observations on serum uric acid levels and the risk of idiopathic Parkinson's disease. *American Journal of Epidemiology*, *144*(5), 480–484.
de Lau, L. M., Koudstaal, P. J., Hofman, A., & Breteler, M. M. (2005). Serum uric acid levels and the risk of Parkinson disease. *Annals of Neurology*, *58*(5), 797–800.
Domino, E. F., Ni, L., & Zhang, H. (1999). Nicotine alone and in combination with L-DOPA methyl ester or the D(2) agonist N-0923 in MPTP-induced chronic hemiparkinsonian monkeys. *Experimental Neurology*, *158*(2), 414–421.
Dunbar, G. C., Inglis, F., Kuchibhatla, R., Sharma, T., Tomlinson, M., & Wamsley, J. (2007). Effect of ispronicline, a neuronal nicotinic acetylcholine receptor partial agonist, in subjects with age associated memory impairment (AAMI). *Journal of Psychopharmacology*, *21*(2), 171–178.
Dunbar, G. C., Kuchibhatla, R. V., & Lee, G. (2011). A randomized double-blind study comparing 25 and 50 mg TC-1734 (AZD3480) with placebo, in older subjects with age-associated memory impairment. *Journal of Psychopharmacology*, *25*(8), 1020–1029.
Duty, S. (2012). Targeting glutamate receptors to tackle the pathogenesis, clinical symptoms and levodopa-induced dyskinesia associated with Parkinson's disease. *CNS Drugs*, *26*(12), 1017–1032.
Duty, S., & Jenner, P. (2011). Animal models of Parkinson's disease: A source of novel treatments and clues to the cause of the disease. *British Journal of Pharmacology*, *164*(4), 1357–1391.
Ebersbach, G., Stock, M., Muller, J., Wenning, G., Wissel, J., & Poewe, W. (1999). Worsening of motor performance in patients with Parkinson's disease following transdermal nicotine administration. *Movement Disorders*, *14*(6), 1011–1013.
Elbaz, A., & Moisan, F. (2008). Update in the epidemiology of Parkinson's disease. *Current Opinion in Neurology*, *21*(4), 454–460.

Fox, S. H., Chuang, R., & Brotchie, J. M. (2009). Serotonin and Parkinson's disease: On movement, mood, and madness. *Movement Disorders, 24,* 1255–1266.

Freedman, R. (2014). α7-Nicotinic acetylcholine receptor agonists for cognitive enhancement in schizophrenia. *Annual Review of Medicine, 65,* 245–261.

Gao, X. M., Cooper, T., Suckow, R. F., & Tamminga, C. A. (2006). Multidose risperidone treatment evaluated in a rodent model of tardive dyskinesia. *Neuropsychopharmacology, 31*(9), 1864–1868.

Gasparini, F., Di Paolo, T., & Gomez-Mancilla, B. (2013). Metabotropic glutamate receptors for Parkinson's disease therapy. *Parkinson's Disease, 2013,* 196028.

Geerts, H. (2012). α7 Nicotinic receptor modulators for cognitive deficits in schizophrenia and Alzheimer's disease. *Expert Opinion on Investigational Drugs, 21*(1), 59–65.

Gershanik, O. S., & Gomez-Arevalo, G. J. (2011). Typical and atypical neuroleptics. *Handbook of Clinical Neurology, 100,* 579–599.

Giorguieff-Chesselet, M. F., Kemel, M. L., Wandscheer, D., & Glowinski, J. (1979). Regulation of dopamine release by presynaptic nicotinic receptors in rat striatal slices: Effect of nicotine in a low concentration. *Life Sciences, 25*(14), 1257–1262.

Grady, S., Marks, M. J., Wonnacott, S., & Collins, A. C. (1992). Characterization of nicotinic receptor-mediated [3H]dopamine release from synaptosomes prepared from mouse striatum. *Journal of Neurochemistry, 59*(3), 848–856.

Graef, S., Schonknecht, P., Sabri, O., & Hegerl, U. (2011). Cholinergic receptor subtypes and their role in cognition, emotion, and vigilance control: An overview of preclinical and clinical findings. *Psychopharmacology, 215*(2), 205–229.

Han, Z. Y., Zoli, M., Cardona, A., Bourgeois, J. P., Changeux, J. P., & Le Novere, N. (2003). Localization of [3H]nicotine, [3H]cytisine, [3H]epibatidine, and [125I]alpha-bungarotoxin binding sites in the brain of Macaca mulatta. *The Journal of Comparative Neurology, 461*(1), 49–60.

Hanagasi, H. A., Lees, A., Johnson, J. O., Singleton, A., & Emre, M. (2007). Smoking-responsive juvenile-onset Parkinsonism. *Movement Disorders, 22*(1), 115–119.

Heumann, R., Moratalla, R., Herrero, M. T., Chakrabarty, K., Drucker-Colin, R., Garcia-Montes, J. R., et al. (2014). Dyskinesia in Parkinson's disease: Mechanisms and current non-pharmacological interventions. *Journal of Neurochemistry, 130*(4), 472–489.

Holmes, A. D., Copland, D. A., Silburn, P. A., & Chenery, H. J. (2011). Acute nicotine enhances strategy-based semantic processing in Parkinson's disease. *The International Journal of Neuropsychopharmacology, 14,* 877–885.

Huang, L. Z., Campos, C., Ly, J., Carroll, F. I., & Quik, M. (2011). Nicotinic receptor agonists decrease L-dopa-induced dyskinesias most effectively in moderately lesioned parkinsonian rats. *Neuropharmacology, 60,* 861–868.

Huang, L., Grady, S. R., & Quik, M. (2011). Nicotine reduces L-dopa-induced dyskinesias by acting at {beta}2 nicotinic receptors. *The Journal of Pharmacology and Experimental Therapeutics, 338,* 932–941.

Huot, P., Johnston, T. H., Koprich, J. B., Fox, S. H., & Brotchie, J. M. (2013). The pharmacology of L-DOPA-induced dyskinesia in Parkinson's disease. *Pharmacological Reviews, 65*(1), 171–222.

Hurst, R., Rollema, H., & Bertrand, D. (2013). Nicotinic acetylcholine receptors: From basic science to therapeutics. *Pharmacology & Therapeutics, 137*(1), 22–54.

Iderberg, H., Francardo, V., & Pioli, E. Y. (2012). Animal models of L-DOPA-induced dyskinesia: An update on the current options. *Neuroscience, 211,* 13–27.

Iravani, M. M., McCreary, A. C., & Jenner, P. (2012). Striatal plasticity in Parkinson's disease and L-DOPA induced dyskinesia. *Parkinsonism & Related Disorders, 18*(Suppl. 1), S123–S125.

Ishikawa, A., & Miyatake, T. (1993). Effects of smoking in patients with early-onset Parkinson's disease. *Journal of the Neurological Sciences, 117*(1–2), 28–32.

Janson, A. M., Fuxe, K., Agnati, L. F., Kitayama, I., Harfstrand, A., Andersson, K., et al. (1988). Chronic nicotine treatment counteracts the disappearance of tyrosine-hydroxylase–immunoreactive nerve cell bodies, dendrites and terminals in the mesostriatal dopamine system of the male rat after partial hemitransection. *Brain Research*, *455*(2), 332–345.

Jenner, P. (2009). Functional models of Parkinson's disease: A valuable tool in the development of novel therapies. *Annals of Neurology*, *64*(S2), S16–S29.

Jeste, D. V., & Caligiuri, M. P. (1993). Tardive dyskinesia. *Schizophrenia Bulletin*, *19*(2), 303–315.

Jeyarasasingam, G., Tompkins, L., & Quik, M. (2002). Stimulation of non-alpha7 nicotinic receptors partially protects dopaminergic neurons from 1-methyl-4-phenylpyridinium-induced toxicity in culture. *Neuroscience*, *109*(2), 275–285.

Johnston, T. H., Huot, P., Fox, S. H., Koprich, J. B., Szeliga, K. T., James, J. W., et al. (2013). TC-8831, a nicotinic acetylcholine receptor agonist, reduces L-DOPA-induced dyskinesia in the MPTP macaque. *Neuropharmacology*, *73*, 337–347.

Kelton, M. C., Kahn, H. J., Conrath, C. L., & Newhouse, P. A. (2000). The effects of nicotine on Parkinson's disease. *Brain and Cognition*, *43*(1–3), 274–282.

Kobylecki, C., Burn, D. J., Kass-Iliyya, L., Kellett, M. W., Crossman, A. R., & Silverdale, M. A. (2014). Randomized clinical trial of topiramate for levodopa-induced dyskinesia in Parkinson's disease. *Parkinsonism & Related Disorders*, *20*(4), 452–455.

Kutlu, M. G., & Gould, T. J. (2015). Nicotinic receptors, memory, and hippocampus. *Current Topics in Behavioral Neurosciences*, *23*, 137–163.

Lendvai, B., Kassai, F., Szajli, A., & Nemethy, Z. (2013). alpha7 nicotinic acetylcholine receptors and their role in cognition. *Brain Research Bulletin*, *93*, 86–96.

Levin, E. D. (2013). Complex relationships of nicotinic receptor actions and cognitive functions. *Biochemical Pharmacology*, *86*(8), 1145–1152.

Levin, E. D., Hall, B. J., & Rezvani, A. H. (2015). Heterogeneity across brain regions and neurotransmitter interactions with nicotinic effects on memory function. *Current Topics in Behavioral Neurosciences*, *23*, 87–101.

Maggio, R., Riva, M., Vaglini, F., Fornai, F., Molteni, R., Armogida, M., et al. (1998). Nicotine prevents experimental parkinsonism in rodents and induces striatal increase of neurotrophic factors. *Journal of Neurochemistry*, *71*(6), 2439–2446.

Maggio, R., Riva, M., Vaglini, F., Fornai, F., Racagni, G., & Corsini, G. U. (1997). Striatal increase of neurotrophic factors as a mechanism of nicotine protection in experimental parkinsonism. *Journal of Neural Transmission*, *104*(10), 1113–1123.

Malone, R. P., & Waheed, A. (2009). The role of antipsychotics in the management of behavioural symptoms in children and adolescents with autism. *Drugs*, *69*(5), 535–548.

Marshall, J., & Schnieden, H. (1966). Effect of adrenaline, noradrenaline, atropine, and nicotine on some types of human tremor. *Journal of Neurology, Neurosurgery, and Psychiatry*, *29*(3), 214–218.

Meissner, W. G., Frasier, M., Gasser, T., Goetz, C. G., Lozano, A., Piccini, P., et al. (2011). Priorities in Parkinson's disease research. *Nature Reviews. Drug Discovery*, *10*(5), 377–393.

Meshul, C. K., Kamel, D., Moore, C., Kay, T. S., & Krentz, L. (2002). Nicotine alters striatal glutamate function and decreases the apomorphine-induced contralateral rotations in 6-OHDA-lesioned rats. *Experimental Neurology*, *175*(1), 257–274.

Mihalak, K. B., Carroll, F. I., & Luetje, C. W. (2006). Varenicline is a partial agonist at alpha4beta2 and a full agonist at alpha7 neuronal nicotinic receptors. *Molecular Pharmacology*, *70*(3), 801–805.

Millar, N. S., & Gotti, C. (2009). Diversity of vertebrate nicotinic acetylcholine receptors. *Neuropharmacology*, *56*(1), 237–246.

Mineur, Y. S., Einstein, E. B., Seymour, P. A., Coe, J. W., O'Neill, B. T., Rollema, H., et al. (2011). alpha4beta2 nicotinic acetylcholine receptor partial agonists with low intrinsic efficacy have antidepressant-like properties. *Behavioural Pharmacology, 22*(4), 291–299.

Mineur, Y. S., & Picciotto, M. R. (2010). Nicotine receptors and depression: Revisiting and revising the cholinergic hypothesis. *Trends in Pharmacological Sciences, 31*, 580–586.

Mitsuoka, T., Kaseda, Y., Yamashita, H., Kohriyama, T., Kawakami, H., Nakamura, S., et al. (2002). Effects of nicotine chewing gum on UPDRS score and P300 in early-onset parkinsonism. *Hiroshima Journal of Medical Sciences, 51*(1), 33–39.

Morelli, M., Carta, A. R., Kachroo, A., & Schwarzschild, M. A. (2010). Pathophysiological roles for purines: Adenosine, caffeine and urate. *Progress in Brain Research, 183*, 183–208.

Mudo, G., Belluardo, N., Mauro, A., & Fuxe, K. (2007). Acute intermittent nicotine treatment induces fibroblast growth factor-2 in the subventricular zone of the adult rat brain and enhances neuronal precursor cell proliferation. *Neuroscience, 145*(2), 470–483.

Noyce, A. J., Bestwick, J. P., Silveira-Moriyama, L., Hawkes, C. H., Giovannoni, G., Lees, A. J., et al. (2012). Meta-analysis of early nonmotor features and risk factors for Parkinson disease. *Annals of Neurology, 72*(6), 893–901.

Obeso, J. A., Rodriguez-Oroz, M. C., Goetz, C. G., Marin, C., Kordower, J. H., Rodriguez, M., et al. (2010). Missing pieces in the Parkinson's disease puzzle. *Nature Medicine, 16*(6), 653–661.

Pahwa, R., & Lyons, K. E. (2014). Treatment of early Parkinson's disease. *Current Opinion in Neurology, 27*(4), 442–449.

Park, H. J., Lee, P. H., Ahn, Y. W., Choi, Y. J., Lee, G., Lee, D. Y., et al. (2007). Neuroprotective effect of nicotine on dopaminergic neurons by anti-inflammatory action. *The European Journal of Neuroscience, 26*(1), 79–89.

Pauly, J. R., Marks, M. J., Gross, S. D., & Collins, A. C. (1991). An autoradiographic analysis of cholinergic receptors in mouse brain after chronic nicotine treatment. *The Journal of Pharmacology and Experimental Therapeutics, 258*(3), 1127–1136.

Perestelo-Perez, L., Rivero-Santana, A., Perez-Ramos, J., Serrano-Perez, P., Panetta, J., & Hilarion, P. (2014). Deep brain stimulation in Parkinson's disease: Meta-analysis of randomized controlled trials. *Journal of Neurology, 261*, 2051–2060.

Philip, N. S., Carpenter, L. L., Tyrka, A. R., & Price, L. H. (2010). Nicotinic acetylcholine receptors and depression: A review of the preclinical and clinical literature. *Psychopharmacology, 212*, 1–12.

Philip, N. S., Carpenter, L. L., Tyrka, A. R., Whiteley, L. B., & Price, L. H. (2009). Varenicline augmentation in depressed smokers: An 8-week, open-label study. *The Journal of Clinical Psychiatry, 70*(7), 1026–1031.

Picciotto, M. R., Lewis, A. S., van Schalkwyk, G. I., & Mineur, Y. S. (2015). Mood and anxiety regulation by nicotinic acetylcholine receptors: A potential pathway to modulate aggression and related behavioral states. *Neuropharmacology, 96*, 235–243.

Pickar, D., Vinik, J., & Bartko, J. J. (2008). Pharmacotherapy of schizophrenic patients: Preponderance of off-label drug use. *PLoS One, 3*(9), e3150.

Poorthuis, R. B., Goriounova, N. A., Couey, J. J., & Mansvelder, H. D. (2009). Nicotinic actions on neuronal networks for cognition: General principles and long-term consequences. *Biochemical Pharmacology, 78*(7), 668–676.

Prediger, R. D. (2010). Effects of caffeine in Parkinson's disease: From neuroprotection to the management of motor and non-motor symptoms. *Journal of Alzheimer's Disease, 20*(Suppl. 1), S205–S220.

Quik, M., Campos, C., Bordia, T., Strachan, J. P., Zhang, J., McIntosh, J. M., et al. (2013). alpha4beta2 nicotinic receptors play a role in the nAChR-mediated decline in L-dopa-induced dyskinesias in parkinsonian rats. *Neuropharmacology, 71*, 191–203.

Quik, M., Campos, C., & Grady, S. (2013). Multiple CNS nicotinic receptors mediate L-dopa-induced dyskinesias: Studies with parkinsonian nicotinic receptor knockout mice. *Biochemical Pharmacology, 86*, 1153–1162.

Quik, M., Cox, H., Parameswaran, N., O'Leary, K., Langston, J. W., & Di Monte, D. (2007). Nicotine reduces levodopa-induced dyskinesias in lesioned monkeys. *Annals of Neurology, 62*, 588–596.

Quik, M., Mallela, A., Chin, M., McIntosh, J. M., Perez, X. A., & Bordia, T. (2013). Nicotine-mediated improvement in L-dopa-induced dyskinesias in MPTP-lesioned monkeys is dependent on dopamine nerve terminal function. *Neurobiology of Disease, 50*, 30–41.

Quik, M., Mallela, A., Ly, J., & Zhang, D. (2013). Nicotine reduces established L-dopa-induced dyskinesias in a monkey model of Parkinson's disease. *Movement Disorders, 28*, 1398–1406.

Quik, M., O'Neill, M., & Perez, X. A. (2007). Nicotine neuroprotection against nigrostriatal damage: Importance of the animal model. *Trends in Pharmacological Sciences, 28*, 229–235.

Quik, M., Parameswaran, N., McCallum, S. E., Bordia, T., Bao, S., McCormack, A., et al. (2006). Chronic oral nicotine treatment protects against striatal degeneration in MPTP-treated primates. *Journal of Neurochemistry, 98*(6), 1866–1875.

Quik, M., Park, K. M., Hrachova, M., Mallela, A., Huang, L. Z., McIntosh, J. M., et al. (2012). Role for alpha6 nicotinic receptors in L-dopa-induced dyskinesias in parkinsonian mice. *Neuropharmacology, 63*(3), 450–459.

Quik, M., Perez, X. A., & Bordia, T. (2012). Nicotine as a potential neuroprotective agent for Parkinson's disease. *Movement Disorders, 27*(8), 947–957.

Quik, M., & Wonnacott, S. (2011). {alpha}6{beta}2* and {alpha}4{beta}2* nicotinic acetylcholine receptors as drug targets for Parkinson's disease. *Pharmacological Reviews, 63*(4), 938–966.

Quik, M., Zhang, D., Perez, X. A., & Bordia, T. (2014a). Role for the nicotinic cholinergic system in movement disorders: Therapeutic implications. *Pharmacology & Therapeutics, 144*, 50–59.

Quik, M., Zhang, D., Perez, X. A., & Bordia, T. (2014b). Role for the nicotinic cholinergic system in movement disorders: Therapeutic implications. *Pharmacology & Therapeutics, 144*(1), 50–59.

Rapier, C., Lunt, G. G., & Wonnacott, S. (1990). Nicotinic modulation of [3H]dopamine release from striatal synaptosomes: Pharmacological characterisation. *Journal of Neurochemistry, 54*(3), 937–945.

Rascol, O., Fox, S., Gasparini, F., Kenney, C., Di Paolo, T., & Gomez-Mancilla, B. (2014). Use of metabotropic glutamate 5-receptor antagonists for treatment of levodopa-induced dyskinesias. *Parkinsonism & Related Disorders, 20*, 947–956.

Rezvani, A. H., Timofeeva, O., Sexton, H. G., DeCuir, D., Xiao, Y., Gordon, C. J., et al. (2012). Effects of sazetidine-A, a selective alpha4beta2* nicotinic receptor desensitizing agent, on body temperature regulation in mice and rats. *European Journal of Pharmacology, 682*(1–3), 110–117.

Rhodes, J. D., Hawk, L. W., Jr., Ashare, R. L., Schlienz, N. J., & Mahoney, M. C. (2012). The effects of varenicline on attention and inhibitory control among treatment-seeking smokers. *Psychopharmacology, 223*(2), 131–138.

Ritz, B., Ascherio, A., Checkoway, H., Marder, K. S., Nelson, L. M., Rocca, W. A., et al. (2007). Pooled analysis of tobacco use and risk of Parkinson disease. *Archives of Neurology, 64*(7), 990–997.

Rizzone, M. G., Fasano, A., Daniele, A., Zibetti, M., Merola, A., Rizzi, L., et al. (2014). Long-term outcome of subthalamic nucleus DBS in Parkinson's disease: From the advanced phase towards the late stage of the disease? *Parkinsonism & Related Disorders, 20*, 376–381.

Rodgman, A., & Perfetti, T. A. (2013). *The chemical components of tobacco and tobacco smoke* (2nd ed.). Boca Raton: CRC Press.

Ryan, R. E., Ross, S. A., Drago, J., & Loiacono, R. E. (2001). Dose-related neuroprotective effects of chronic nicotine in 6-hydroxydopamine treated rats, and loss of neuroprotection in alpha4 nicotinic receptor subunit knockout mice. *British Journal of Pharmacology, 132*(8), 1650–1656.

Rylander, D. (2012). The serotonin system: A potential target for anti-dyskinetic treatments and biomarker discovery. *Parkinsonism & Related Disorders, 18*(Suppl. 1), S126–S128.

Sarter, M., Parikh, V., & Howe, W. M. (2009). nAChR agonist-induced cognition enhancement: Integration of cognitive and neuronal mechanisms. *Biochemical Pharmacology, 78*(7), 658–667.

Schaeffer, E., Pilotto, A., & Berg, D. (2014). Pharmacological strategies for the management of levodopa-induced dyskinesia in patients with Parkinson's disease. *CNS Drugs, 28*, 1155–1184.

Schapira, A. H. (2009). Neurobiology and treatment of Parkinson's disease. *Trends in Pharmacological Sciences, 30*(1), 41–47.

Schapira, A. H., & Jenner, P. (2011). Etiology and pathogenesis of Parkinson's disease. *Movement Disorders, 26*(6), 1049–1055.

Schapira, A. H., Olanow, C. W., Greenamyre, J. T., & Bezard, E. (2014). Slowing of neurodegeneration in Parkinson's disease and Huntington's disease: Future therapeutic perspectives. *Lancet, 384*(9942), 545–555.

Schneider, J. S., Van Velson, M., Menzaghi, F., & Lloyd, G. K. (1998). Effects of the nicotinic acetylcholine receptor agonist SIB-1508Y on object retrieval performance in MPTP-treated monkeys: Comparison with levodopa treatment. *Annals of Neurology, 43*(3), 311–317.

Schwarzschild, M. A., Schwid, S. R., Marek, K., Watts, A., Lang, A. E., Oakes, D., et al. (2008). Serum urate as a predictor of clinical and radiographic progression in Parkinson disease. *Archives of Neurology, 65*(6), 716–723.

Searles Nielsen, S., Gallagher, L. G., Lundin, J. I., Longstreth, W. T., Jr., Smith-Weller, T., Franklin, G. M., et al. (2012). Environmental tobacco smoke and Parkinson's disease. *Movement Disorders, 27*(2), 293–296.

Seeman, P. (2010). Dopamine D2 receptors as treatment targets in schizophrenia. *Clinical Schizophrenia & Related Psychoses, 4*(1), 56–73.

Sgambato-Faure, V., & Cenci, M. A. (2012). Glutamatergic mechanisms in the dyskinesias induced by pharmacological dopamine replacement and deep brain stimulation for the treatment of Parkinson's disease. *Progress in Neurobiology, 96*(1), 69–86.

Shim, J. C., Jung, D. U., Jung, S. S., Seo, Y. S., Cho, D. M., Lee, J. H., et al. (2012). Adjunctive varenicline treatment with antipsychotic medications for cognitive impairments in people with schizophrenia: A randomized double-blind placebo-controlled trial. *Neuropsychopharmacology, 37*(3), 660–668.

Shimohama, S. (2009). Nicotinic receptor-mediated neuroprotection in neurodegenerative disease models. *Biological & Pharmaceutical Bulletin, 32*(3), 332–336.

Shoulson, I. (2006). Randomized placebo-controlled study of the nicotinic agonist SIB-1508Y in Parkinson disease. *Neurology, 66*(3), 408–410.

Silver, A. A., Shytle, R. D., Sheehan, K. H., Sheehan, D. V., Ramos, A., & Sanberg, P. R. (2001). Multicenter, double-blind, placebo-controlled study of mecamylamine monotherapy for Tourette's disorder. *Journal of the American Academy of Child and Adolescent Psychiatry, 40*(9), 1103–1110.

Stayte, S., & Vissel, B. (2014). Advances in non-dopaminergic treatments for Parkinson's disease. *Frontiers in Neuroscience, 8*, 113.

Stuckenholz, V., Bacher, M., Balzer-Geldsetzer, M., Alvarez-Fischer, D., Oertel, W. H., Dodel, R. C., et al. (2013). The alpha7 nAChR agonist PNU-282987 reduces inflammation and MPTP-induced nigral dopaminergic cell loss in mice. *Journal of Parkinson's Disease, 3*(2), 161–172.

Suh, D. C., Pahwa, R., & Mallya, U. (2012). Treatment patterns and associated costs with Parkinson's disease levodopa induced dyskinesia. *Journal of the Neurological Sciences*, *319*(1–2), 24–31.

Suzuki, S., Kawamata, J., Matsushita, T., Matsumura, A., Hisahara, S., Takata, K., et al. (2013). 3-[(2,4-Dimethoxy)benzylidene]-anabaseine dihydrochloride protects against 6-hydroxydopamine-induced parkinsonian neurodegeneration through alpha7 nicotinic acetylcholine receptor stimulation in rats. *Journal of Neuroscience Research*, *91*(3), 462–471.

Szabo, N., Kincses, Z. T., & Vecsei, L. (2011). Novel therapy in Parkinson's disease: Adenosine A(2A) receptor antagonists. *Expert Opinion on Drug Metabolism & Toxicology*, 7(4), 441–455.

Tanner, C. M. (2010). Advances in environmental epidemiology. *Movement Disorders*, *25*(Suppl. 1), S58–S62.

Tanner, C. M., Goldman, S. M., Aston, D. A., Ottman, R., Ellenberg, J., Mayeux, R., et al. (2002). Smoking and Parkinson's disease in twins. *Neurology*, *58*(4), 581–588.

Tarsy, D., Lungu, C., & Baldessarini, R. J. (2011). Epidemiology of tardive dyskinesia before and during the era of modern antipsychotic drugs. *Handbook of Clinical Neurology*, *100*, 601–616.

Turrone, P., Remington, G., & Nobrega, J. N. (2002). The vacuous chewing movement (VCM) model of tardive dyskinesia revisited: Is there a relationship to dopamine D(2) receptor occupancy? *Neuroscience and Biobehavioral Reviews*, *26*(3), 361–380.

Vieregge, A., Sieberer, M., Jacobs, H., Hagenah, J. M., & Vieregge, P. (2001). Transdermal nicotine in PD: A randomized, double-blind, placebo-controlled study. *Neurology*, *57*(6), 1032–1035.

Villafane, G., Cesaro, P., Rialland, A., Baloul, S., Azimi, S., Bourdet, C., et al. (2007). Chronic high dose transdermal nicotine in Parkinson's disease: An open trial. *European Journal of Neurology*, *14*, 1313–1316.

Waddington, J. L. (1990). Spontaneous orofacial movements induced in rodents by very long-term neuroleptic drug administration: Phenomenology, pathophysiology and putative relationship to tardive dyskinesia. *Psychopharmacology*, *101*(4), 431–447.

Ween, H., Thorin-Hagene, K., Andersen, E., Gronlien, J. H., Lee, C. H., Gopalakrishnan, M., et al. (2010). alpha3* and alpha7 nAChR-mediated Ca(2+) transient generation in IMR-32 neuroblastoma cells. *Neurochemistry International*, *57*, 269–277.

Weisskopf, M. G., O'Reilly, E., Chen, H., Schwarzschild, M. A., & Ascherio, A. (2007). Plasma urate and risk of Parkinson's disease. *American Journal of Epidemiology*, *166*(5), 561–567.

Wichmann, T., DeLong, M. R., Guridi, J., & Obeso, J. A. (2011). Milestones in research on the pathophysiology of Parkinson's disease. *Movement Disorders*, *26*(6), 1032–1041.

Wirdefeldt, K., Adami, H. O., Cole, P., Trichopoulos, D., & Mandel, J. (2011). Epidemiology and etiology of Parkinson's disease: A review of the evidence. *European Journal of Epidemiology*, *26*(Suppl. 1), S1–S58.

Zhang, D., Bordia, T., McGregor, M., McIntosh, J. M., Decker, M. W., & Quik, M. (2014). ABT-089 and ABT-894 reduce levodopa-induced dyskinesias in a monkey model of Parkinson's disease. *Movement Disorders*, *29*(4), 508–517.

Zhang, D., Mallela, A., Sohn, D., Carroll, F. I., Bencherif, M., Letchworth, S., et al. (2013). Nicotinic receptor agonists reduce L-DOPA-induced dyskinesias in a monkey model of Parkinson's disease. *The Journal of Pharmacology and Experimental Therapeutics*, *347*(1), 225–234.

Zhang, D., McGregor, M., Decker, M. W., & Quik, M. (2014). The alpha7 nicotinic receptor agonist ABT-107 decreases L-Dopa-induced dyskinesias in parkinsonian monkeys. *The Journal of Pharmacology and Experimental Therapeutics*, *351*(1), 25–32.

Zupancic, M. L. (2011). Role of atypical antipsychotics in rapid cycling bipolar disorder: A review of the literature. *Annals of Clinical Psychiatry*, *23*(2), 141–149.
Zurkovsky, L., Taylor, W. D., & Newhouse, P. A. (2013). Cognition as a therapeutic target in late-life depression: Potential for nicotinic therapeutics. *Biochemical Pharmacology*, *86*(8), 1133–1144.

PART 5

Alzheimer's disease

CHAPTER TEN

Nicotinic Cholinergic Mechanisms in Alzheimer's Disease

Jianxin Shen*, Jie Wu*,†,1

*Department of Physiology, Shantou University Medical College, Shantou, Guangdong, China

†Divisions of Neurology and Neurobiology, Barrow Neurological Institute, St. Joseph's Hospital and Medical Center, Phoenix Arizona, USA

[1]Corresponding author: e-mail address: jie.wu@dignityhealth.org

Contents

Abstract

Alzheimer's disease (AD) is a neurodegenerative condition characterized by increased accumulation of Aβ and degeneration of cholinergic signaling between basal forebrain and hippocampus. Nicotinic acetylcholine receptors (nAChRs) are important mediators of cholinergic signaling in modulation of learning and memory function. Accumulating lines of evidence indicate that a nAChR subtype, α7 receptor (α7-nAChR), plays an important role in modulations of excitatory neurotransmitter release, improvement of learning and memory ability, and enhancement of cognitive function. Importantly, the expression and function of α7-nAChRs is altered in the brain of AD animal models and AD patients, suggesting that this nAChR subtype participates in AD pathogenesis and may serve as a novel therapeutic target for AD treatment. However, the mechanisms underlying the role of α7-nAChRs in AD pathogenesis are very complex, and either neuroprotective effects or neurotoxic effects may occur through the α7-nAChRs. These effects depend on the levels of α7-nAChR expression and function, disease stages, or the use of α7-nAChR agonists, antagonists, or allosteric modulators. In this chapter, we summarize recent progresses in the roles of α7-nAChRs played in AD pathogenesis and therapy.

International Review of Neurobiology, Volume 124
ISSN 0074-7742
http://dx.doi.org/10.1016/bs.irn.2015.08.002

1. HOMOMERIC AND HETEROMERIC α7-nAChRs

Nicotinic acetylcholine receptors (nAChRs) are prototypical members of the ligand-gated ion channel superfamily of neurotransmitter receptors. nAChRs represent both classic and contemporary models for the establishment of concepts pertaining to mechanisms of drug action, synaptic transmission, and structural/functional diversity of transmembrane signaling molecules (Albuquerque et al., 1997; Gotti, Carbonnelle, Moretti, Zwart, & Clementi, 2000; Lindstrom, 1996; Lukas et al., 1999). nAChRs are found throughout the body (e.g., in muscle, autonomic and sensory ganglia, and the central nervous system, CNS) (Albuquerque et al., 1997; Gotti et al., 2000; Lindstrom, 1996; Lukas et al., 1999). They exist as multiple, diverse subtypes composed as pentamers of unique combinations from a family of at least 17 (α1–α10, β1–β4, γ, δ, ε) similar, but genetically distinct, subunits (Champtiaux et al., 2003; Dajas-Bailador & Wonnacott, 2004; Fucile, 2004; Shen & Yakel, 2009). There are two major subtypes of nAChRs in the CNS. One subtype is the heteropentamers containing two or more different α or β subunit, for example, α4β2-nAChRs (Cooper, Harkness, Baker, & Millar, 1999; Nashmi et al., 2003; Scholze et al., 2011). Another is the homomeric, neurotoxin-binding, functionally versatile α7-nAChR. α7-nAChRs have been known to exist for many years based on their ability to bind the curare-mimetic neurotoxin, α-bungarotoxin (Bgt; Clarke, 1992; Sargent, Bryan, Streichert, & Garrett, 1991; Sargent & Garrett, 1995; Schmidt & Freeman, 1980). They have long been shown to exhibit many of the biochemical and pharmacological features of true nAChRs, to have brain distributions sub- or perisynaptic to cholinergic terminals, to have levels of expression sensitive to chronic nicotine exposure and/or modification of cholinergic inputs, and to reveal hints of functional significance in electrophysiological studies (Clarke, 1992; Sargent et al., 1991; Sargent & Garrett, 1995; Schmidt & Freeman, 1980). However, their physiological relevance was elusive and their functional study was confounded until heterologous expression studies of α7-nAChR homomers revealed unusually rapid, agonist-induced, calcium ion-permeable channel opening and inactivation (Couturier et al., 1990; Seguela, Wadiche, Dineley-Miller, Dani, & Patrick, 1993). Subsequently, renewed searches for functions of natural Bgt-binding nAChRs uncovered short-lived, nicotine-gated, toxin-sensitive, inward currents and/or elevations of intracellular Ca^{2+} in chick autonomic neurons (Franceschini et al., 2002; Liu, Xie, Emadi,

Sierks, & Wu, 2015; Thomsen et al., 2015; Vijayaraghavan, Pugh, Zhang, Rathouz, & Berg, 1992), in human ganglionic neuron-like clonal cells (Puchacz, Buisson, Bertrand, & Lukas, 1994), and in rat CNS neurons (Albuquerque et al., 1995; Alkondon & Albuquerque, 1991, 1993; Durazzo, Mattsson, Weiner, & Alzheimer's Disease Neuroimaging, 2014; McGehee & Role, 1995; Wu, George, et al., 2004; Zorumski, Thio, Isenberg, & Clifford, 1992). Many of the cell types naturally expressing Bgt-sensitive, functional nAChRs have been shown to express CHRNA7 as well as some native form of Bgt-binding nAChR (Albuquerque et al., 1995; Alkondon & Albuquerque, 1991, 1993; Lukas et al., 1999; McGehee & Role, 1995; Wu, George, et al., 2004; Zorumski et al., 1992). Knockout of the α7 gene leads to absence of Bgt-binding nAChR in cell lines or in mice (Orr-Urtreger et al., 1997).

Although it is well known that most α7-nAChRs are formed as homomeric pentamer (Fig. 1A), emerging lines of evidence demonstrate that nAChR α7 and β2 subunits can be coexpressed in most forebrain cholinergic neurons (Azam, Winzer-Serhan, & Leslie, 2003). The heterologous expression work indicates that nAChR α7 and β2 subunits can assemble together to form heteromeric, functional channels in *Xenopus* oocytes (Khiroug et al., 2002). We found that in mouse basal forebrain cholinergic neurons, nAChR α7 and β2 subunits are coexpressed and coassemble to form heteromeric, functional α7β2-nAChRs (Liu et al., 2009). More importantly, we found that heteromeric α7β2-nAChRs exhibit high sensitivity to pathologically relevant concentrations (e.g., 1 nM) of amyloid beta-peptide (Aβ). Thereafter, a series of reports demonstrated expression, assembling, function, and stoichiometry of this novel α7β2-nAChR (Murray et al., 2009, 2012; Onaivi, Ishiguro, Gu, & Liu, 2012). The α7β2-nAChR type is also likely to exist in human brainstem, hippocampus,

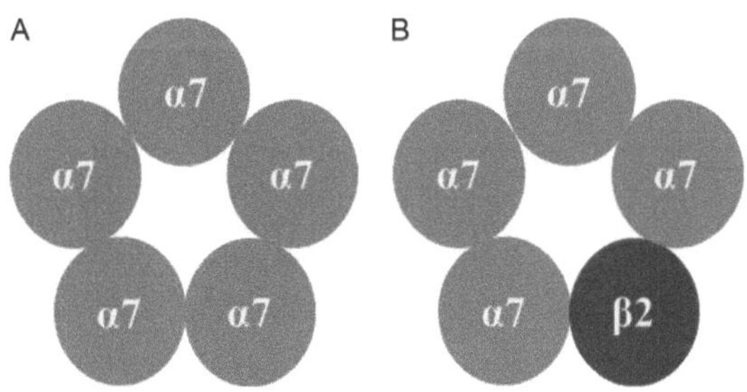

Figure 1 Homomeric and heteromeric α7-nAChRs. (A) Homomeric α7-nAChRs. (B) Heteromeric α7β2-nAChRs.

basal forebrain, and cerebral cortex (Machaalani, Kashi, & Waters, 2010; Moretti et al., 2014; Thomsen et al., 2015). Collectively, this novel subtype of α7-containing nAChRs (α7*-nAChRs) is expressed in the CNS and may play an important role in AD pathogenesis and therapeutic approaches (Fig. 1B).

2. ALTERATIONS OF α7-nAChR EXPRESSIONS IN AD ANIMAL MODELS AND AD PATIENTS

Epidemiological investigations demonstrate a complicated relationship between cigarette smoking and AD incidence. The early report suggests a negative (Fratiglioni & Wang, 2000), but a recent report suggests a positive relationship between cigarette smoking and AD incidence (Durazzo et al., 2014). Nevertheless, cigarette smoking likely affects AD pathogenesis and process, and nAChRs play an important role in mediation of this effect. For example, a significant loss of radioligand binding sites corresponding to α4β2-nAChR has been consistently observed at autopsy in a number of neocortical areas and the hippocampi of patients with AD (Burghaus et al., 2000; Nordberg, 2001). Losses in α7*-nAChR-like radioligand binding sites have been also reported in several brain regions of AD patients (Burghaus et al., 2000) in an early study. Decreases in numbers of radioligand binding sites corresponding to α7*-nAChRs are among the earliest events detected in AD, preceding changes in cholinergic markers and neuronal loss (Burghaus et al., 2000). Anticholinergic activity is known to impair memory while nicotine exposure improves cognitive function in AD patients (Levin & Rezvani, 2002), supporting crucial roles for cholinergic signaling and α7*-nAChRs in cognitive function. Activation of α7*-nAChRs might moderate Aβ toxicity. However, several studies showed higher levels of α7-nAChR mRNA and protein expression in amyloid precursor protein (APP) AD animal models and in AD patients (Counts et al., 2007; Dineley, Xia, Bui, Sweatt, & Zheng, 2002; Hellstrom-Lindahl, Court, et al., 2004, Hellstrom-Lindahl, Mousavi, Ravid, & Nordberg, 2004; Hellstrom-Lindahl, Mousavi, Zhang, Ravid, & Nordberg, 1999; Ikonomovic et al., 2009; Jones et al., 2006), and functionally, α7-nAChR-mediated currents are comparable between APP transgenic and wild-type mice at adult (7 months) ages (Spencer et al., 2006). In addition, chronic exposure to Aβ increased α7-nAChR expression in glial cells (Liu, Xie, Lukas, St John, & Wu, 2013; Xiu, Nordberg, Zhang, & Guan, 2005; Yu, Guan, Bogdanovic, & Nordberg, 2005). One interpretation is that the inhibition of α7-nAChR by Aβ may trigger receptor upregulation (Molinari et al., 1998), and these upregulated α7*-nAChRs may play a toxic role in AD pathogenesis (Liu et al., 2015, 2013).

3. INTERACTION OF AMYLOID BETA PEPTIDES (Aβ) AND α7-nAChRs

Aβ accumulation and aggregation in neuritic or senile plaques and severe, selective cholinergic neuronal deficits are characteristic hallmarks of AD (Palop & Mucke, 2010; Selkoe, 1999). The extent of learning and memory deficits in AD is proportional to the degree of forebrain cholinergic neuronal degeneration, and the extent of Aβ deposition is used to characterize disease severity (Selkoe, 1999). Processes such as impairment of neurotrophic support and disorders in glucose metabolism have been implicated in cholinergic neuronal loss and AD (Dolezal & Kasparova, 2003). However, clear, neurotoxic effects of Aβ across a range of *in vivo* or *in vitro* models suggest that Aβ plays roles in cholinergic neuronal degeneration and consequent learning and memory deficits (Parri, Hernandez, & Dineley, 2011; Selkoe, 1999). APP is a 39- to 43-amino acid transmembrane fragment of a larger precursor molecule (Selkoe, 1999). Insertion of APP into the cell plasmalemma affects membrane fluidity (Avdulov et al., 1997; Kanfer, Sorrentino, & Sitar, 1999; Muller, Koch, Eckert, Hartmann, & Scheuer, 1995). APP also affects intracellular Ca^{2+} homeostasis through either production of cation ionophores or activation of ligand- and/or voltage-gated channels (Arispe, Pollard, & Rojas, 1994; Sanderson, Butler, & Ingram, 1997). Other evidence exists revealing the harmful effects of Aβ on cellular Ca^{2+} homeostasis, neurotransmission, neuronal signaling, and receptor/ion channel function (Fraser, Suh, & Djamgoz, 1997). However, most of the early, relevant studies were performed using Aβ at concentrations ranging between 100 nM and 10 μM, which are much higher than Aβ concentrations (<5 nM) found in AD model brain. In addition, those studies did not utilize cholinergic neurons.

In studies suggesting possible pathophysiological links between Aβ, nAChRs, and AD, Wang et al. (Wang, Lee, D'Andrea, et al., 2000; Wang, Lee, Davis, & Shank, 2000) reported very high affinity binding of $A\beta_{1-40}$ and $A\beta_{1-42}$ to α7-nAChRs and lower affinity binding of Aβ to α4β2-nAChRs. Two other groups reported direct and functionally relevant interactions of $A\beta_{1-42}$ with α7-nAChRs (Liu, Kawai, & Berg, 2001; Pettit, Shao, & Yakel, 2001). Thereafter, several groups, including ours, reported the effects of Aβ on a variety of nAChRs (Dougherty, Wu, & Nichols, 2003; Liu & Wu, 2006; Wu, Kuo, et al., 2004). Most *in vitro* studies show an inhibitory effect of acute application of Aβ to neural preparations (single cells or brain slices), but some reports demonstrate an activating effect of Aβ toward heterologously transfected and native nAChRs. A summary of the interactions between Aβ and nAChRs has been published in recent reviews (Dineley, 2007; Dineley, Pandya, & Yakel, 2015; Hernandez & Dineley, 2012; Liu & Wu, 2006; Lombardo & Maskos, 2015; Parri et al., 2011). It is now apparent that the interaction between Aβ and nAChRs is very complex depending on nAChR subunit composition and stoichiometry, neuronal phenotype and brain location, as well as Aβ nature, concentrations, forms, length of exposure period, and aggregative status. It is clear that the structure and aggregative status confer Aβ unique functional properties such as the ability to perturb cholinergic signaling and impair learning and memory. Specific natural Aβ oligomers (dimers, trimers, and a 56 kDa dodecamer aggregate) have been linked to disruption of cognitive function and impaired memory (Cleary et al., 2005; Lesne et al., 2006). For freshly prepared Aβ, Dineley et al. found that their preparations predominantly contained monomer, trimer, and hexamer oligomers based on molecular weight estimates, but it is unknown which subset of these structures is responsible for Aβ's effects (Dineley, Bell, Bui, & Sweatt, 2002; Dineley et al., 2001).

4. COINCIDENT ROLES OF α7-nAChRs IN Aβ TOXICITY

Accumulating evidence links α7-nAChRs with Aβ deposition and AD pathogenesis. Many recent studies reported that Aβ directly modulates α7-nAChR function (Dineley, Bell, et al., 2002; Dineley, Xia, et al., 2002; Lamb, Melton, & Yakel, 2005; Liu et al., 2009, 2001; Pettit et al., 2001; Wu, Kuo, et al., 2004). Most of these findings, including ours, suggest that acute Aβ directly inhibits α7-nAChR function (Liu et al., 2009, 2001; Pettit et al., 2001). On the other end, activation of α7-nAChR appeared to reduce Aβ toxicity and protect targeted neurons. For instance, nAChR activation inhibits amyloid plaque formation *in vitro* and *in vivo* (Geerts, 2005), activates α-secretase cleavage of APP (Lahiri et al., 2002), increases acetylcholine (ACh) release, facilitates Aβ internalization (Nagele, D'Andrea, Anderson, & Wang, 2002), inhibits activity of the MAPK/NF-M/c-myc pathway (Liu et al., 2007), reduces Aβ production, and attenuates tau phosphorylation (Sadot et al., 1996). These findings suggest that signaling through α7-nAChRs not only is involved in cognitive function but also could protect against the kind of insults associated with AD. This concept has led to develop α7-nAChR agonists (or partial agonists) as a therapeutic strategy for treating AD (Valles, Borroni, & Barrantes, 2014).

Although Aβ inhibits α7-nAChRs acutely, chronic exposure to Aβ increased α7-nAChR expression (Liu et al., 2013; Xiu et al., 2005; Yu et al., 2005), and high expression levels of α7-nAChRs occur in both AD patients and preclinical AD models (Counts et al., 2007; Dineley, Xia, et al., 2002; Hellstrom-Lindahl, Court, et al., 2004; Hellstrom-Lindahl, Mousavi, et al., 2004; Hellstrom-Lindahl et al., 1999; Ikonomovic et al., 2009; Jones et al., 2006). As already mentioned, a possible interpretation of these apparently contradicting facts is that inhibition of α7-nAChR by Aβ may trigger receptor upregulation (Molinari et al., 1998). These data suggest a new concept, namely, that α7-nAChR may exhibit hyper- rather than hypoexpression and function in basal forebrain and hippocampal neurons in animal models of AD or even in AD patients. Considering the feature of high Ca^{2+} permeability in α7-nAChRs (Shen & Yakel, 2009; Skok, 2002; Szabo, Zelles, & Lendvai, 2008; Uteshev, 2010), and that these receptors are predominantly expressed in glutamatergic presynaptic terminals (Grilli et al., 2006; Jones & Wonnacott, 2004; Lin et al., 2010), higher expression of α7-nAChRs may lead to toxicity and participate in AD pathogenesis. Experiments that support this concept include the significantly high incidence of epilepsy in AD patients and APP animal models of AD, neuronal hyperexcitation in hippocampal neurons after chronic Aβ exposures (Liu et al., 2013), and the fact that APP-α7-nAChR$^{-/-}$ mice were shown to have less synaptic functional impairment and learning and memory behavioral deficits (Dziewczapolski, Glogowski, Masliah, & Heinemann, 2009). Currently, we further confirm that chronic exposure pathologically relevant concentration of Aβ aggregates to differentiated human neuroblastoma (SH-SY5Y) cells with cholinergic characteristics results in the significant neuronal toxicity (Liu et al., 2015).

5. ROLES OF α7-nAChRs IN NEURONAL HYPEREXCITATION IN AD

Epileptic seizures have long been recognized as a complication of the clinical syndrome of AD, especially in the early stage of disease, but have hitherto been viewed essentially as epiphenomena of the neurodegenerative process. Progress with animal models of AD has suggested that this view may be incorrect, and that seizures may be a reflection of pathophysiological processes similar to or overlapping with those responsible for cognitive decline in AD (Larner, 2010). Accumulating evidence indicates that in AD patients, the incidence of epilepsy is significantly higher than that of age-matched, non-AD subjects (Amatniek et al., 2006; Hesdorffer, Hauser, Annegers, Kokmen, & Rocca, 1996; Hommet, Mondon, Camus, De Toffol, & Constans, 2008; Lozsadi & Larner, 2006; Mendez & Lim, 2003). More importantly, the risk of seizure activity is especially high at an early stage of dementia onset with a reported 87-fold enhancement of seizure incidence in early-stage AD patients compared to an age-matched reference population (Amatniek et al., 2006; Mendez, 1994). Rabinowicz, Starkstein, Leiguarda, and Coleman (2000) reported that the epileptiform discharges were associated with transient episodes of amnestic wandering and disorientation in AD. Interestingly, the relationship between epilepsy and AD is even tighter in autosomal dominant early-onset AD linked to the mutations in presenilin-1, in presenilin-2, and/or APP (Edwards-Lee, Cook, Fairbanks, Leuchter, & Cummings, 2000; Marcon et al., 2004; Snider et al., 2005). Seizures in AD patients are mostly of partial origin, with both complex partial and secondary generalized seizures. Seizure pathophysiology may relate to increased amyloid beta-peptide production, structural alterations in neurons related to cytoskeletal dysfunction, cerebrovascular changes, neurotransmitter dysfunction, or combinations thereof. Together, these data suggest that epileptic seizures are part of the AD phenotype (Westmark, Westmark, Beard, Hildebrandt, & Malter, 2008), and that neuronal hyperexcitation and seizures may be one of the important mechanisms for Aβ toxicity and AD pathogenesis (Leonard & McNamara, 2007).

Emerging lines of evidence indicate that high levels of Aβ peptide induce neuronal or neurocircuit hyperexcitation. For example, chronic exposure to high levels of Aβ causes certain neuronal networks more susceptible to hyperexcitation (Del Vecchio, Gold, Novick, Wong, & Hyde, 2004). In animals that overexpressed Aβ, the high levels of Aβ peptide caused epileptiform activity within the entorhinal–hippocampal circuitry (Palop et al., 2007). Westmark et al. (2008) compared seizure threshold (test response to pentylenetetrazol) between an AD animal model (Tg2576) and wild-type mice and found a reduction of seizure threshold in the AD animal model. By studying the mechanisms of neuronal hyperexcitability in AD animal models, Busche et al. (2008) found that impaired synaptic inhibition was responsible for the phenomenon rather than intracellular Ca^{2+} release from intracellular stores or increased intrinsic neuronal firing. Furthermore, Aβ elevates presynaptic Ca^{2+} levels in an α7-nAChR-dependent manner (Dougherty et al., 2003), and Aβ also increases presynaptic glutamate release via activation of α7-nAChRs in forebrain cholinergic neurons (Chin, Ma, MacTavish, & Jhamandas, 2007). These lines of evidence suggest that Aβ induces neuronal hyperexcitation through an alteration of synaptic function, in which functional α7-nAChRs likely play a role in mediating Aβ-induced neuronal hyperexcitability.

Activation of α7-nAChRs enhances cognitive function, and the concept that α7-nAChR agonists can be used to treat AD has gained some favor. Thus far, several clinical trials are developing that are targeting nAChRs for mild to moderate AD, e.g., EVP-6124 (FORUM Pharmaceuticals), also known as MT-4666 (Mitsubishi Tanabe Pharma Corporation), an α7-nAChR partial agonist, varenicline (Pfizer), an α4β2* nAChR partial agonist, and AZD-3480 (AstraZeneca). However, recent evidence has challenged this concept. For example, chronic exposure to Aβ upregulates α7-nAChRs in glial cells (Xiu et al., 2005; Yu et al., 2005). Therefore, α7-nAChRs may exhibit high rather than low expression and function in hippocampal neurons in AD animal models or even in AD patients. In addition, emerging evidence has indicated the roles for functional α7-nAChRs in the modulation of neuronal excitability and firing synchronization. For instance, studies using hippocampal slices show that nAChR α7 KO mice exhibit lower seizure susceptibility after exposure to Aβ than found for WT controls (Minkeviciene et al., 2009). Pharmacological experiments demonstrate a positive correlation between the number of α-bungarotoxin (Bgt) binding sites in the hippocampus and sensitivity to nicotine-induced seizures (Miner & Collins, 1989; Miner, Marks, & Collins, 1984, 1985). Pharmacological study indicates that α7-nAChRs are important in mediating nicotine-induced seizures (Damaj, Glassco, Dukat, & Martin, 1999). However, the nicotine-induced seizures are presented in the α7-nAChR KO mice (Franceschini et al., 2002). In addition, mutant mice harboring an α7 subunit gain-of-function mutation (L250T) have increased sensitivity to both spontaneous, and nicotine-induced, seizures (Broide et al., 2002; Gil, Sack, Kedmi, Harmelin, & Orr-Urtreger, 2002). Moreover, a selective α7-nAChR partial agonist, JN403, exhibits anticonvulsant properties at high doses, probably through desensitization of α7-nAChRs (Feuerbach et al., 2009). Collectively, given features of α7-nAChRs such as high Ca^{2+} permeability and modulation of excitatory neurotransmitter release, these receptors may play a critical role in modulating neuronal excitability and synchronization.

6. COINCIDENT ROLES OF α7-nAChRs IN AD PATHOGENESIS

A key question is the real role of α7 nAChRs in AD pathogenesis. To address this critical question, a mouse model was generated by crossing the Tg2576 APP mouse model of AD with α7-nAChR KO mice to produce APP mice without α7-nAChRs ($APP^{+/-}\alpha7^{-/-}$ mice). Those mice could be examined for various pathological alterations, including Aβ levels and deposition, cell apoptosis, synaptic plasticity, and learning memory behavior, and compared to WT mice. Dr. Dineley's lab was one of the groups that analyzed the $APP^{+/-}\alpha7^{-/-}$ mice (Hernandez, Kayed, Zheng, Sweatt, & Dineley, 2010). They found that the learning and memory deficits seen in 5-month-old APP mice are more severe in the $APP^{+/-}\alpha7^{-/-}$ mice. Analyses of animals in early-stage preplaque cognitive decline revealed signs of neurodegeneration in $APP^{+/-}\alpha7^{-/-}$ hippocampus as well as loss of cholinergic functionality in the basal forebrain and hippocampus. Based on these findings, Dinley and colleagues hypothesized that increased soluble oligomeric Aβ may underlie the additional impairment of cognitive function in $APP^{+/-}\alpha7^{-/-}$. Thus, α7-nAChRs in a mouse model for early-stage AD appear to serve a neuroprotective role through maintenance of the septohippocampal cholinergic system and preservation of hippocampal integrity, possibly through influences on Aβ accumulation and oligomerization. However, studies from Dr. Heinemann's lab showed an opposite role of the α7-nAChRs in AD pathogenesis (Dziewczapolski et al., 2009). Using a new line of transgenic mice obtained by mating two established transgenic lines, PDAPP (J9) (Hsia et al., 1999) and α7-nAChR-null (Orr-Urtreger et al., 1997), they demonstrated that despite the presence of high amounts of APP and amyloid deposits, deletion of the α7-nAChR subunit leads to a protection from the dysfunction in synaptic integrity (pathology and plasticity) and learning and memory behavior. Based on their findings, the authors concluded that the α7-nAChR plays a role in AD pathogenesis, and perturbing α7-nAChR function could be beneficial in the treatment of AD.

7. CONCLUSIONS

Collectively, the α7-nAChR plays an important role in AD pathogenesis and has the potential to become an important therapeutic target. However, the mechanisms underlying the influence of α7-nAChRs on AD pathology are very complex. Both neuroprotective and toxic effects may occur through α7-nAChRs, depending on α7-nAChR expression and function levels, disease stage, or the use of α7-nAChR agonists, antagonists, or allosteric modulators. Fully understanding the impact of α7-nAChRs on AD pathogenesis and therapeutic outcomes will significantly improve the cholinergic-based diagnosis and treatment of AD.

REFERENCES

Albuquerque, E. X., Alkondon, M., Pereira, E. F., Castro, N. G., Schrattenholz, A., Barbosa, C. T., et al. (1997). Properties of neuronal nicotinic acetylcholine receptors: Pharmacological characterization and modulation of synaptic function. *The Journal of Pharmacology and Experimental Therapeutics*, *280*(3), 1117–1136.

Albuquerque, E. X., Pereira, E. F., Castro, N. G., Alkondon, M., Reinhardt, S., Schroder, H., et al. (1995). Nicotinic receptor function in the mammalian central nervous system. *Annals of the New York Academy of Sciences*, *757*, 48–72.

Alkondon, M., & Albuquerque, E. X. (1991). Initial characterization of the nicotinic acetylcholine receptors in rat hippocampal neurons. *Journal of Receptor Research*, *11*(6), 1001–1021.

Alkondon, M., & Albuquerque, E. X. (1993). Diversity of nicotinic acetylcholine receptors in rat hippocampal neurons. I. Pharmacological and functional evidence for distinct structural subtypes. *The Journal of Pharmacology and Experimental Therapeutics*, *265*(3), 1455–1473.

Amatniek, J. C., Hauser, W. A., DelCastillo-Castaneda, C., Jacobs, D. M., Marder, K., Bell, K., et al. (2006). Incidence and predictors of seizures in patients with Alzheimer's disease. *Epilepsia*, *47*(5), 867–872.

Arispe, N., Pollard, H. B., & Rojas, E. (1994). beta-Amyloid Ca(2+)-channel hypothesis for neuronal death in Alzheimer disease. *Molecular and Cellular Biochemistry*, *140*(2), 119–125.

Avdulov, N. A., Chochina, S. V., Igbavboa, U., O'Hare, E. O., Schroeder, F., Cleary, J. P., et al. (1997). Amyloid beta-peptides increase annular and bulk fluidity and induce lipid peroxidation in brain synaptic plasma membranes. *Journal of Neurochemistry*, *68*(5), 2086–2091.

Azam, L., Winzer-Serhan, U., & Leslie, F. M. (2003). Co-expression of alpha7 and beta2 nicotinic acetylcholine receptor subunit mRNAs within rat brain cholinergic neurons. *Neuroscience*, *119*(4), 965–977.

Broide, R. S., Salas, R., Ji, D., Paylor, R., Patrick, J. W., Dani, J. A., et al. (2002). Increased sensitivity to nicotine-induced seizures in mice expressing the L250T alpha 7 nicotinic acetylcholine receptor mutation. *Molecular Pharmacology*, *61*(3), 695–705.

Burghaus, L., Schutz, U., Krempel, U., de Vos, R. A., Jansen Steur, E. N., Wevers, A., et al. (2000). Quantitative assessment of nicotinic acetylcholine receptor proteins in the cerebral cortex of Alzheimer patients. *Brain Research. Molecular Brain Research*, *76*(2), 385–388.

Busche, M. A., Eichhoff, G., Adelsberger, H., Abramowski, D., Wiederhold, K. H., Haass, C., et al. (2008). Clusters of hyperactive neurons near amyloid plaques in a mouse model of Alzheimer's disease. *Science*, *321*(5896), 1686–1689.

Champtiaux, N., Gotti, C., Cordero-Erausquin, M., David, D. J., Przybylski, C., Lena, C., et al. (2003). Subunit composition of functional nicotinic receptors in dopaminergic neurons investigated with knock-out mice. *The Journal of Neuroscience*, *23*(21), 7820–7829.

Chin, J. H., Ma, L., MacTavish, D., & Jhamandas, J. H. (2007). Amyloid beta protein modulates glutamate-mediated neurotransmission in the rat basal forebrain: Involvement of presynaptic neuronal nicotinic acetylcholine and metabotropic glutamate receptors. *The Journal of Neuroscience*, *27*(35), 9262–9269. http://dx.doi.org/10.1523/JNEUROSCI.1843-07.2007.

Clarke, P. B. (1992). The fall and rise of neuronal alpha bungarotoxin binding proteins. *Trends in Pharmacological Sciences*, *13*(11), 407–413.

Cleary, J. P., Walsh, D. M., Hofmeister, J. J., Shankar, G. M., Kuskowski, M. A., Selkoe, D. J., et al. (2005). Natural oligomers of the amyloid-beta protein specifically disrupt cognitive function. *Nature Neuroscience*, *8*(1), 79–84.

Cooper, S. T., Harkness, P. C., Baker, E. R., & Millar, N. S. (1999). Up-regulation of cell-surface alpha4beta2 neuronal nicotinic receptors by lower temperature and expression of chimeric subunits. *The Journal of Biological Chemistry*, *274*(38), 27145–27152.

Counts, S. E., He, B., Che, S., Ikonomovic, M. D., DeKosky, S. T., Ginsberg, S. D., et al. (2007). Alpha7 nicotinic receptor up-regulation in cholinergic basal forebrain neurons in Alzheimer disease. *Archives of Neurology*, *64*(12), 1771–1776. http://dx.doi.org/10.1001/archneur.64.12.1771.

Couturier, S., Bertrand, D., Matter, J. M., Hernandez, M. C., Bertrand, S., Millar, N., et al. (1990). A neuronal nicotinic acetylcholine receptor subunit (alpha 7) is developmentally regulated and forms a homo-oligomeric channel blocked by alpha-BTX. *Neuron*, *5*(6), 847–856.

Dajas-Bailador, F. A., & Wonnacott, S. (2004). Nicotinic acetylcholine receptors and the regulation of neuronal signalling. *Trends in Pharmacological Sciences*, *25*(6), 317–324.

Damaj, M. I., Glassco, W., Dukat, M., & Martin, B. R. (1999). Pharmacological characterization of nicotine-induced seizures in mice. *The Journal of Pharmacology and Experimental Therapeutics*, *291*(3), 1284–1291.

Del Vecchio, R. A., Gold, L. H., Novick, S. J., Wong, G., & Hyde, L. A. (2004). Increased seizure threshold and severity in young transgenic CRND8 mice. *Neuroscience Letters*, *367*(2), 164–167.

Dineley, K. T. (2007). Beta-amyloid peptide–nicotinic acetylcholine receptor interaction: The two faces of health and disease. *Frontiers in Bioscience*, *12*, 5030–5038.

Dineley, K. T., Bell, K. A., Bui, D., & Sweatt, J. D. (2002). Beta-amyloid peptide activates alpha 7 nicotinic acetylcholine receptors expressed in Xenopus oocytes. *The Journal of Biological Chemistry*, 277(28), 25056–25061.

Dineley, K. T., Pandya, A. A., & Yakel, J. L. (2015). Nicotinic ACh receptors as therapeutic targets in CNS disorders. *Trends in Pharmacological Sciences*, *36*(2), 96–108.

Dineley, K. T., Westerman, M., Bui, D., Bell, K., Ashe, K. H., & Sweatt, J. D. (2001). Beta-amyloid activates the mitogen-activated protein kinase cascade via hippocampal alpha7 nicotinic acetylcholine receptors: In vitro and in vivo mechanisms related to Alzheimer's disease. *The Journal of Neuroscience*, *21*(12), 4125–4133.

Dineley, K. T., Xia, X., Bui, D., Sweatt, J. D., & Zheng, H. (2002). Accelerated plaque accumulation, associative learning deficits, and up-regulation of alpha 7 nicotinic receptor protein in transgenic mice co-expressing mutant human presenilin 1 and amyloid precursor proteins. *The Journal of Biological Chemistry*, *277*(25), 22768–22780.

Dolezal, V., & Kasparova, J. (2003). Beta-amyloid and cholinergic neurons. *Neurochemical Research*, *28*(3–4), 499–506.

Dougherty, J. J., Wu, J., & Nichols, R. A. (2003). Beta-amyloid regulation of presynaptic nicotinic receptors in rat hippocampus and neocortex. *The Journal of Neuroscience*, *23*(17), 6740–6747.

Durazzo, T. C., Mattsson, N., Weiner, M. W., & Alzheimer's Disease Neuroimaging, I. (2014). Smoking and increased Alzheimer's disease risk: A review of potential mechanisms. *Alzheimers Dement*, *10*(3 Suppl.), S122–S145.

Dziewczapolski, G., Glogowski, C. M., Masliah, E., & Heinemann, S. F. (2009). Deletion of the alpha 7 nicotinic acetylcholine receptor gene improves cognitive deficits and synaptic pathology in a mouse model of Alzheimer's disease. *The Journal of Neuroscience, 29*(27), 8805–8815.

Edwards-Lee, T., Cook, I., Fairbanks, L., Leuchter, A., & Cummings, J. L. (2000). Quantitative electroencephalographic correlates of psychosis in Alzheimer disease. *Neuropsychiatry, Neuropsychology, and Behavioral Neurology, 13*(3), 163–170.

Feuerbach, D., Lingenhoehl, K., Olpe, H. R., Vassout, A., Gentsch, C., Chaperon, F., et al. (2009). The selective nicotinic acetylcholine receptor alpha7 agonist JN403 is active in animal models of cognition, sensory gating, epilepsy and pain. *Neuropharmacology, 56*(1), 254–263.

Franceschini, D., Paylor, R., Broide, R., Salas, R., Bassetto, L., Gotti, C., et al. (2002). Absence of alpha7-containing neuronal nicotinic acetylcholine receptors does not prevent nicotine-induced seizures. *Brain Research. Molecular Brain Research, 98*(1–2), 29–40.

Fraser, S. P., Suh, Y. H., & Djamgoz, M. B. (1997). Ionic effects of the Alzheimer's disease beta-amyloid precursor protein and its metabolic fragments. *Trends in Neurosciences, 20*(2), 67–72.

Fratiglioni, L., & Wang, H. X. (2000). Smoking and Parkinson's and Alzheimer's disease: Review of the epidemiological studies. *Behavioural Brain Research, 113*(1–2), 117–120.

Fucile, S. (2004). Ca^{2+} permeability of nicotinic acetylcholine receptors. *Cell Calcium, 35*(1), 1–8. Journal article review.

Geerts, H. (2005). Indicators of neuroprotection with galantamine. *Brain Research Bulletin, 64*(6), 519–524.

Gil, Z., Sack, R. A., Kedmi, M., Harmelin, A., & Orr-Urtreger, A. (2002). Increased sensitivity to nicotine-induced seizures in mice heterozygous for the L250T mutation in the alpha7 nicotinic acetylcholine receptor. *Neuroreport, 13*(2), 191–196.

Gotti, C., Carbonnelle, E., Moretti, M., Zwart, R., & Clementi, F. (2000). Drugs selective for nicotinic receptor subtypes: A real possibility or a dream? *Behavioural Brain Research, 113*(1–2), 183–192.

Grilli, M., Raiteri, L., Patti, L., Parodi, M., Robino, F., Raiteri, M., et al. (2006). Modulation of the function of presynaptic alpha7 and non-alpha7 nicotinic receptors by the tryptophan metabolites, 5-hydroxyindole and kynurenate in mouse brain. *British Journal of Pharmacology, 149*(6), 724–732.

Hellstrom-Lindahl, E., Court, J., Keverne, J., Svedberg, M., Lee, M., Marutle, A., et al. (2004). Nicotine reduces A beta in the brain and cerebral vessels of APPsw mice. *The European Journal of Neuroscience, 19*(10), 2703–2710.

Hellstrom-Lindahl, E., Mousavi, M., Ravid, R., & Nordberg, A. (2004). Reduced levels of Abeta 40 and Abeta 42 in brains of smoking controls and Alzheimer's patients. *Neurobiology of Disease, 15*(2), 351–360.

Hellstrom-Lindahl, E., Mousavi, M., Zhang, X., Ravid, R., & Nordberg, A. (1999). Regional distribution of nicotinic receptor subunit mRNAs in human brain: Comparison between Alzheimer and normal brain. *Brain Research. Molecular Brain Research, 66*(1–2), 94–103.

Hernandez, C. M., & Dineley, K. T. (2012). alpha7 nicotinic acetylcholine receptors in Alzheimer's disease: Neuroprotective, neurotrophic or both? *Current Drug Targets, 13*(5), 613–622.

Hernandez, C. M., Kayed, R., Zheng, H., Sweatt, J. D., & Dineley, K. T. (2010). Loss of alpha7 nicotinic receptors enhances beta-amyloid oligomer accumulation, exacerbating early-stage cognitive decline and septohippocampal pathology in a mouse model of Alzheimer's disease. *The Journal of Neuroscience, 30*(7), 2442–2453.

Hesdorffer, D. C., Hauser, W. A., Annegers, J. F., Kokmen, E., & Rocca, W. A. (1996). Dementia and adult-onset unprovoked seizures. *Neurology, 46*(3), 727–730.

Hommet, C., Mondon, K., Camus, V., De Toffol, B., & Constans, T. (2008). Epilepsy and dementia in the elderly. *Dementia and Geriatric Cognitive Disorders*, *25*(4), 293–300.

Hsia, A. Y., Masliah, E., McConlogue, L., Yu, G. Q., Tatsuno, G., Hu, K., et al. (1999). Plaque-independent disruption of neural circuits in Alzheimer's disease mouse models. *Proceedings of the National Academy of Sciences of the United States of America*, *96*(6), 3228–3233.

Ikonomovic, M. D., Wecker, L., Abrahamson, E. E., Wuu, J., Counts, S. E., Ginsberg, S. D., et al. (2009). Cortical alpha7 nicotinic acetylcholine receptor and beta-amyloid levels in early Alzheimer disease. *Archives of Neurology*, *66*(5), 646–651.

Jones, I. W., Westmacott, A., Chan, E., Jones, R. W., Dineley, K., O'Neill, M. J., et al. (2006). Alpha7 nicotinic acetylcholine receptor expression in Alzheimer's disease: Receptor densities in brain regions of the APP(SWE) mouse model and in human peripheral blood lymphocytes. *Journal of Molecular Neuroscience*, *30*(1–2), 83–84.

Jones, I. W., & Wonnacott, S. (2004). Precise localization of alpha7 nicotinic acetylcholine receptors on glutamatergic axon terminals in the rat ventral tegmental area. *The Journal of Neuroscience*, *24*(50), 11244–11252.

Kanfer, J. N., Sorrentino, G., & Sitar, D. S. (1999). Amyloid beta peptide membrane perturbation is the basis for its biological effects. *Neurochemical Research*, *24*(12), 1621–1630.

Khiroug, S. S., Harkness, P. C., Lamb, P. W., Sudweeks, S. N., Khiroug, L., Millar, N. S., et al. (2002). Rat nicotinic ACh receptor alpha7 and beta2 subunits co-assemble to form functional heteromeric nicotinic receptor channels. *The Journal of Physiology*, *540*(Pt. 2), 425–434.

Lahiri, D. K., Utsuki, T., Chen, D., Farlow, M. R., Shoaib, M., Ingram, D. K., et al. (2002). Nicotine reduces the secretion of Alzheimer's beta-amyloid precursor protein containing beta-amyloid peptide in the rat without altering synaptic proteins. *Annals of the New York Academy of Sciences*, *965*, 364–372.

Lamb, P. W., Melton, M. A., & Yakel, J. L. (2005). Inhibition of neuronal nicotinic acetylcholine receptor channels expressed in Xenopus oocytes by beta-amyloid1-42 peptide. *Journal of Molecular Neuroscience*, *27*(1), 13–21.

Larner, A. J. (2010). Epileptic seizures in AD patients. *Neuromolecular Medicine*, *12*(1), 71–77.

Leonard, A. S., & McNamara, J. O. (2007). Does epileptiform activity contribute to cognitive impairment in Alzheimer's disease? *Neuron*, *55*(5), 677–678.

Lesne, S., Koh, M. T., Kotilinek, L., Kayed, R., Glabe, C. G., Yang, A., et al. (2006). A specific amyloid-beta protein assembly in the brain impairs memory. *Nature*, *440*(7082), 352–357. http://dx.doi.org/10.1038/nature04533.

Levin, E. D., & Rezvani, A. H. (2002). Nicotinic treatment for cognitive dysfunction. *Current Drug Targets. CNS and Neurological Disorders*, *1*(4), 423–431.

Lin, H., Vicini, S., Hsu, F. C., Doshi, S., Takano, H., Coulter, D. A., et al. (2010). Axonal alpha7 nicotinic ACh receptors modulate presynaptic NMDA receptor expression and structural plasticity of glutamatergic presynaptic boutons. *Proceedings of the National Academy of Sciences of the United States of America*, *107*(38), 16661–16666.

Lindstrom, J. (1996). Neuronal nicotinic acetylcholine receptors. *Ion Channels*, *4*, 377–450.

Liu, Q., Huang, Y., Xue, F., Simard, A., DeChon, J., Li, G., et al. (2009). A novel nicotinic acetylcholine receptor subtype in basal forebrain cholinergic neurons with high sensitivity to amyloid peptides. *The Journal of Neuroscience*, *29*(4), 918–929.

Liu, Q., Kawai, H., & Berg, D. K. (2001). Beta-amyloid peptide blocks the response of alpha 7-containing nicotinic receptors on hippocampal neurons. *Proceedings of the National Academy of Sciences of the United States of America*, *98*(8), 4734–4739.

Liu, Q., & Wu, J. (2006). Neuronal nicotinic acetylcholine receptors serve as sensitive targets that mediate beta-amyloid neurotoxicity. *Acta Pharmacologica Sinica*, *27*(10), 1277–1286.

Liu, Q., Xie, X., Emadi, S., Sierks, M. R., & Wu, J. (2015). A novel nicotinic mechanism underlies beta-amyloid-induced neurotoxicity. *Neuropharmacology*, *97*, 457–463.

Liu, Q., Xie, X., Lukas, R. J., St John, P. A., & Wu, J. (2013). A novel nicotinic mechanism underlies beta-amyloid-induced neuronal hyperexcitation. *The Journal of Neuroscience, 33*(17), 7253–7263.

Liu, Q., Zhang, J., Zhu, H., Qin, C., Chen, Q., & Zhao, B. (2007). Dissecting the signaling pathway of nicotine-mediated neuroprotection in a mouse Alzheimer disease model. *The FASEB Journal, 21*(1), 61–73.

Lombardo, S., & Maskos, U. (2015). Role of the nicotinic acetylcholine receptor in Alzheimer's disease pathology and treatment. *Neuropharmacology, 96*(Pt. B), 255–262.

Lozsadi, D. A., & Larner, A. J. (2006). Prevalence and causes of seizures at the time of diagnosis of probable Alzheimer's disease. *Dementia and Geriatric Cognitive Disorders, 22*(2), 121–124.

Lukas, R. J., Changeux, J. P., Le Novère, N., Albuquerque, E. X., Balfour, D. J., Berg, D. K., et al. (1999). International Union of Pharmacology. XX. Current status of the nomenclature for nicotinic acetylcholine receptors and their subunits. *Pharmacological Reviews, 51*(2), 397–401.

Machaalani, R., Kashi, P. K., & Waters, K. A. (2010). Distribution of nicotinic acetylcholine receptor subunits alpha7 and beta2 in the human brainstem and hippocampal formation. *Journal of Chemical Neuroanatomy, 40*(3), 223–231.

Marcon, G., Giaccone, G., Cupidi, C., Balestrieri, M., Beltrami, C. A., Finato, N., et al. (2004). Neuropathological and clinical phenotype of an Italian Alzheimer family with M239V mutation of presenilin 2 gene. *Journal of Neuropathology and Experimental Neurology, 63*(3), 199–209.

McGehee, D. S., & Role, L. W. (1995). Physiological diversity of nicotinic acetylcholine receptors expressed by vertebrate neurons. *Annual Review of Physiology, 57*, 521–546.

Mendez, M. F. (1994). Huntington's disease: Update and review of neuropsychiatric aspects. *International Journal of Psychiatry in Medicine, 24*(3), 189–208.

Mendez, M., & Lim, G. (2003). Seizures in elderly patients with dementia: Epidemiology and management. *Drugs & Aging, 20*(11), 791–803.

Miner, L. L., & Collins, A. C. (1989). Strain comparison of nicotine-induced seizure sensitivity and nicotinic receptors. *Pharmacology, Biochemistry, and Behavior, 33*(2), 469–475.

Miner, L. L., Marks, M. J., & Collins, A. C. (1984). Classical genetic analysis of nicotine-induced seizures and nicotinic receptors. *The Journal of Pharmacology and Experimental Therapeutics, 231*(3), 545–554.

Miner, L. L., Marks, M. J., & Collins, A. C. (1985). Relationship between nicotine-induced seizures and hippocampal nicotinic receptors. *Life Sciences, 37*(1), 75–83.

Minkeviciene, R., Rheims, S., Dobszay, M. B., Zilberter, M., Hartikainen, J., Fulop, L., et al. (2009). Amyloid beta-induced neuronal hyperexcitability triggers progressive epilepsy. *The Journal of Neuroscience, 29*(11), 3453–3462.

Molinari, E. J., Delbono, O., Messi, M. L., Renganathan, M., Arneric, S. P., Sullivan, J. P., et al. (1998). Up-regulation of human alpha7 nicotinic receptors by chronic treatment with activator and antagonist ligands. *European Journal of Pharmacology, 347*(1), 131–139.

Moretti, M., Zoli, M., George, A. A., Lukas, R. J., Pistillo, F., Maskos, U., et al. (2014). The novel alpha7beta2-nicotinic acetylcholine receptor subtype is expressed in mouse and human basal forebrain: Biochemical and pharmacological characterization. *Molecular Pharmacology, 86*(3), 306–317.

Muller, W. E., Koch, S., Eckert, A., Hartmann, H., & Scheuer, K. (1995). beta-Amyloid peptide decreases membrane fluidity. *Brain Research, 674*(1), 133–136. 0006-8993(94)01463-R [pii].

Murray, T. A., Bertrand, D., Papke, R. L., George, A. A., Pantoja, R., Srinivasan, R., et al. (2012). alpha7beta2 nicotinic acetylcholine receptors assemble, function, and are activated primarily via their alpha7-alpha7 interfaces. *Molecular Pharmacology, 81*(2), 175–188.

Murray, T., Srinivasan, R., Liu, Q., Maxwell, J., Pantoja, R., Whiteaker, P., et al. (2009). Coassembly of fluorescently-tagged α7 and β2 nicotinic acetylcholine receptor subunits in mammalian cell line detected by FRET. In *2009 Neuroscience meeting planner*, Chicago, IL: Society for Neuroscience. Online, Program No. 227.12/C8.

Nagele, R. G., D'Andrea, M. R., Anderson, W. J., & Wang, H. Y. (2002). Intracellular accumulation of beta-amyloid(1–42) in neurons is facilitated by the alpha 7 nicotinic acetylcholine receptor in Alzheimer's disease. *Neuroscience, 110*(2), 199–211.

Nashmi, R., Dickinson, M. E., McKinney, S., Jareb, M., Labarca, C., Fraser, S. E., et al. (2003). Assembly of alpha4beta2 nicotinic acetylcholine receptors assessed with functional fluorescently labeled subunits: Effects of localization, trafficking, and nicotine-induced upregulation in clonal mammalian cells and in cultured midbrain neurons. *The Journal of Neuroscience, 23*(37), 11554–11567.

Nordberg, A. (2001). Nicotinic receptor abnormalities of Alzheimer's disease: Therapeutic implications. *Biological Psychiatry, 49*(3), 200–210. S0006322300011252 [pii].

Onaivi, E. S., Ishiguro, H., Gu, S., & Liu, Q. R. (2012). CNS effects of CB2 cannabinoid receptors: Beyond neuro-immuno-cannabinoid activity. *Journal of Psychopharmacology, 26*(1), 92–103.

Orr-Urtreger, A., Goldner, F. M., Saeki, M., Lorenzo, I., Goldberg, L., De Biasi, M., et al. (1997). Mice deficient in the alpha7 neuronal nicotinic acetylcholine receptor lack alpha-bungarotoxin binding sites and hippocampal fast nicotinic currents. *The Journal of Neuroscience, 17*(23), 9165–9171.

Palop, J. J., Chin, J., Roberson, E. D., Wang, J., Thwin, M. T., Bien-Ly, N., et al. (2007). Aberrant excitatory neuronal activity and compensatory remodeling of inhibitory hippocampal circuits in mouse models of Alzheimer's disease. *Neuron, 55*(5), 697–711.

Palop, J. J., & Mucke, L. (2010). Amyloid-beta-induced neuronal dysfunction in Alzheimer's disease: From synapses toward neural networks. *Nature Neuroscience, 13*(7), 812–818.

Parri, H. R., Hernandez, C. M., & Dineley, K. T. (2011). Research update: Alpha7 nicotinic acetylcholine receptor mechanisms in Alzheimer's disease. *Biochemical Pharmacology, 82*(8), 931–942.

Pettit, D. L., Shao, Z., & Yakel, J. L. (2001). beta-Amyloid(1–42) peptide directly modulates nicotinic receptors in the rat hippocampal slice. *The Journal of Neuroscience, 21*(1), Rc120.

Puchacz, E., Buisson, B., Bertrand, D., & Lukas, R. J. (1994). Functional expression of nicotinic acetylcholine receptors containing rat alpha 7 subunits in human SH-SY5Y neuroblastoma cells. *FEBS Letters, 354*(2), 155–159.

Rabinowicz, A. L., Starkstein, S. E., Leiguarda, R. C., & Coleman, A. E. (2000). Transient epileptic amnesia in dementia: A treatable unrecognized cause of episodic amnestic wandering. *Alzheimer Disease and Associated Disorders, 14*(4), 231–233.

Sadot, E., Gurwitz, D., Barg, J., Behar, L., Ginzburg, I., & Fisher, A. (1996). Activation of m1 muscarinic acetylcholine receptor regulates tau phosphorylation in transfected PC12 cells. *Journal of Neurochemistry, 66*(2), 877–880.

Sanderson, K. L., Butler, L., & Ingram, V. M. (1997). Aggregates of a beta-amyloid peptide are required to induce calcium currents in neuron-like human teratocarcinoma cells: Relation to Alzheimer's disease. *Brain Research, 744*(1), 7–14.

Sargent, P. B., Bryan, G. K., Streichert, L. C., & Garrett, E. N. (1991). Denervation does not alter the number of neuronal bungarotoxin binding sites on autonomic neurons in the frog cardiac ganglion. *The Journal of Neuroscience, 11*(11), 3610–3623.

Sargent, P. B., & Garrett, E. N. (1995). The characterization of alpha-bungarotoxin receptors on the surface of parasympathetic neurons in the frog heart. *Brain Research, 680*(1–2), 99–107.

Schmidt, J. T., & Freeman, J. A. (1980). Electrophysiologic evidence that retinotectal synaptic transmission in the goldfish is nicotinic cholinergic. *Brain Research, 187*(1), 129–142.

Scholze, P., Ciuraszkiewicz, A., Groessl, F., Orr-Urtreger, A., McIntosh, J. M., & Huck, S. (2011). alpha4beta2 nicotinic acetylcholine receptors in the early postnatal mouse superior cervical ganglion. *Developmental Neurobiology*, *71*(5), 390–399.

Seguela, P., Wadiche, J., Dineley-Miller, K., Dani, J. A., & Patrick, J. W. (1993). Molecular cloning, functional properties, and distribution of rat brain alpha 7: A nicotinic cation channel highly permeable to calcium. *The Journal of Neuroscience*, *13*(2), 596–604.

Selkoe, D. J. (1999). Translating cell biology into therapeutic advances in Alzheimer's disease. *Nature*, *399*(6738 Suppl.), A23–A31.

Shen, J. X., & Yakel, J. L. (2009). Nicotinic acetylcholine receptor-mediated calcium signaling in the nervous system. *Acta Pharmacologica Sinica*, *30*(6), 673–680.

Skok, V. I. (2002). Nicotinic acetylcholine receptors in autonomic ganglia. *Autonomic Neuroscience*, *97*(1), 1–11.

Snider, B. J., Norton, J., Coats, M. A., Chakraverty, S., Hou, C. E., Jervis, R., et al. (2005). Novel presenilin 1 mutation (S170F) causing Alzheimer disease with Lewy bodies in the third decade of life. *Archives of Neurology*, *62*(12), 1821–1830.

Spencer, J. P., Weil, A., Hill, K., Hussain, I., Richardson, J. C., Cusdin, F. S., et al. (2006). Transgenic mice over-expressing human beta-amyloid have functional nicotinic alpha 7 receptors. *Neuroscience*, *137*(3), 795–805. http://dx.doi.org/10.1016/j.neuroscience.2005.10.007.

Szabo, S. I., Zelles, T., & Lendvai, B. (2008). Intracellular Ca2+ dynamics of hippocampal interneurons following nicotinic acetylcholine receptor activation. *Neurochemistry International*, *52*(1–2), 135–141.

Thomsen, M. S., Zwart, R., Ursu, D., Jensen, M. M., Pinborg, L. H., Gilmour, G., et al. (2015). alpha7 and beta2 nicotinic acetylcholine receptor subunits form heteromeric receptor complexes that are expressed in the human cortex and display distinct pharmacological properties. *PLoS One*, *10*(6), e0130572.

Uteshev, V. V. (2010). Evaluation of Ca2+ permeability of nicotinic acetylcholine receptors in hypothalamic histaminergic neurons. *Acta Pharmacologica Sinica*, *42*(1), 8–20.

Valles, A. S., Borroni, M. V., & Barrantes, F. J. (2014). Targeting brain alpha7 nicotinic acetylcholine receptors in Alzheimer's disease: Rationale and current status. *CNS Drugs*, *28*(11), 975–987.

Vijayaraghavan, S., Pugh, P. C., Zhang, Z. W., Rathouz, M. M., & Berg, D. K. (1992). Nicotinic receptors that bind alpha-bungarotoxin on neurons raise intracellular free Ca^{2+}. *Neuron*, *8*(2), 353–362. Journal article.

Wang, H. Y., Lee, D. H., D'Andrea, M. R., Peterson, P. A., Shank, R. P., & Reitz, A. B. (2000). beta-Amyloid(1–42) binds to alpha7 nicotinic acetylcholine receptor with high affinity. Implications for Alzheimer's disease pathology. *The Journal of Biological Chemistry*, *275*(8), 5626–5632.

Wang, H. Y., Lee, D. H., Davis, C. B., & Shank, R. P. (2000). Amyloid peptide Abeta(1–42) binds selectively and with picomolar affinity to alpha7 nicotinic acetylcholine receptors. *Journal of Neurochemistry*, *75*(3), 1155–1161.

Westmark, C. J., Westmark, P. R., Beard, A. M., Hildebrandt, S. M., & Malter, J. S. (2008). Seizure susceptibility and mortality in mice that over-express amyloid precursor protein. *International Journal of Clinical and Experimental Pathology*, *1*(2), 157–168.

Wu, J., George, A. A., Schroeder, K. M., Xu, L., Marxer-Miller, S., Lucero, L., et al. (2004). Electrophysiological, pharmacological, and molecular evidence for alpha7-nicotinic acetylcholine receptors in rat midbrain dopamine neurons. *The Journal of Pharmacology and Experimental Therapeutics*, *311*(1), 80–91.

Wu, J., Kuo, Y. P., George, A. A., Xu, L., Hu, J., & Lukas, R. J. (2004). beta-Amyloid directly inhibits human alpha4beta2-nicotinic acetylcholine receptors heterologously expressed in human SH-EP1 cells. *The Journal of Biological Chemistry*, *279*(36), 37842–37851.

Xiu, J., Nordberg, A., Zhang, J. T., & Guan, Z. Z. (2005). Expression of nicotinic receptors on primary cultures of rat astrocytes and up-regulation of the alpha7, alpha4 and beta2 subunits in response to nanomolar concentrations of the beta-amyloid peptide(1–42). *Neurochemistry International*, *47*(4), 281–290.

Yu, W. F., Guan, Z. Z., Bogdanovic, N., & Nordberg, A. (2005). High selective expression of alpha7 nicotinic receptors on astrocytes in the brains of patients with sporadic Alzheimer's disease and patients carrying Swedish APP 670/671 mutation: A possible association with neuritic plaques. *Experimental Neurology*, *192*(1), 215–225.

Zorumski, C. F., Thio, L. L., Isenberg, K. E., & Clifford, D. B. (1992). Nicotinic acetylcholine currents in cultured postnatal rat hippocampal neurons. *Molecular Pharmacology*, *41*(5), 931–936.

INDEX

Note: Page numbers followed by "*f*" indicate figures and "*t*" indicate tables.

H

L

M

N

CONTENTS OF RECENT VOLUMES

Volume 42

Volume 43

Volume 44

Volume 45

Volume 46

Volume 47

Volume 48

Volume 49

Volume 50

Volume 53

Volume 54

Volume 55

Volume 56

Volume 57

Volume 58

Volume 59

Volume 60

Volume 61

Volume 62

Volume 63

Volume 64

Volume 65

Volume 66

Volume 67

Volume 68

Volume 69

Volume 70

Volume 71

Volume 72

Volume 73

Volume 74

Volume 75

Volume 76

Volume 77

Volume 78

Volume 79

Volume 80

Volume 81

Volume 82

Volume 83

Volume 84

Volume 85

Volume 86

Volume 87

Volume 88

Volume 89

Volume 90

Volume 91

Volume 92

Volume 93

Volume 94

Volume 95

Volume 96

Volume 97

Volume 98

Volume 99

Volume 100

Volume 101

Volume 102

Volume 103

Volume 104

Volume 105

Volume 106

Volume 107

Volume 108

Volume 109

Volume 110

Volume 111

Volume 112

Volume 113

Volume 114

Volume 115

Volume 119

Volume 120

Volume 121

Volume 122

Volume 123